Unconventional Co[m...]
2007

Andrew Adamatzky, Benjamin De Lacy Costello,
Larry Bull, Susan Stepney, Christof Teuscher
Editors

Unconventional Computing
2007

Luniver Press
2007

Published by Luniver Press
Frome BA11 6TT United Kingdom

British Library Cataloguing-in-Publication Data
A catalogue record for this book is available from the British Library

Unconventional Computing 2007

ISBN-10: 1-905986-05-X
ISBN-13: 978-1-905986-05-7

Editorial

Unconventional computing is the quest for groundbreaking new algorithms and physical implementations of novel and ultimately — compared to classical approaches – more powerful computing paradigms and machines based on and inspired by principles of information processing in physical, chemical and biological systems.

The book offers a timely collection of papers presented at the 2007 Unconventional Computing 2007 conference which was held in Bristol, July 12-14, 2007. The event was financially supported by the Engineering and Physical Sciences Research Council (EPSRC), the UK Government's leading funding agency for research and training in engineering and the physical sciences.

The Unconventional Computing 2007 conference is a successor of several previous events: The Grand Challenges in Non-Classical Computation International Workshop (York, UK, 2005), the Workshop on Unconventional Computing at ECAL2005 (Kent, UK, 2005), the workshop From Utopian to Genuine Unconventional Computers at UC2006 (York, UK, 2006), and Unconventional Computation: Quo Vadis? conference (Santa Fe, USA, 2007).

The conference aims to bring together world-leading scientists whose research focuses on non-traditional theoretical machines, experimental prototypes and genuine implementations of non-classical computing devices, who try to revisit existing approaches in unconventional computing, provide scientists and engineers with blueprints of realisable computing devices, and take a critical glance at the design of novel and emergent computing systems to point out failures and shortcomings of both theoretical and experimental approaches.

This book includes cutting-edge theoretical work on quantum and kinematic Turing machines, computational complexity of physical systems, molecular and chemical computation, processing incomplete information, physical hypercomputation, automata networks and swarms. They are complemented by recent results on experimental implementations of logical and arithmetical circuits in a domino substrate, DNA computers, and self-assembly.

We hope the book will encourage interdisciplinary interactions in the field of emergent computing paradigms and architectures; develop a common interface between computer science, biology, mathematics, chemistry, electronics engineering, and physics; create new research communities in non-classical computation; and promote the transfer of knowledge between different research communities.

Andrew Adamatzky, Larry Bull, Benjamin De Lacy Costello,
Susan Stepney, Christof Teuscher,
The Conference Organizers
Bristol (UK), York (UK), Los Alamos (USA)
May 2007

Contents

Unconventional Computing 2007

Unentangling nuclear magnetic resonance computing

Matthias Bechmann[1], John A. Clark[2], Angelika Sebald[1], Susan Stepney[2]

[1] Department of Chemistry, University of York, YO10 5DD, UK
[2] Department of Computer Science, University of York, UK, YO10 5DD

Abstract Nuclear magnetic resonance (NMR) is typically thought of as a possible technology for quantum computation. Here we instead outline how commercially available NMR spectrometers could be used to perform non-quantum computation: from addressable 3D memory, to a programmable 3D reaction-diffusion computer.

1 Introduction

Recently, exciting border-crossing joint activities between computer sciences and nuclear magnetic resonance (NMR) spectroscopy have emerged. From the point of view of NMR spectroscopy, one would probably first think of recent applications of genetic algorithms for finding improved pulse sequences in solution- and solid-state NMR [8] [9]. In the context of computer science, probably NMR first comes to mind as a tool for implementing and testing ideas in the area of quantum computing [5] [7]. NMR can play this role due to its uniquely well defined quantum mechanical nature and the relative ease of (some) experimental implementations. Commercial NMR spectrometers are widely available, for example providing the prime analytical technique in synthetic organic chemistry, or as a slightly more sophisticated method for the study of molecular structures in biochemistry.

Here we describe some ideas for mutually beneficial interplay between NMR and unconventional computation other than quantum computing. We first give a brief summary of the most basic features and concepts encountered in (solution-state) NMR spectroscopy. Then we propose various computational approaches, and outline how such concepts could be implemented in a real NMR spectrometer.

2   Background: The physics of NMR

Here we give a brief and qualitative sketch of the basic principles involved in NMR; we refer the reader to the literature [11] for further reading about the physical background, the theory and quantum mechanical description of NMR.

Nuclear magnetic resonance exploits an intrinsic property of most isotopes in the period table of the elements: the magnetic moments of atomic nuclei. Associated with this magnetic moment is a property called spin. For the sake of simplicity we consider only the case of isotopes with spin quantum number $I = 1/2$. In addition, we initially assume that we have a homogeneous sample with only one type of spins present. The sample is placed in a very strong and homogeneous external magnetic field B_0 along the z-direction of the laboratory coordinate system (see figure 1). The spins start precessing around the B_0 direction at the so-called Larmor frequency (which is isotope specific) and a net magnetisation of the spins in the sample builds up along the z-direction. Adopting a reference frame rotating at the Larmor frequency around the z-direction yields a description where the spin dynamics appear static. Therefore, we can now easily inspect the effect of a radio frequency (r.f.) pulse along the x-direction of the rotating frame as the effect of an additional static magnetic field B_1 in the x-direction of the rotating frame on the magnetisation vector. During the pulse, the spins precess around the direction of the applied field B_1: for example, an r.f. pulse of suitable amplitude and duration at the Larmor frequency tips the magnetisation vector in the $-y$-direction (a so-called $(\pi/2)_x$-pulse, where the subscript indicates the direction/phase of the r.f. pulse), giving the maximum possible signal in a detection coil aligned in the y-direction. (See figure 2 for examples of the effects of various pulses.)

After switching off the r.f. pulse, the signal is observed in the time domain in the form of a free induction decay (FID) in the receiver coil (see figure 1). Commonly, this time domain signal is transformed to the frequency domain by applying a complex Fourier transformation. By varying the phase, amplitude and/or duration of the r.f. pulse, the magnetisation vector can be tipped in any direction – and the corresponding NMR spectrum carries frequency as well as phase information. This is the simplest form of a common, one-dimensional (1D) NMR spectrum.

The 'burst of radiation' in the form of the r.f. pulse leaves the system in a non-equilibrium state. Accordingly, after the pulse the system returns to equilibrium z-magnetisation. This happens with a characteristic time constant T_1 and measures the rate at which z-magnetisation is re-built. Another effect, with a characteristic time constant T_2, concerns the xy-plane: initially, right after an r.f. pulse, all the spins are in phase.

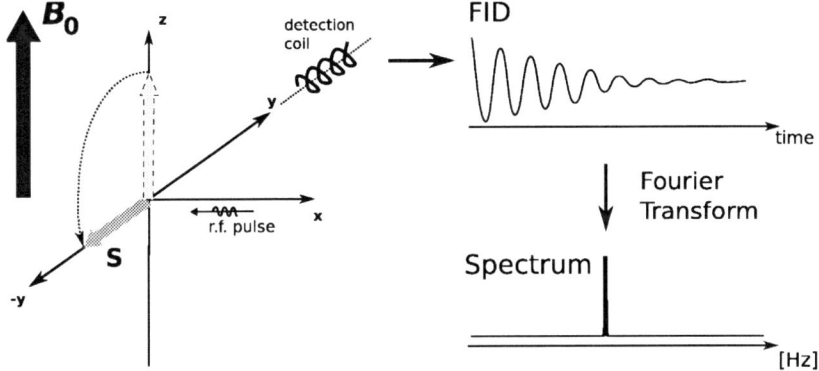

Figure 1. Larmor precession in the rotating frame, resonance, FID detection and frequency-domain spectrum.

Small fluctuating local fields cause a dephasing in the xy-plane at a rate measured by T_2. In non-viscous liquids at ambient conditions usually $T_1 \sim T_2$. As T_1 and T_2 are of the order of seconds, NMR resonances under these conditions tend to be very narrow, giving very high spectral resolution.

Now we add another feature. Suppose we add magnetic field gradients (in the x,y,z directions, either static or pulsed gradients) to the external static magnetic field. Such gradients alter the resonance frequencies of spins in different locations (voxels) of the sample. In our simple system of a homogeneous liquid sample with only one type of spins present, different voxels of the sample are now represented by different resonance frequencies – as the spins in different regions of the sample feel different local magnetic fields, depending on the gradients used. Thus, gradients add space (and further phase) encoding to the scenario. If we now take advantage of selective r.f. pulses, we have enormous degrees of freedom to select slices and specific regions in the sample for all kinds of 'spin gymnastics'. A practical, and probably the most widely known, application of these principles is magnetic resonance imaging (MRI) in the medical context where NMR signals of ^1H nuclei in fat and water, our predominant soft body materials, deliver three-dimensional images. Although keeping these principles in mind, we do not aim at using them for image production, but rather aim to exploit them in a sense of localised spectroscopy.

Having generally available now frequency, phase and space encoding we need to expand the 1D NMR spectrum to include a second dimension. 2D NMR needs the application of a sequence of pulses and/or

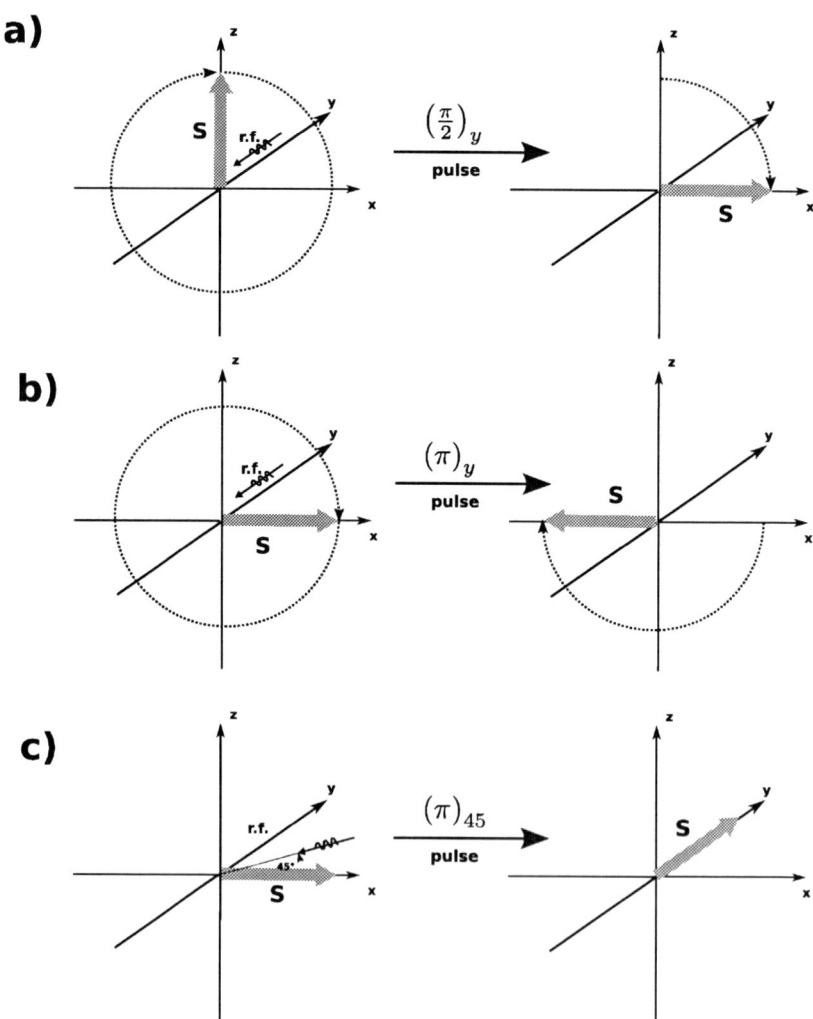

Figure 2. Illustration of the the effects of different r.f. pulses on magnetisation vectors S: (a) a $(\pi/2)_y$-pulse acting on the $+z$-magnetisation vector; (b) a $(\pi)_y$-pulse acting on the $+x$-magnetisation vector; (c) a $(\pi)_{45}$-pulse acting on the $+x$-magnetisation vector.

events in order to induce dependency of the NMR signal(s) on two time variables. These two time variables, along with different kinds of pulse sequences, are able to deliver 2D NMR spectra displaying a wide range of e.g. correlations amongst spins, after complex Fourier transformation of the time-domain data with regard to two time variables. In that respect, 2D NMR is not just a simple expansion of 1D NMR but adds a fundamentally important additional feature. (Note that here the '2D' refers to two *temporal* dimensions, not to spatial dimensions.)

If, in gedankenexperiments, we now add more complexity to our samples (e.g. by including samples with more than one spin species present, by including interactions amongst spins by dipolar coupling interactions, by including samples with anisotropic properties leading to anisotropic signatures in NMR spectra) plus allow for virtually unlimited complexity of pulse (and gradient switching) sequences, it takes little imagination to see that we have a highly controllable, extremely well defined – and highly complex – system at our disposal.

We claim that this system can, and should, be exploited for further experiments, beyond the implementation of quantum-computing model systems. As noted earlier, NMR is one proposed approach to quantum computation [5] [7]. There, the interactions between nuclear spins (couplings) are used to encode qubits, and entangling those spins gives qubits. However, like all quantum computation implementations, there are the usual engineering problems of maintaining coherence and scaling up to large numbers of qubits. Here we take a step back, and investigate using the nuclear spins to encode *classical* bits, and use liquid-state NMR spectroscopy to do classical (as in non-quantum) massively parallel computation.

3 Encoding bits

The liquid sample is divided into spatial regions, or *voxels*, by the field gradients and associated resonant frequencies: reading spins at discretised frequencies corresponds to a discretised space. Each voxel thus defined can encode a bit (or multiple bits from different spins in a complex molecule).

As described above, the basic ingredients for encoding bits are the frequency, duration/amplitude and phase of r.f. pulses, together with the magnetisation vector in the absence or presence of further r.f. pulses. We need to distinguish two different scenarios: a 'quasi-static' case which, in NMR terms would be referred to as a 'on resonance' condition, and a 'quasi-dynamic' case which, in NMR terms would be called a 'off resonance' case. The former condition permits a time-independent operation

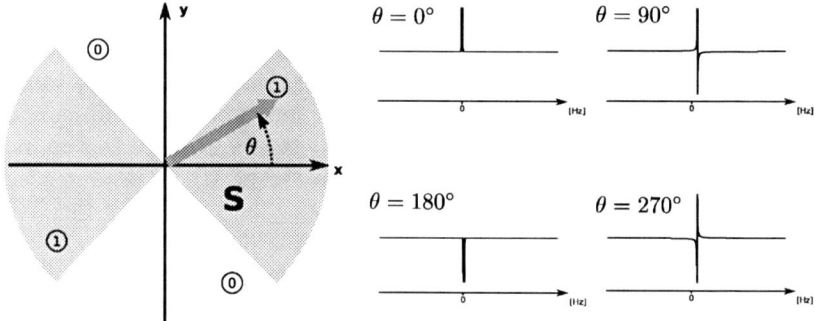

Figure 3. Definition of the '1' and '0' regions in the xy-plane (left) and some spectra corresponding to various different $\pi/2$ preparation r.f. pulses (right).

whereas the latter condition is more complicated in that time becomes an integral part of the logic.

3.1   The on-resonance, quasi-static condition

We start from equilibrium magnetisation (along the $+z$-direction) with a $(\pi/2)$-preparation pulse, for example (see figure 3) a $(\pi/2)_y$-pulse giving $+x$-magnetisation, or a $(\pi/2)_x$-pulse giving $-y$-magnetisation. We define the areas in the xy-plane covering $\pm 45°$ around the $+x$- and $-x$-directions as representing 1, the remaining areas as 0 (see figure 3; this choice of areas will become obvious in the following).

Thus, the phase and the duration/amplitude of the preparation $(\pi/2)$ r.f. pulse permits us to prepare a 1 or a 0 state. Here we have chosen the r.f. frequency such that it is *precisely* on resonance, with the consequence that the magnetisation vectors in the xy-plane, again, appear static ('they stay where we put them', loosely speaking). We can now introduce the second bit (see figure 4): a $(\pi)_{45}$-pulse P (1), or no pulse (0).

Application of the $(\pi)_{45}$-pulse changes the first bit from 1 to 0, whereas in the absence of this pulse the first bit remains in its 1 state. If the presence or absence of the r.f. pulse is controlled by the value of another bit, the result is a CNOT gate on those two bits (see table 1). Currently this requires some computation external to medium itself, to physically control the r.f. pulse; in section 5.3 we also discuss computation intrinsic to the medium.

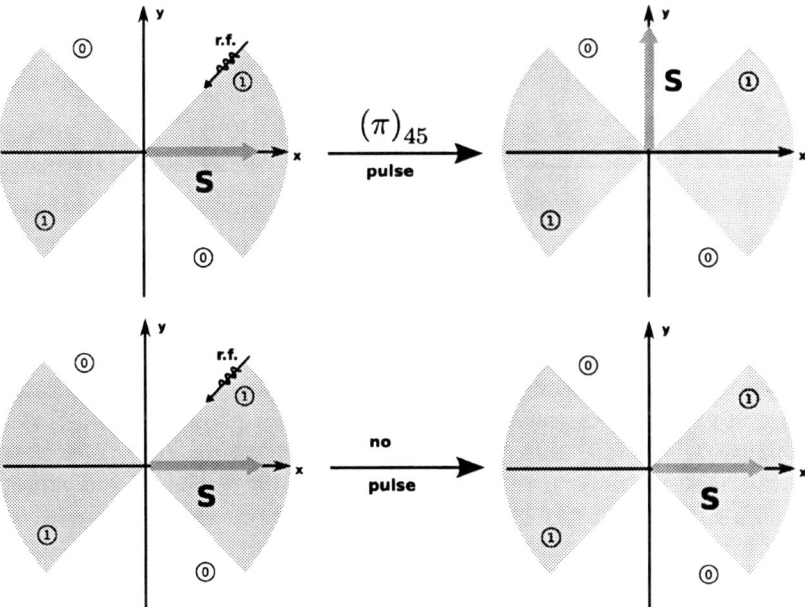

Figure 4. The effect of presence (top) and absence (bottom) of the $(\pi)_{45}$-pulse on the $+x$-magnetisation vector.

pulse (control bit)	target bit (spin)	final spin
off (0)	0	0
off (0)	1	1
on (1)	0	1
on (1)	1	0

Table 1. Components of a CNOT gate.

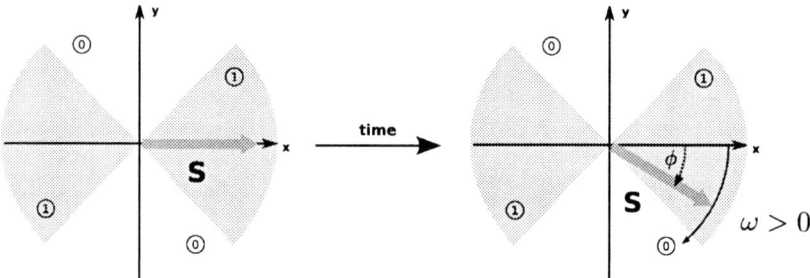

Figure 5. The effect of an off-resonance preparation pulse as f(time).

3.2   The off-resonance, quasi-dynamic condition

We start again with equilibrium magnetisation and a $(\pi/2)$-preparation pulse, flipping the magnetisation vector to the xy-plane, as is illustrated in figure 5. The only difference to the previous on-resonance condition is that now we chose a r.f. pulse not exactly on resonance but strong enough to flip or invert the magnetisation vector. The consequence of this frequency mismatch between Larmor frequency and r.f. frequency is the loss of time-independence in the rotating-frame picture (see figure 5): now the magnetisation vector appears either 'slower' or 'faster' than the reference frame, depending on the choice of r.f. frequency.

The amount of frequency mismatch obviously determines how fast the magnetisation vector rotates in the xy-plane. As a function of time, the magnetisation vector now periodically travels through all the '0' and '1' regions.

If, for example, we record the NMR response after a period τ sufficient for the vector to rotate by $90°$, a previously 1 bit will have become a 0 bit, and vice versa. Hence, this time shift (clock shift) corresponds to a NOT operation. From the bottom part of figure 6 we can further see that at specific points in time we can create exactly those conditions we had previously considered in the on-resonance case, and in full analogy to that situation we can again apply (or omit) $(\pi)_{45}$-pulses in order to selectively negate bits – but not only at these specific points in time.

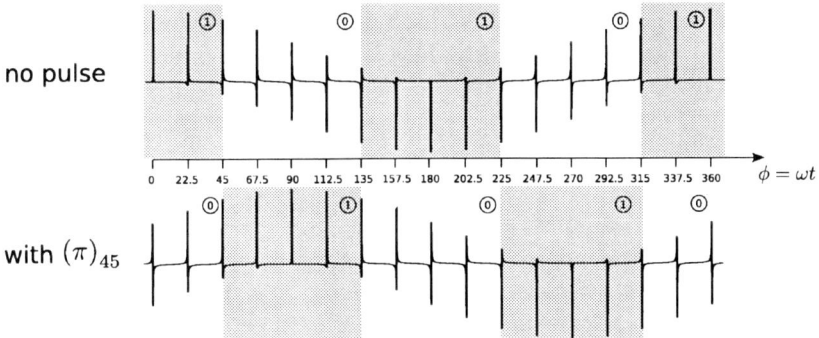

Figure 6. The events in the off-resonance scenario as a f(time) without (top) and with (bottom) application of a $(\pi)_{45}$-pulse.

The off-resonance situation adds an element of choice (time / clock) to the previous on-resonance condition, in addition to a further degree of (potential) complication as the rotation of the magnetisation vector in the xy-plane can be of either positive or negative sign, and the sign is retained after a $(\pi)_{45}$-pulse.

We need to take the off-resonance situation into account for a variety of reasons. We might have a sample with more than one spin species present – in that case, only one spin species can be in the on-resonance condition at a time. Alternatively, we may have a sample with only one spin species present but may alter the resonance frequency in different voxels of the sample by applying field gradients. Or we may be interested in implementing logics which require time dependence of the spin dynamics in a voxel.

3.3 Implementation

Any of the above operations and bits based on phases of pulses and magnetisation vectors are straightforward to implement on contemporary NMR spectrometers; they can be carried out on liquid as well as on solid samples; and the necessity of having isolated spin species available does not impose any severe restrictions on the choice of suitable samples for practical implementations.

4 3D Memory

Conventional solid state memory chips have a limitation: one can extract memory values from the physical edges only, so memory values stored in the middle of the chip need to be moved to the edges before they can be read. This limits i/o bandwidth. This is a well-know limitation of CCD chips, for example.

Mechanisms that can directly address 'internal' memory bits usually involve electromagnetic radiation (such as light), as this requires no wires. For example, the bacteriorhodopsin protein is being developed as a three dimensional optical storage memory [3] [4] [16]. This protein has two stable states that can be set by pulses of different frequency laser light, with a switching time of ~ 0.5 ns. So a protein molecule can store a bit: in practice several thousand are used per bit. The theoretical storage density is also limited by the wavelength of the laser light. The quoted theoretical limit is 10^{18} bits m^{-3}, with practical limits being a few orders of magnitude lower. Each bit can be addressed directly by suitably focussing the lasers. Parallel access to a 2D slice of the memory is achieved by arrays of read/write lasers.

The NMR spin states in voxels could similarly be used as a directly 3D addressable RAM. The molecules involved are smaller and simpler than the bacteriorhodopsin protein. The spins equilibrate on the (long) T_1 timescale, and so the memory would need to be refreshed periodically.

The potential data density is determined by the NMR voxel size, determined by the spectroscopy resolution available. Current resolution is on the order of 1 mm^3 for medical imaging applications (MRI). Localised spectroscopy can achieve resolutions 1 μm^3; current sample sizes of ~ 100mm^3 could store $\sim 10^8$ bits at these resolutions. The total sample size is limited by the volume over which the magnetic field B_0 can be made homogeneous. This resolution implies a data density of 10^{18} bits m^{-3}, comparable to the quoted bacteriorhodopsin figures. Currently it is significantly less than that, of course: the sample that actually stores the data is dwarfed by the vast size of the surrounding superconducting magnet (~ 1m^3) and r.f. pulse generator (~ 1m^3). (Smaller non-cryogenic NMR technologies also exist, such as the portable "NMR mouse" [6], which is moved over a surface to analyse its composition, providing "mobile NMR".)

Memory voxels are written by r.f. pulses, which can be delivered every $1-5$ μs. The number of pulses required to write multiple voxels depends on the complexity of the data being written: very regular patterns of data can be written with a single pulse, more complex patterns require a pulse sequence. The writing time scales with the data complexity (limited by the ability to generate the complex r.f. pulse sequences), not with the sample size (except for the increased power requirements of the pulse). The reading can be performed in a constant time (~ 1 ms) per slice of memory. The read/write r.f. system can be easily modulated to address only parts of the memory, with consequent gains in speed.

5 Computation

Beyond storing data in the medium, we would like to be able to perform computations on that data using the physical properties of the medium. This requires the data encoded in different bits to interact in a data-dependent manner. For interactions between different bits encoded in a single voxel (for example, interactions between spins in a complex molecule) this can be done using the same encodings as NMR quantum computation (see for example [15]), but without quantum entanglement being used to give superpositions of multiple configurations.

We would also like the the data in different voxels to interact in a data-dependent manner. We have seen an example earlier (the CNOT gate), where the value of one bit controlled the application of an r.f. pulse on another bit. That design requires some external decision making (read one bit, and on the basis of its value, choose whether or not to fire the r.f. pulse at another bit.) We would also like an *intrinsic* mechanism for the bits to interact directly within the medium. This is difficult to do in general (hence the dearth of, say, NMR-based quantum cellular automata). However, there is a variety of methods for allowing data in adjacent voxels to interact through non-linear interactions of adjacent phases. These include modulating the logical grid, and physical movement (diffusion) of the molecules.

5.1 Grid modulation

The size and positions of the voxels in the grid can be programmatically changed by modulating the magnetic field.

One can alter the size of the grid cells between timesteps, alternately enlarging the cells (thereby combining the different information contents of several smaller cells) and reducing them (thereby spreading the information across several (smaller) cells (see figure 7).

One can alter the position of the grid between timesteps, changing the offsets. For example, in classical cellular automata (CAs) the Margolus neighbourhood [14] is used to simulate particle dynamics. Here the entire grid is partitioned into neighbourhoods of 2×2 cells. The CA rule is applied to each neighbourhood. Then, on the following timestep, the neighbourhoods are offset by $(1, 1)$, allowing a different set of four cells to interact (see figure 8).

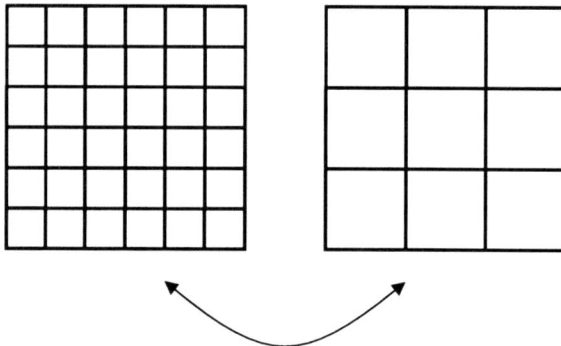

Figure 7. Modulating the voxel cell size between steps

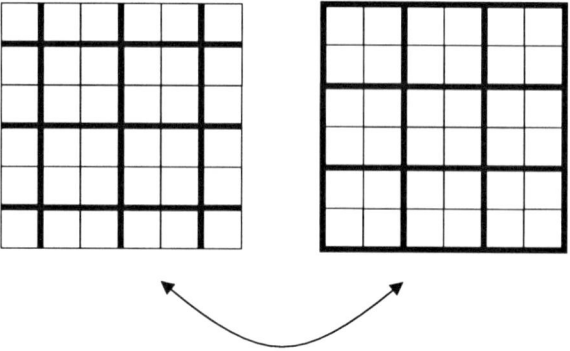

Figure 8. Modulating the voxel cell position between steps

Given each NMR voxel is bulk matter, one could employ an analogous process with half-integer grid movements, allowing corners of adjacent cells in one timestep to become members of the same cell in subsequent timesteps. The resulting bulk spin (and hence bit value) would depend on some combined or average property of the individual spins.

Then combinations of size and position changes can be used to provide programmatic control over information interactions.

5.2 Field gradients as parameters

Varying the gradient of the magnetic field allows vertical slices to experience a different value of the field. This can be used to provide a parameter value, and perform an array of 2D calculations in parallel, with identical computations, but different parameter values.

5.3 Diffusion

In solids, spins of neighbouring molecules interact by dipolar coupling, with the result that the spin state diffuses through the system. Also, ions can physically diffuse through crystal lattices, at a much slower rate than the spin diffusion rate. The resultant bulk spin associated with the voxel is a non-linear combination of the original spin, the diffusing spin states, and the diffusing molecules.

The result is a **programmable 3D reaction-diffusion computer**. The 'reaction' is that of spins in combination, so is of a strictly limited kind, unlike chemical reaction-diffusion (RD) computers, which have a wide range of non-linear chemical reactions available [2]. Chemical RD computers have relatively limited interactivity: they tend to be programmed through the initial distribution of chemical reagents, with some opportunity of adding spots of chemicals as the reaction proceeds, or (usually passively) illuminating patterned areas over the light-sensitive reactions [10] [13]. The NMR 3DRD computer has the potential for detailed programmability: a 'program' consists of a sequence of r.f. pulses; each timestep a programmed r.f. pulse can manipulate each of the bits in the medium. The choice of paths through the program can be a function of the current state of the system (read out in the receiver coil). Clearly, if this function of current state is being calculated out of the medium, in the conventional computer controlling the r.f. pulse program, it needs to be a relatively simple calculation, such as a property of a few voxels, or certain bulk properties of the entire medium. The computation performed is an emergent property of all these actions. See figure 9.

One particular use of the programmability would be to isolate certain volumes by the use of continually zeroed voxels while part of a

program steps (RF pulses)

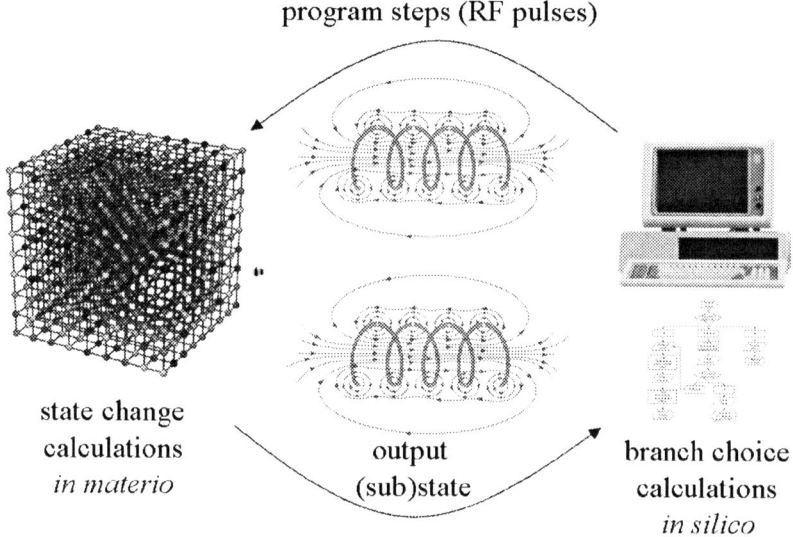

state change
calculations
in materio

output
(sub)state

branch choice
calculations
in silico

Figure 9. Schematic architecture of the programmable 3D NMR reaction-diffusion computer

computation was being performed, then reprogramming these voxels to allow interactions to occur between the regions in another part of the computation.

6   Variations on a theme

6.1   Other materials

So far we have restricted most of the discussion to isotropic liquids with simple spin states (for example, the ^{1}H spins in water), and simple solids.

Molecules with more complex spin states exist, particularly the ones used for quantum NMR computing, where the different spin states per molecule are used to encode different qubits. However, even in a relatively simple two-spin state molecule, the Larmor frequencies are different, and the spins interact in a complex non-linear manner that could potentially be exploited for computation.

Inhomogeneous materials with complex structures, for example, layers with different properties, can be constructed, with gels or solids. These can have anisotropic spin diffusion rates, that again could be exploited for different kinds of computation.

6.2   Dynamically programmable matter

The consideration of complex materials leads on to the possibility of dynamically programmable matter. One requires a material where the the spin state changes some interesting properties of the material. Clearly it changes the r.f. properties, as this is what is used to read out the current state: there may be other properties that are also usefully changed.

One could then program the medium to have the desired properties to perform some task, with those properties being a function of time. That is, the matter could be programmed to react, or adapt, to changing circumstances. For example, consider adaptive optics. These are used in astronomy to remove the twinkling caused by atmospheric fluctuations, which otherwise produce blurred images from ground-based observations. This is achieved by computer controlled flexible mirrors, or variable refraction (for an introduction, see for example [12]). The longer the wavelength, the easier this is to do, so the r.f. domain may be an ideal application.

Other programmable matter applications will become apparent as the capability of the NMR computer, particularly with complex materials, is explored further.

6.3   Breaking the model

One of the major advantages of the proposed approach is the availability of commercial off the shelf NMR spectrometers, and the associated laboratory expertise. However, some of the design properties of those NMR spectrometers are to enable conventional NMR spectroscopy experiments to be performed effectively in a homogeneous environment. These design constraints might not be suitable for moving to full NMR computers that could exploit information processing capabilities of as inhomogeneous environment, such as inhomogeneous magnetic fields (there is a whole sub-branch of NMR dealing with inhomogeneous magnetic fields, applied to, for example, oil exploration), or space and time dependent temperature fluctuations in the medium. Much of the time, effort, and expense, of current instruments goes into controlling these properties: if NMR computers can operate (and operate more effectively) in less homogeneous environments, then there is even greater potential for smaller, lower cost devices.

In our initial proposal, we are looking at discretised 3D space (voxels) and time (Larmor period clocking). However, we will also need to look at the possibilities offered by considering the medium to be a continuum existing in continuous time. This could potentially offer a simpler design of the r.f. pulse structure. However, all the usual problems of classical analogue computing will also become apparent, particularly the sensitivity to noise. So initially we are concentrating on the discretised spatial and temporal domain.

7   Next Steps

It is important to recognise that we are not proposing the NMR computer as a *universal* computer in the Turing sense. We want to understand what the physics of NMR offers computation: what the computation physical system can do in a 'natural' manner. We want to exploit its physical properties, not to shoe-horn it into an unnatural (for it) computational paradigm. Hence we have *not* done the usual first step in a novel computational domain of demonstrating how the device can be made to emulate a variety of logic gates. The exception here is the NOT gate implementation (section 3), since there are very natural ways to achieve this in the NMR setting.

Given that we are not striving for universality, what is needed is a classification of the computational capabilities of an NMR computer, and the application areas for which it is best suited.

NMR is an incredibly rich domain. We propose to start simply, with the following prototypes in simple media:

1. prototype an implementation of a 3D memory in water, and measure:
   - the memory capacity (achievable spatial resolution)
   - the write rate (r.f. pulse programmability, complexity and frequency)
   - the read rate (read cycles, exploitation of phase)
   - the required refresh rate (relaxation timescale)
   - the error rates
2. prototype an implementation of a programmable 3D reaction-diffusion computer in a simple solid, and measure:
   - the spin diffusion rates in various media
   - the ion diffusion rates in various media
   - the logical operations implemented by these diffusions, with different grid modulation options
   - the constraints on the computation in the controlling r.f. pulse generator
   - the computational tradeoffs between the restricted spin reaction types, and the flexible data programmability
3. prototype an implementation of a hybrid NMR-silicon computer, balancing the computation in the medium, and in the external computer, exploiting the strengths of each approach

4. investigate suitable programming styles to encompass these various uses (including the use of evolutionary algorithms to program an unfamiliar and complex device)
5. prototype a range of applications to help determine computational and application classes (based on existing chemical RD examples [2], such as Voronoi diagram calculation, and on amorphous computing [1] applications, and how these can be extended to exploit the 3D and programmability capabilities of the NMR computer)

8 Conclusions

We have described how a commercially available NMR spectrometer could be used to perform (non-quantum) computation: from addressable 3D memory, to a programmable 3D reaction-diffusion computer. We make no claims as to its Turing universality, or otherwise: rather we seek to exploit the computational abilities naturally occurring in interacting (but not necessarily entangled) switchable protons spins in bulk media.

Acknowledgements

Financial support of our work (MB and AS) by the Deutsche Forschungsgemeinschaft is gratefully acknowledged.

References

[1] Harold Abelson, Don Allen, Daniel Coore, Chris Hanson, George Homsy, Thomas F. Knight, Radhika Nagpal, Erik Rauch, Gerald J. Sussman, and Ron Weiss. Amorphous computing. *CACM*, 43(5):74–82, 2001.

[2] Andrew Adamatzky, Benjamin De Lacy Costello, and Tetsuya Asai. *Reaction-Diffusion Computers*. Elsevier, 2005.

[3] Robert R. Burge. Protein-based optical computing and memories. *Computer*, 25(11):56–67, November 1992.

[4] Robert R. Burge. Protein-based computers. *Scientific American*, 272(3):66–71, March 1995.

[5] D. G. Cory, A. F. Fahmy, and T. F. Havel. Ensemble quantum computing by NMR spectroscopy. *Proc. Natl. Acad. Sci. USA*, 94:1634–1639, 1997.

[6] G. Eidmann, R. Savelsberg, P. Blümler, and B. Blümich. The NMR MOUSE®: A mobile universal surface explorer. *J. Magn. Reson. A*, 122:104–109, 1996.

[7] N. A. Gershenfeld and I. L. Chuang. Bulk spin-resonance quantum computation. *Science*, 275:350–356, 1997.

[8] C. Kehlet, T. Vosegaard, N. Khaneja, S. J. Glaser, and N. C. Nielsen. Low-power dipolar recoupling in solid-state NMR developed using optimal control theory. *Chem. Phys. Lett.*, 414:204–209, 2005.

[9] K. Kobzar, T. Skinner, N. Khaneja, S. J. Glaser, and B. Luy. Exploring the limits of broadband excitation and inversion pulses. *J. Magn. Reson.*, 170:236–243, 2004.

[10] L. Kuhnert, K. I. Agladze, and V. I. Krinsky. Image processing using light-sensitive chemical waves. *Nature*, 337:244–247, 1989.

[11] Malcolm H. Levitt. *Spin Dynamics: Basics of Nuclear Magnetic Resonance*. Wiley, 2001.

[12] Jason Porter, Hope Queener, Julianna Lin, and Karen Thorn. *Adaptive Optics for Vision Science: Principles, Practices, Design and Applications*. Wiley, 2006.

[13] N. G. Rambidi and D. Yakovenchuk. Chemical reaction-diffusion implementation of finding the shortest paths in a labyrinth. *Phys.Rev.E*, 63:026607, 2001.

[14] Tommaso Toffoli and Norman H. Margolus. *Cellular Automata Machines: a new environment for modeling*. MIT Press, 1987.

[15] Lieven M. K. Vandersypen, Costantino S. Yannoni, and Isaac L. Chuang. Liquid state NMR quantum computing. In David M. Grant and Robin K. Harris, editors, *Encyclopedia of Nuclear Magnetic Resonance, Volume 9: Advances in NMR*, pages 687–697. Wiley, 2002.

[16] B. Xi, Kevin J. Wise, Jeffrey A. Stuart, and Robert R. Birge. Bacteriorhodopsin-based 3D optical memory. In V. Renugopalakrishnan and Randolph V. Lewis, editors, *Bionanotechnology: Proteins to Nanodevices*, pages 39–59. Springer, 2006.

Implementation of logical operations on a domino substrate

Simon O'Keefe

Department of Computer Science, University of York, York, UK YO10 5DD
sok@cs.york.ac.uk

Abstract This paper presents some simple ideas about domino gates. Domino gates are implementations of logical operations using toppling dominos to represent the movement of information (bits). The actively toppling domino represents a logic one, and absence of toppling represents logic zero. The domino model can be used as a representation of single-shot wave gates in unconstrained media, such as sub-excitable Belousov-Zhabotinsky reactors. On a practical note, they are slightly easier to set up and use than collision logic models using 'billiard balls'.

1 Introduction

Recent developments in the manipulation of chemical waves have introduced the possibility of computation performed with chemical 'wavelets', in which a chemical reaction proceeds at a specific point in a thin-layer reactor rather than everywhere at once. For a discussion of these reaction-diffusion computers, see [2]. The reactor typically contains a Belousov-Zhabotinsky mixture [3]. The locus of the reaction is in the form of a wave front that spreads through the reactor, visible as a change in colour as the catalyst changes state. In the most recent research [1, 4], the wavelets are dynamic but limited in extent – they move about the reactor as the reaction progresses, but they do not spread out as a 'normal' wave does. Using these wavelets, we can adopt a dynamic or 'collision based' approach to computing using mobile patterns.

These patterns are "self-localisations" [1] of excitation in the medium, and they represent quanta of information. These quanta travel in space and their interactions may be interpreted as logical operations. Given the right conditions, they do not pass through each other, and unlike trigger waves in excitable media they do not always annihilate when they meet each other. Instead, they sometimes act more like physical objects, and can be made to bounce off each other like colliding balls.

In this system, truth values are represented by the presence or absence of travelling quanta. Because they are localised (they do not spread out) they do not require canalisation of the medium to control their propagation. The system is thus said to be 'architecture-free' – there is no predetermined architecture, a trajectory of a wave fragment is an instantaneous wire.

The gates are instantaneously realised by the collision of two wave fragments. A typical gate would have two inputs (wave trajectories) and at least three outputs – the two original trajectories, which are the paths along which either wave will continue in the absence of the other, and the trajectory or trajectories taken by the waves as the result of the collision between waves.

The basis for architecture-free computational systems lies in physical models that describe the behaviour of large number of particles. With the right interpretation, these may be seen as models of computation, based on elastic collisions between hard spheres - Fredkin's Billiard Ball Model [5, 6].

The Billiard Ball Model consists of hard spheres moving about in space (for simplicity, on a plane). If the centre of a sphere is at a point in space at a given time, we say there is a 1 there, otherwise a there is a 0. Thus the 1's move from place to place, as the billiard balls are moved around, but the number of 1's is conserved.

The key insight is that "every place where a collision of finite-diameter hard spheres might occur can be viewed as a boolean logic gate" [5]. So the loci of potential collisions forms a network of gates. From the point of view of the ball, the path a ball takes depends on whether it hits anything. In other words, it makes a decision about its future trajectory based on the presence or absence of another bit of information.

To make this feasible, we need to constrain the position and the momentum of the balls. All initial and final positions are on a grid (with the grid spacing related to the diameter of the balls), and colliding balls have perpendicular momentum vectors. These constraints make actual implementation of a billiard ball computer problematic.

We introduce here a physical system that has combines the physical simplicity of the billiard ball model with the wave-based computational approach of the chemical system – the *domino race* or *domino gate*.

Domino gates are implementations of logical operations using toppling dominos to represent the movement of information (bits). The actively toppling domino represents a logic one, and absence of toppling represents logic zero. The domino model can be used as a representation of single-shot wave gates in unconstrained media, such as sub-excitable Belousov-Zhabotinsky reactors.

Unlike the billiard ball model, which has a strict physical basis, the domino model is based on a relatively weak analogy between chemical-wave based systems and propgating disturbance in an arrangement of balanced dominoes. However, they are slightly easier to set up and use than collision logic models using 'billiard' balls.

2   Simple gates

The movement of information around the circuit is represented by the toppling of the dominos. The inputs to the ciruit are external forces that cause the dominos to start toppling at specific points in the circuit. If a particular input to the circuit is a '1', then the exposed domino on that input is toppled. If the input is a '0', then it is not toppled. The notion of timing is therefore essential, as it is possible for the circuit to produce no change in its output elements, and therefore the time at which the change (or lack of it) is to be observed must be known. This may be challenging for circuits in general, as the signal propagation times along different paths may vary greatly.

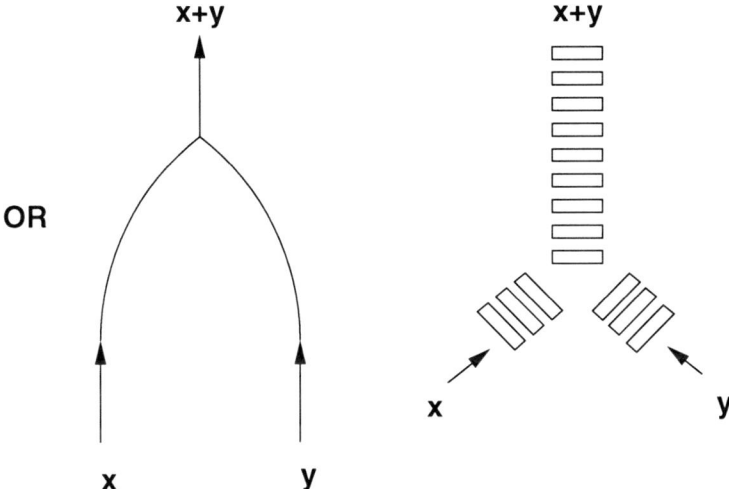

Figure 1. Implementation of the domino OR logic gate.

The behaviour of the circuits is best illustrated with an example. In figure 1 an OR gate is shown. The two inputs are on the bottom of the diagram and the output is at the top. If a domino is toppled at either of the inputs, then the domino at the output will topple. If both inputs are '1', the output is still '1', and if neither input is '1' then the output is zero (it does not topple).

Implementation of simple, single gates using dominos is straightforward as the interactions between the inputs may be considered only between adjacent lines. Figure 2 illustrates schematically the layouts required for the NOT, OR, NOR and XOR gates (an implementation for AND is discussed in the next section).

The NOT gate requires an auxilliary signal (indicated by the 1 input). This auxilliary signal is propagated if, and only if, the input x is zero (no signal propagated). If the input is a '1', the toppling domino line follows the curve around and topples dominos in the line from the auxilliary input. Since the input line is curved around, the toppling is not propagated towards the output. This illustrates the nature of the relative timings of signals required. Because a domino system does not typically recover after a signal has been propagated, the effects of a signal at a gate may be effectively latched. Thus in the case of the NOT, the auxilliary input may be propagated at any time after the actual input, so long as there is sufficient delay to ensure that the input reaches the point where the two input lines are joined.

The NOR gate is a simple extension of the NOT gate. The output is only '1' if none of the inputs have been a '1', since if any of the inputs is a '1', the toppling dominos block the line for the auxilliary input.

The XOR gate is slightly more interesting. A single input '1' must pass through to the output, but if both inputs are '1' the output must remain at '0'. A convenient way to arrange this is to have the two inputs interfere with each other, by directing the input pulses in opposite directions along the centre section joining the two inputs.

3   A half-adder circuit

The implementation of a half-adder circuit is more complex than the simple gates discussed in the preceding section. The main complication is that the usual left-to-right or bottom-to-top layout for circuit elements requires the signals to physically cross. This can be achieved in a domino circuit by using bridges to carry one line of dominos over another line.

Give two inputs x and y, the outputs from a half-adder circuit are $x \oplus y$ and $x.y$, where $\oplus$ means 'exclusive OR'. The circuit for $x \oplus y$ has been given in the preceding section. The circuit for $x.y$ requires some

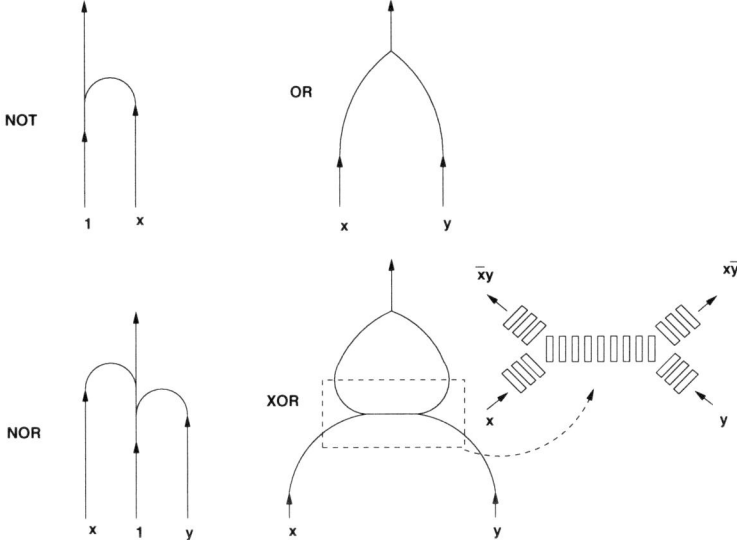

Figure 2. Simple implementations of other domino logic gates.

work. Firstly, the inputs x and y need to be duplicated, and carried outside the layout for the XOR ciruit. This involves the use of one bridge, if we going to produce a layout which has inputs and outputs that are exposed and could be chained with other circuits. Secondly, we need to calculate $x.y$ given the inputs. One way to do this is to use the identity $x.y = \overline{\overline{x} + \overline{y}}$. If we negate the inputs x and y, we can then use an OR gate followed by a further negation to compute the AND. An outline of a circuit to compute the half-adder outputs is shown in figure 3. It should be noted that the three auxilliary inputs could be produced from a single input. It should be appreciated that the timing constraints in the circuit are crucial – x and y must be input within a short interval so as to arrive at the centre of the XOR gate at the same time, and they must also arrive at the NOT gates on the auxilliary inputs before the auxilliary inputs do. The third auxilliary input that drives the final AND output must be timed to arrive at the final inverter later than the inverter's control input.

Figure 4 shows the implementation of the circuit using wooden dominos 43mm x 22mm x 7mm. Spacing between dominos along straight runs is approximately 40mm, but spacings are smaller where runs are curved. The time between applying inputs and observing outputs is of the order of 5 seconds. The left picture in figure 4 shows the initialised circuit, the

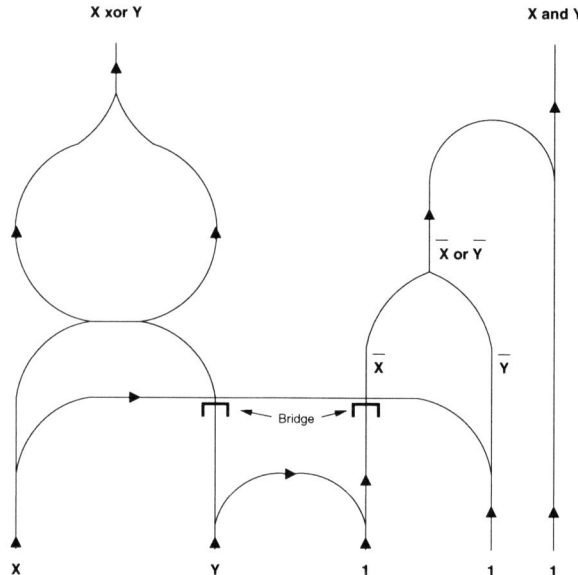

Figure 3. Sketch of the implementation of a domino half adder circuit.

right picture shows the circuit after applying inputs $x = 1$ and $y = 1$. Reset time before reuse of the circuit is of the order of 200 seconds.

4   Alternative implementation

Initial experiments have been conducted with varous types of dominos, composed of plastic and of wood. The results of this macroscopic implementation are discussed in preceding sections.

An alternative to the macroscopic implementation is to engineer 'domino' circuits at the nanoscale. The primary difficulty for nanoscale use is the initialisation of the circuit elements – getting all the dominos to stand up. If we assume the existence of nanobots capable of contructing the initial domino layout on a nanoscale-smooth surface, by the placement of anchored or attached nano-dominos, then initialisation of the circuit becomes the relatively straightforward task of using a nanobot to crawl the circuit and 'energise' each element individually.

More challenging still is the construction of nanoscale elements which may be reused within the circuit without resetting the whole circuit. In other words, we would like the circuit elements to reset themselves automatically. One possibility is to use an arrangement similar to a relaxation oscillator in reverse – when a domino topples, it makes a contact

 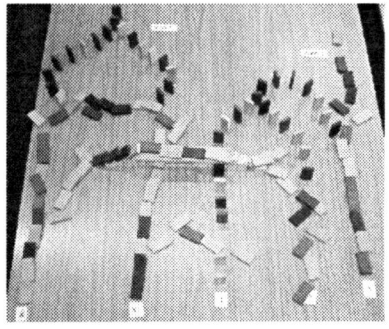

(a)   (b)

Figure 4. Implementation of a domino half adder circuit. *(a)* The initialised circuit. Inputs are along the bottom of the image, in the order, x, y, auxilliary inputs. Outputs are at the top of the image, $x \oplus y$ on the left and $x.y$ on the right. *(b)* The circuit after applying inputs $x = 1$ and $y = 1$. The output show $x \oplus y = 0$ and $x.y = 1$, which is the correct output.

and a sudden charging of a capacitive element places enough positive charge to attract the negatively-charged base of the domino and pull it upright. Once upright, the contact is broken and the charge then leaks through a resistance to ground. A small region immediately below the base of the domino is insulated from the main capacitive element and normally remains charged. This provides sufficient attraction to make the domino metastable – small disturbances to the physical equilibrium are insufficient to topple the domino, but larger disturbances move its base sufficiently far away from the charged element that gravity may take over. Figure 5 shows schematically a row of four dominos and the arrangement of the charging circuit via the domino hinge and the discharge circuit. The geometry is fairly crucial – each domino must have sufficient leeway to fall and topple the next in line, but then come to rest closing the charging circuit.

5 Conclusion

As we can implement a universal set using dominoes (since we have implemented a NAND gate as part of the half-adder) we can in theory produce a universal computer. The fact that the gates are single use (unless the dominoes are 'energised' again by standing them upright) should not matter as we can duplicate the gates as many times as necessary.

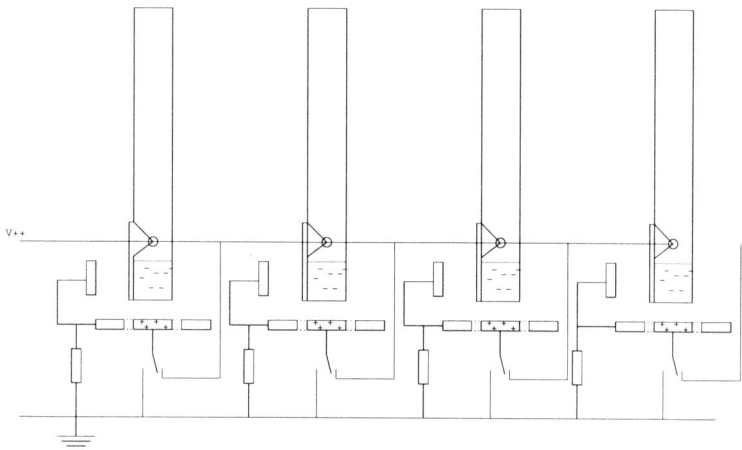

Figure 5. Simple implementations of domino logic gates.

A more serious constraint is due to the geometry. If the implementation is restricted to the (pseudo) two-dimensional case, in which we do not have bridges to allow one line of dominos to pass over another line, then the passage of a wave along a line of dominos creates a barrier of 'de-energised' dominos across which we cannot pass information, so the movement of information must be constructed very carefully. Alternatively we can allow the implementation to protrude into the third dimension by the use of bridges which allow one line of dominos to cross another.

6   Acknowledgement

Thanks are due to Susan Stepney for useful discussion about nano-scale implementation.

References

[1] A. Adamatzky.   Collision-based computing in Belousov-Zhabotinsky medium. *Chaos, Solitons and Fractals*, 21:1259–1264, 2004.

[2] B. Adamatzky, A. and de Lacy Costello and T Asai. *Reaction-diffusion computers*. Elsevier, Amsterdam, 2005.

[3] R. Aliev. Mechanism of the BZ reaction [online], [Accessed April 2006]. Available at: http://people.musc.edu/ alievr/BZ/BZexplain.html.

[4] B. de Lacy Costello and A. Adamatzky. Experimental implementation of collision-based gates in Belousov-Zhabotinsky medium. *Chaos, Solitons and Fractals*, 25:535–544, 2005.

[5] F. Fredkin and T. Toffoli. Conservative logic. *Int J Theor Phys*, 21:219–253, 1982.

[6] N. Margolus. Physics-like models of computation. *Physica D*, 10:81–95, 1984.

A kinematic Turing machine

William M. Stevens

Department of Physics and Astronomy, Open University, Walton Hall,
Milton Keynes, MK7 6AA, UK
william@stevens93.fsnet.co.uk,
WWW home page: http://www.srm.org.uk

Abstract A Turing Machine in a three dimensional discrete
space environment containing movable cubic parts is described.
All of the cubic parts are identical in function. The only function
that a part performs is to move a neighbouring part by one
unit. Parts can be connected to neighbouring parts. When one
part moves, parts that it is connected to also move. An example
program for the Turing Machine is given.

1 Introduction

The earliest computers were mechanical. Babbage's analytical engine [2]
consisted of sets of cogged wheels which stored decimal numbers and
which could be engaged by a mechanism able to perform arithmetical
operations on the numbers represented by the angular position of the
wheels. Züse used interlocking sliding bars to implement boolean operations in his mechanical computers [14].

Recently, there has been a revival of interest in mechanical computing, partly out of a desire to explore the relationship between computation and physics, and partly because it may be possible to build
nanoscale computing devices using mechanical logic. Fredkin and Toffoli
described a model of computation based on elastic collisions between
identical balls [4]. Merkle described a model based on sliding, interlocking rods [7].

This paper is motivated by a desire to understand what kinds of mechanical interactions are required in mechanical computers. An attempt
is made to model the kinematic behaviour of a simple system without
modelling any mechanism that gives rise to the kinematic behaviour in
question. The model used can be classified as a 'kinematic automaton'.
It is based in a 3D cellular space, where each cell can contain a cubic
part. Collections of connected parts can move together like a rigid body.
From such collections, mechanisms can be made.

A Turing Machine can be implemented in this system, and is presented in this paper. The Turing Machine presented has some similarities to that described by Laing in [6]. The most notable similarity is the way in which state transitions are implemented. In a state diagram, state transitions can be represented as arrows that lead from one state to another. In [6] and in the system described here, state transitions are embodied as paths that run from one state to another that get followed by the Turing Machine.

All mechanisms are made from identical component parts. This scheme arose during research into self-replication and automatic construction. If one is interested in simplifying the process of automatically constructing a computer from a set of component parts, then designing that computer using a single type of component part is advantageous. Making a computer that operates on the same type of component from which it is made is of interest from the perspective of self-replication because a self-replicating machine must be able to process the same kinds of parts that it is built from.

It is advised that this paper be read in conjunction with the supplementary material that is available for simulating the Turing Machine, as described in section 7.

2   A kinematic simulation environment

A three dimensional discrete space simulation environment that supports moveable cubic parts is used in this paper. The system is called CBlocks3D. A part called the 'SlideOn' part is used to implement a computing scheme. The SlideOn part is shown in figure 1.

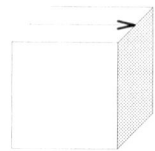

Figure 1. The SlideOn part.

The SlideOn part will move any part positioned above it one unit in the direction that the arrow is pointing. The term 'active face' is used to refer to the face of the part that has the arrow on it.

A universe evolves through a series of states. The deterministic rule described in subsection 2.2 determines how a universe evolves from one state to another as parts act upon other parts, giving rise to movement.

2.1   Describing a state of the universe

We define six direction vectors

$$EAST = (1,0,0) \quad WEST = (-1,0,0)$$
$$NORTH = (0,1,0) \quad SOUTH = (0,-1,0)$$
$$FRONT = (0,0,1) \quad BACK = (0,0,-1)$$

Let D denote the set of these vectors

$$D = \{EAST, WEST, NORTH, SOUTH, FRONT, BACK\}$$

The primary axis of a part is used to refer to a vector that lies on the line from the centre of the part to the centre of the active face of the part. The secondary axis of a part is the direction in which the arrow on the active face points. E.g., in figure 1 the primary axis points up the page ($NORTH$) and the secondary axis points to the right of the page ($EAST$).

We define the function

$$\text{opposite}((x, y, z)) = (-x, -y, -z)$$

The 3-tuple $((x, y, z), A, B)$ completely describes a part.
Here

$$x, y, z \in \mathbb{Z}$$
$$A \in D$$
$$B \in D \setminus \{A, \text{opposite}(A)\}$$

(x, y, z) is a position vector that specifies the location of a part.

The orientation of a part is described by specifying the direction of the main axis A and the direction of the secondary axis B. Since the main axis and the secondary axis are perpendicular, $A \cdot B = 0$.

The following defines what is meant when a direction vector V is added to a part p. If $p = ((x, y, z), A, B)$ and $V = (x', y', z') \in D$ then $p + V = ((x + x', y + y', z + z'), A, B)$.

The following notation is used to refer to elements in a 3-tuple p,

$p.position$ is the position vector (x, y, z) of p.
$p.main$ is the direction of the main axis of p.
$p.secondary$ is the direction of the secondary axis of p.

We define the neighbour relation $\|$ for parts p_1, p_2

$$p_1 \parallel p_2 \text{ if and only if } p_1.position - p_2.position \in D$$

And the joined relation $\bowtie_C$ for parts p_1, p_2 in a set of parts C

$$p_1 \bowtie_C p_2 \text{ if and only if } p_1 \parallel p_2$$
$$\text{or there exists } p_3 \in C \text{ such that}$$
$$p_1 \parallel p_3 \text{ and } p_3 \bowtie_C p_2$$

It is also useful to have a neighbour predicate which is true if p_2 is a neighbour of p_1 in direction A

$$\text{neighbour}(p_1, p_2, A) \text{ if and only if } p_2.position - p_1.position = A$$

We wish the CBlocks3D system to support collections of joined parts that move together when any one of them is pushed. Such a collection is called a 'construct' and a construct C must satisfy the following conditions. No two parts can have the same coordinates

$$\text{for all } p_1, p_2 \in C, \text{ if } p_1 \neq p_2 \text{ then } p_1.position \neq p_2.position \quad (1)$$

Any pair of parts in C must be joined

$$\text{for all } p_1, p_2 \in C, p_1 \bowtie_C p_2 \quad (2)$$

The active face of a part p_1 in a construct C must not lie against another part p_2 in C

$$\text{for all } p_1 \in C \text{ there does not exist } p_2 \in C$$
$$\text{such that } p_2.position = p_1.position + p_1.main \quad (3)$$

We can define a neighbour relation and a neighbour predicate for constructs C and E based on the neighbour relation and predicate for parts.

$$C \parallel E \text{ if and only if there exists } p_1 \in C, p_2 \in E \text{ such that } p_1 \parallel p_2$$

$$\text{neighbour}(C, E, A) \text{ if and only if there exists}$$
$$p_1 \in C, p_2 \in E \text{ such that } \text{neighbour}(p_1, p_2, A)$$

For constructs, it is also useful to have a joined predicate for constructs C and E in a set of constructs S

$$\text{joined}_S(C, E, A) \text{ if and only if } \text{neighbour}(C, E, A) \text{ or there exists}$$
$$F \in S \text{ such that } \text{neighbour}(C, F, A) \text{ and } \text{joined}_S(F, E, A)$$

A state S is a set of constructs that satisfies the following conditions. Constructs may not overlap.

$$\text{for all } C, E \in S, \text{ if } C \neq E \text{ then there do not exist } p_1 \in C, p_2 \in E \\ \text{such that } p_1.position = p_2.position \tag{4}$$

If the active faces of two or more parts lie against parts in a construct C, the secondary axis of those parts must point in the same direction. This ensures that no attempt is made to slide a construct in two different directions at once.

$$\text{for all } p_1 \in E, p_2 \in F, p_3 \in C, p_4 \in C, \\ p_1.main = p_3.position - p_1.position \text{ and} \\ p_2.main = p_4.position - p_2.position \text{ implies} \\ p_1.secondary = p_2.secondary \tag{5}$$

Note that in (5) it is possible that $E = F$.

2.2 Evolution of a universe

A universe evolves through an infinite sequence of states $S_0, S_1, S_2, \ldots$

S_0 completely determines subsequent states. For $n \geq 0$, S_{n+1} can be determined from S_n using the the algorithm given below. The algorithm can be subdivided into three steps:

Step 1 Work out whether the active face of a part in one construct C comes against a part in another construct E. If so, E is to be moved.

Step 2 Work out whether any constructs will be pushed by E when E moves.

Step 3 Move every construct that is to be moved.

A formal description of this algorithm follows.

Associated with every construct C in S_n is a movement set $M_{C,n} \subseteq D$

Step 1
For each construct E in S_n
    For each part p_1 in E
        If there exists p_2 in a construct $C \in S_n$ such that
        $p_2.position - p_1.position = p_1.main$ then $p_1.secondary \in M_{C,n}$

Step 2
For each construct C in S_n
  If $M_{C,n}$ contains an member A then
    for all constructs D satisfying joined(C, D, A), $A \in M_{D,n}$

It is illegal for any $M_{C,n}$ to have more than one member. This situation would arise if (5) were violated, and also if an attempt were made to displace a construct in two different directions.

Step 3
For each construct C in S_n
  If $M_{C,n}$ is empty, then $C \in S_{n+1}$
    otherwise, if $M_{C,n}$ has a member A then $C + A \in S_{n+1}$

The result of the operation of adding the vector $A \in D$ to the construct C in step 3 is a construct defined by:

$$p \in C \text{ if and only if } p + A \in C + A \qquad (6)$$

It is illegal for (4) to be violated at this point in the algorithm. It is also illegal to have $|S_{n+1}| < |S_n|$. (Without the latter constraint, two identically shaped constructs could end up occupying the same coordinates).

  The 'illegal' conditions described above exist to prevent having to deal with conflict situations that would arise for example, if an attempt were made to slide a construct in two different directions at the same time or if an attempt were made to move two parts into the same location. Of course, rules could be written to resolve such conflict situations (as Arbib does in [1]), but since such situations do not arise in any of the mechanisms described in this paper, and since a description of these conflict resolution rules would be lengthy, we simply define these situations as illegal and rule that no mechanism should depend upon illegal states.

  Note that the Turing Machine described later in this paper does not make use of the ability of constructs to shunt neighbouring constructs along (i.e. Step 2 above).

2.3 Relationship with cellular automata environments

Cellular automata have been widely studied and several results regarding the computational capabilities of simple CA are known. Rendell showed that Conway's Game of Life is computation-universal [9]. Cook showed that 1D cellular automaton rule 110 can be used to implement a cyclic tag system [3].

Some kinematic automata, such as that described in [10], have a direct cellular automata implementation. CBlocks3D does not because of the possibility of connecting parts together so that they move together when pushed.

In common with cellular automata environments, CBlocks3D is based in a discrete space, divided up into identical cells.

The differences between CBlocks3D and cellular automata are listed below.

- In CBlocks3D all parts are identical and do not change with time. In CA cells can be in one of a number of states. The state of a cell can change with time.
- In CBlocks3D parts can move to neighbouring locations. In CA cells are fixed in position. Motion can be simulated by shifting a state from one cell to a neighbour.
- In CBlocks3D a part is directly affected by neighbouring parts, and can also be affected by parts that it is joined to, no matter how far away. In CA, a cell is only directly affected by the states of a finite and predetermined set of its neighbours.
- In CBlocks3D the number of parts is conserved. In CA the number of cells in a particular state may vary.

Cellular automata are excellent tools for studying the logical behaviour of a system and for exploring geometric constraints imposed by particular spaces, but are not the right tool for studying the interactions of rigid structures. The environment described here provides an extension to the cellular automata model that allows the computational properties of interacting rigid structures to be studied. A similar environment was used by Goel and Thompson in [5] to study biological self-assembly.

It is possible to imagine environments that combine the properties of kinematic automata with those of cellular automata. For example, parts in a kinematic automata could be in any one of a number of states and the state of a part could be affected by the states of neighbouring parts. A hybrid environment of this kind is used in [11], [1]. Such systems can be usefully classified as 'kinematic cellular automata'. (A term first used by Toth-Fejel in [12]).

3   Notation and Petri net diagrams

Using the definitions given in the previous section it would be possible to use set notation to completely describe the states that a system goes through as it evolves. Such a description would give little insight into the behaviour of a system and would be a cumbersome tool to use as a means of proving a system's capabilities. In this section, a notation is introduced that can be used to describe the behaviour of a system at a higher level of abstraction than that used in the previous section.

Figures 2 and 3 show successive states S_n and S_{n+1} of a universe.

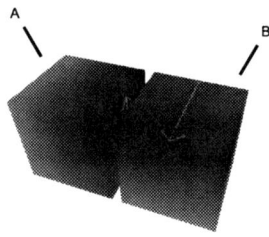

Figure 2. A simple mechanism in state S_n. Constructs A and B are labelled.

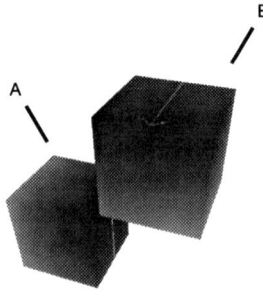

Figure 3. A simple mechanism one time unit later in state S_{n+1}.

Figure 4 is a state diagram describing the system. S_n and S_{n+1} are represented by circles. The transition between S_n and S_{n+1} is represented by an arrow, labelled with the interaction between constructs that causes the state to change. In this case, the interaction is $A, B \rightarrow B[n]$. Meaning 'An interaction between constructs A and B causes B to move North'. The whole diagram means 'When the state of the system is S_n, an interaction between A and B causes B to move North, and the system ends up in state S_{n+1}.

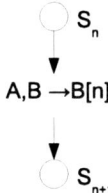

Figure 4. State diagram for a simple mechanism.

For systems containing several interacting mechanisms, a method for describing interacting state diagrams is required. Petri Nets [8] are suitable for this purpose.

Consider the interacting mechanisms shown in figure 5. The behaviour of the mechanism labelled CYC can be described by the state diagram shown in figure 6. This mechanism is not influenced in any way by its neighbouring mechanism, FLIP. On the other hand, construct D in mechanism FLIP is influenced by construct B in mechanism CYC. A Petri Net for the complete system is shown in figure 7. The Petri Net can be thought of as being two state diagrams joined together, with dots to indicate the current state in each diagram. Dots move from one state to another via a transition whenever all arrows leading to a transition start from states with dots in. For more about Petri Nets see [13].

4 Turing Machines — definitions

The following formal description of a Turing Machine is used.

Set of symbols $\Sigma = \{0, 1\}$
Set of states Q
Transition function $\delta : Q \times \Sigma \to Q \times \Sigma \times \{-1, 1\}$
Starting state q_0
Halting state q_h

The state of a Turing Machine at a particular step n in its evolution is called a configuration C_n, defined as follows:

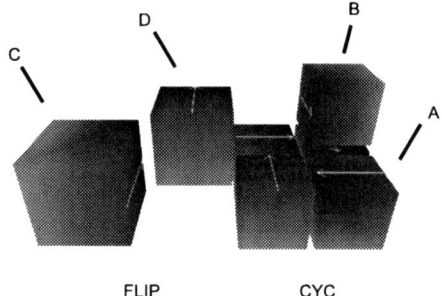

Figure 5. Two interacting mechanisms.

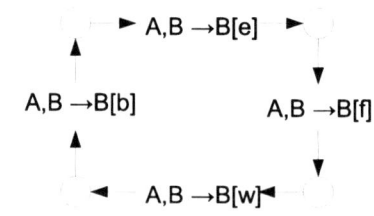

Figure 6. State diagram for CYC mechanism.

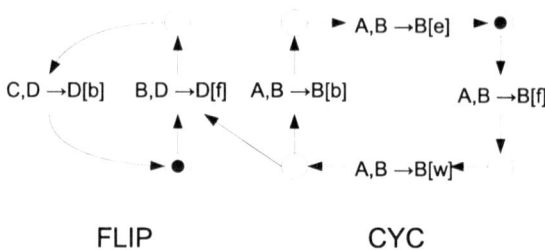

Figure 7. Petri Net diagram for CYC and FLIP mechanisms.

$C_n.q \in Q$ is the state of the machine at step n.
$C_n.a_1...a_t \in \Sigma^t$ are the contents of the tape at step n.
$C_n.i$ is the position of the head on the tape at step n.
$C_n.a_i$ is used as an abbreviation for $C_n.a_{C_n.i}$.

The initial configuration of the machine is:

$C_0.q = q_0$
$C_0.i = 1$
$C_0.a_1...a_t$ are the initial contents of the tape.

The evolution of a Turing Machine over time is given by equation 7. Any contents of the data tape not changed by equation 7 remain unchanged from step n to step $n + 1$.

$$(C_{n+1}.q, C_{n+1}.a_{c_n.i}, C_{n+1}.i - C_n.i) = \delta(C_n.q, C_n.a_{c_n.i}) \qquad (7)$$

$\delta(q, a).state$, $\delta(q, a).symbol$ and $\delta(q, a).direction$ are used when it is necessary to refer to individual elements from the tuple $\delta(q, a)$.

The Turing Machine halts when $C_n.q = q_h$, and C_n is the final configuration of the machine.

5   A Turing Machine in the CBlocks3D environment made from SlideOn parts

5.1   Overview

Figure 8 gives an overview of the Turing Machine presented in this paper.

The Turing Machine contains the following mechanisms, each of which is described in detail later on:

- The data tape mechanism DT: This implements a sequence of bits corresponding to the Turing Machine's tape.
- The program plane construct PP: This implements the transition function δ. For every $q \in Q$ there is a corresponding position of the program plane PP. Let $F : Q \to \mathbb{Z} \times \mathbb{Z}$ be the function that maps Turing Machine states to program plane positions. A state transition is carried out by moving PP from one position to another. Information encoded on PP tells the machine how to respond when a 0 bit or a 1 bit is encountered on the data tape DT. PP also contains parts that trigger the sequencer SQ via the trigger mechanism TR when a state transition has finished.

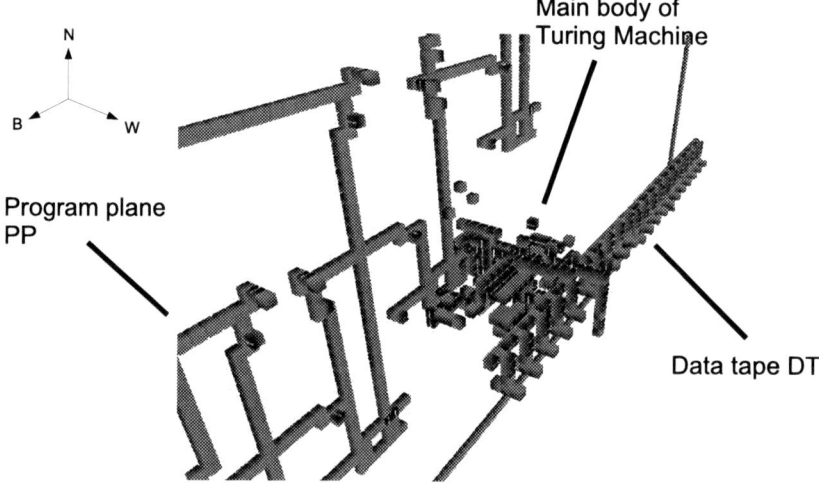

Figure 8. Turing Machine - Overview

- The bit-reading mechanism BR: Interrogates the current position on the data tape DT and activates the selection mechanism SL if a 1 bit is encountered.
- The selection mechanism SL: This moves the program plane PP so as to select the second of two sets of information encoded for a particular state on PP.
- The condition-action mechanism CA: Interrogates the program plane PP and performs bit-writing and/or tape-moving actions according to information associated with the current state on PP.
- The conditional-tape-moving mechanism CTM: Moves the data tape DT six units backward if activated by the condition-action mechanism CA.
- The unconditional-tape-moving mechanism UTM: Moves the data tape DT three units forward when activated by the sequencer SQ.
- The sequencer mechanism SQ: Triggers various parts of the machine in turn in order to perform a cycle of operation. Triggers UTM, then BR, then CA, then PM, then waits to be reset by the program plane PP after a state transition has occurred.
- The plane-moving mechanism PM: Follows tracks on the program plane PP to effect a transition from one state to another by moving PP.

– The trigger mechanism TR: Activated by the program plane PP
when a state transition has finished in order to reset the sequencer
SQ.

Section 6 contains illustrations for all of these mechanisms, along
with Petri Net diagrams which concisely describe the operation of each
mechanism.

5.2   Operation

Suppose that the Turing Machine is in a state corresponding to the
configuration C_n. Then the z coordinate of the data tape DT is related
to $C_n.i$ by $DT.z = 3 \times C_n.i + A$ where A is a constant that depends on
the absolute location of the Turing Machine in space, and bit $C_n.a_p$ is
represented by pin $DT.B_p$.

The main body of the Turing Machine is made from mechanisms BR,
SL, BW, CTM, PM, UTM, SQ and TR and always remains stationary
(apart from movements of internal mechanisms). Only DT and PP move
around.

The operation of the machine is as follows:

1. The machine is in a state corresponding to a configuration C_n.
2. SQ triggers BR, CA, PM and UTM in turn.
3. BR reads the current symbol from DT. If the symbol is 0, SL does
   not move PP, otherwise SL moves PP four units backward.
4. CA first examines PP to determine $\delta(C_n.q, C_n.a_i).symbol$, and then
   sets the current symbol on DT to this.
5. CA then examines PP to determine $\delta(C_n.q, C_n.a_i).direction$ and
   hence which direction to move DT in. It will either leave DT as
   it is, or move DT six units backward.
6. PM causes PP to move along a path from $F(C_n.q)$ to $F(\delta(C_n.q, C_n.a_i)$
   $.state)$
7. When PP has moved to this new position, PP resets SQ.
8. SQ triggers UTM and UTM moves DT three units forward.
9. The machine is now in a state corresponding to configuration C_{n+1}.
   Operation continues at step 1.

So, by the end of a cycle of operation, the following has happened:

Depending on the current location of PP (corresponding to $C_n.q$) and
the position of the pin $DT.B_{C_n.i}$ at the current location on DT (corresponding to $C_n.a_i$), a symbol corresponding to $\delta(C_n.q, C_n.a_i).symbol$ has
been written to DT, DT has been moved 3 places forwards or backwards,
corresdponding to $\delta(C_n.q, C_n.a_i).direction$, and PP has been moved to

a new position corresponding to $\delta(C_n.q, C_n.a_i).state$. This action corresponds to the abstract operation described by equation 7.

The machine can be made to halt by having a path on PP that leads nowhere. When the machine tries to move PP to a new state by following this path, the operation of the machine will cease.

6   Detailed description of mechanisms

6.1   Data tape DT

Figure 9 shows part of the mechanism DT for representing a sequence of binary digits. Figure 10 shows the same mechanism viewed from the opposite direction. The mechanism consists of a long rod A situated against a series of pins B_x that can be individually raised or lowered with respect to the rod to represent binary 1 and 0 digits respectively. The mechanism is designed so that when the rod A is moved east or west, the pins get moved along with it so that the representation is not disturbed.

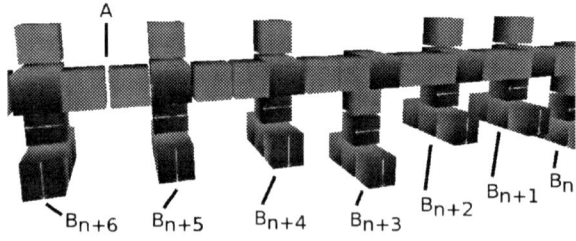

Figure 9. The DT mechanism for representing a binary string. (View 1)

Figures 11 and 12 shows Petri Nets for the data tape DT. Note that DT has a very large number of possible states, equal to the number of possible z-coordinates of DT multiplied by the number of integers that all of the bits on DT can represent. Because of this, figures 11 and 12 use a pair of integers i, j to represent any one of a large set of possible states, where i corresponds to the number represented by all of the bits on DT (and so setting or resetting a bit corresponds to setting $i := i + 2^n$ or $i := i - 2^n$), and where j corresponds to the z-coordinate of DT.

The large, dashed-outline circles in figures 11 and 12 represent states in other Petri Nets. For example, in figure 11, the dashed circle labelled

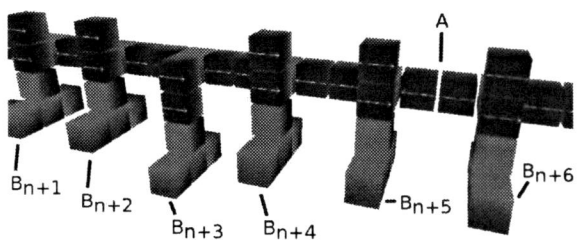

Figure 10. The DT mechanism for representing a binary string. (View 2)

with UTM_1 represents the state labelled 1 in the Petri Net for UTM (figure 24).

6.2 Bit reading mechanism BR

Figure 13 shows a mechanism for reading a bit on the data tape. Figure 14 shows an exploded view of the same mechanism in which the different constructs that make up the mechanism can be distinguished.

Figure 15 shows the Petri Net for the Bit Reading mechanism BR, which explains its operation. Note that this Petri Net uses an 'inhibit' arc (denoted by a line terminated with a circle). A transition that has an inhibit arc as an input cannot take place if the state from which the inhibit arc emanates is marked.

In the Petri Net for BR, the inhibit arc is used so that the network 'chooses' between the path which corresponds to a 1 bit being read from DT, and the path which corresponds to a 0 bit being read from DT.

6.3 Selection mechanism SL

Figure 16 shows the mechanism that selects the second of two positions for a given state on the program plane, if it is activated by BR. Consider the sentence: 'If the Turing Machine is in state q and encounters a 0, then do ..., otherwise if it encounters a 1 then do ...'. The SL mechanism is the implementation of the word 'otherwise' in this sentence.

Figure 17 shows the Petri Net for the SL mechanism.

6.4 Condition action mechanism CA

After the Selection Mechanism SL has ensured that the correct part of the current state on the program plane PP is in the right place, the CA

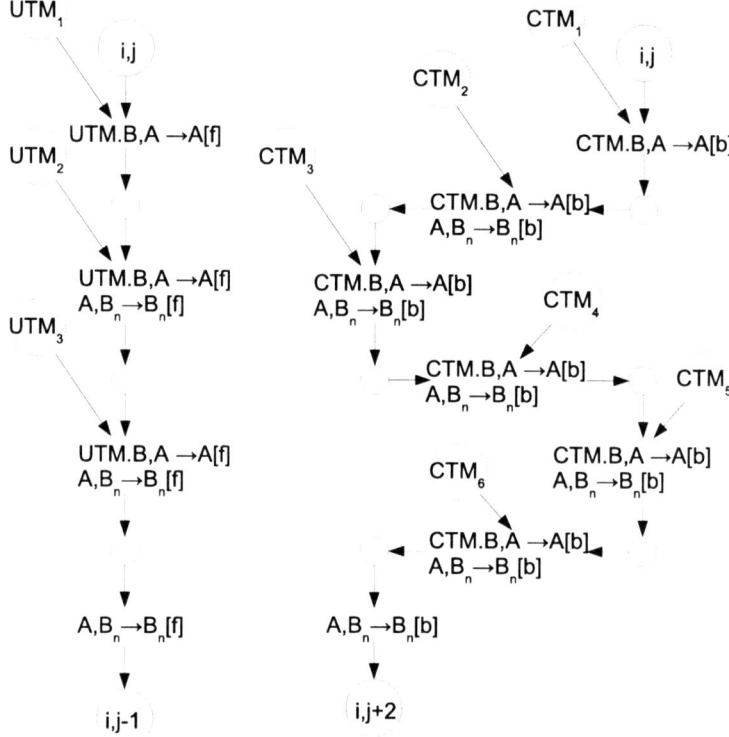

Figure 11. Petri Net for DT. Moving back and forth.

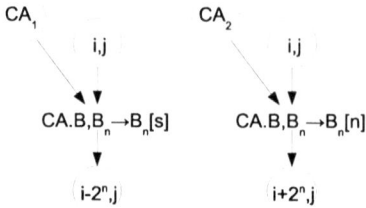

Figure 12. Petri Net for DT. Resetting and setting a bit.

mechanism interrogates that place on PP to work out whether to write a 1 or a 0 onto the data tape DT, and whether to move DT backwards or forwards.

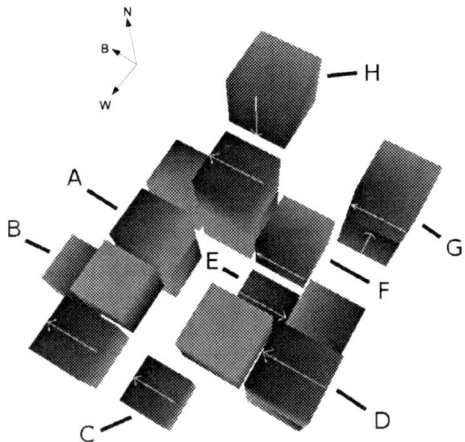

Figure 13. The BR mechanism for detecting the state of a single bit on a data tape.

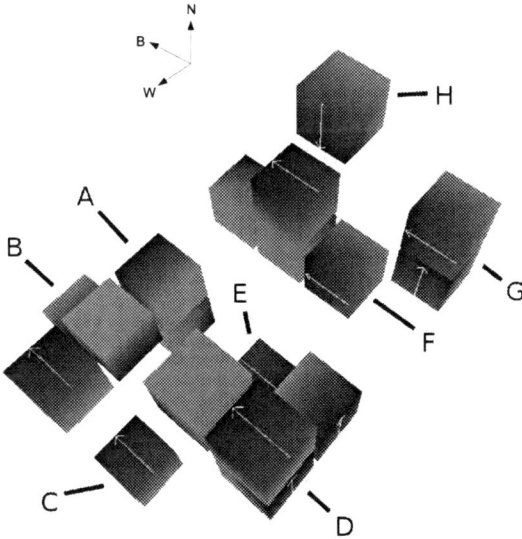

Figure 14. Exploded view of figure 13.

Figures 18 and 19 show the CA mechanism. Basically it consists of two rods, one rod (labelled B) for determining the state of the 'Write a

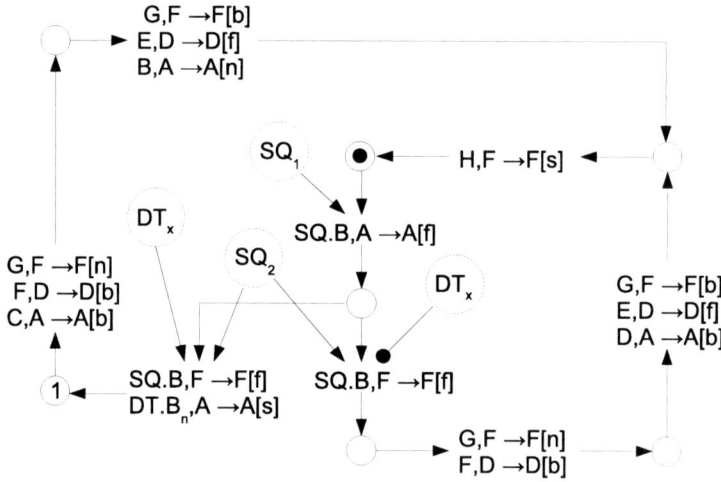

Figure 15. Petri Net for BR.

1' bit on the program plane. The other rod (labelled A) for determining the state of the 'Advance the tape' bit on the program plane. Each rod is pressed against the program plane, and the corresponding action is triggered.

Figure 20 shows the Petri Net for the CA. The Petri Net is complex because the mechanism consists of three main constructs - the two rods A and B, and a construct G that is responsible for moving rod B back to its initial position once PP has been interrogated. Some arcs on the Petri Net that would show dependencies between the three loops in figure 20 have been omitted to avoid clutter, since they do not affect the behaviour of the net.

6.5   Conditional tape moving mechanism CTM

Figure 21 shows the CTM. This mechanism is triggered by the CA when the program plane PP indicates that the data tape DT is to be moved. The CTM moves DT 6 units backwards, so that between them CTM and UTM (see the next subsection) move the DT either 3 units backwards or 3 units forwards.

Figure 22 shows the Petri Net for the CTM.

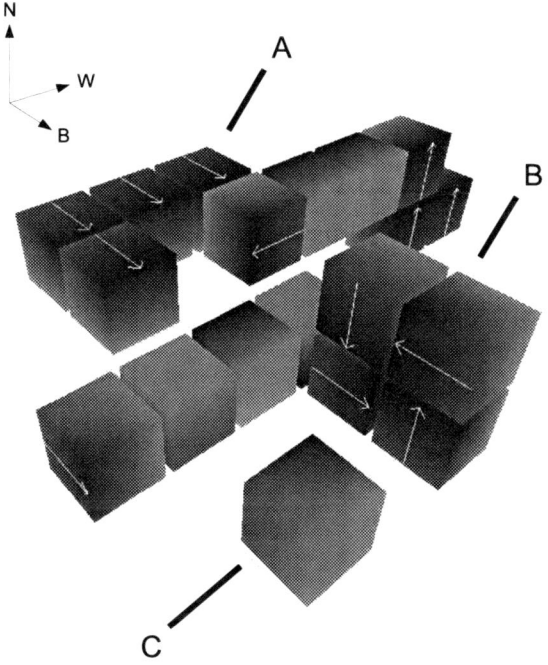

Figure 16. The SL mechanism for conditionally moving the program plane.

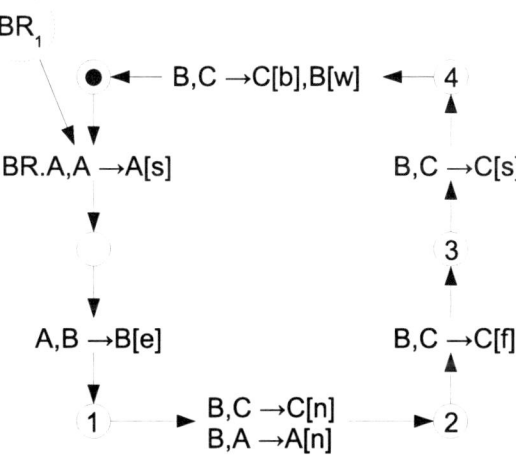

Figure 17. Petri Net for SL.

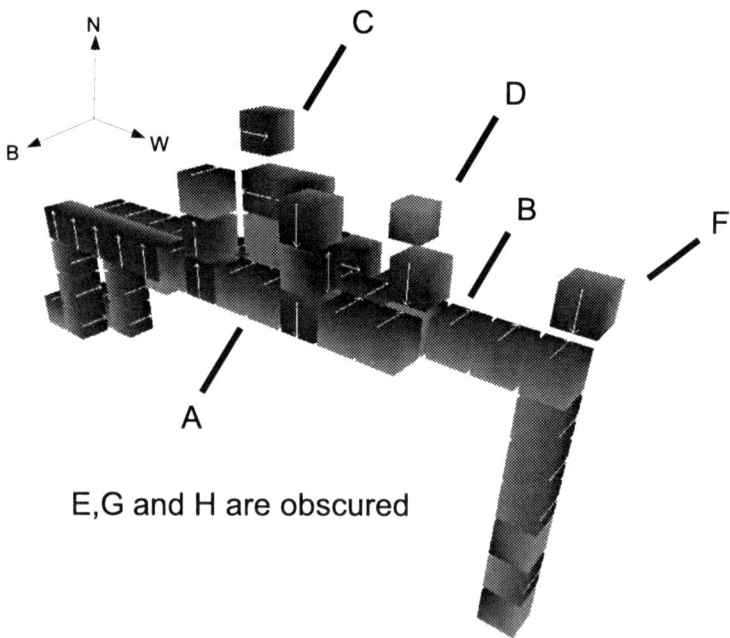

Figure 18. The CA mechanism that performs actions depending on information encoded on the program plane.

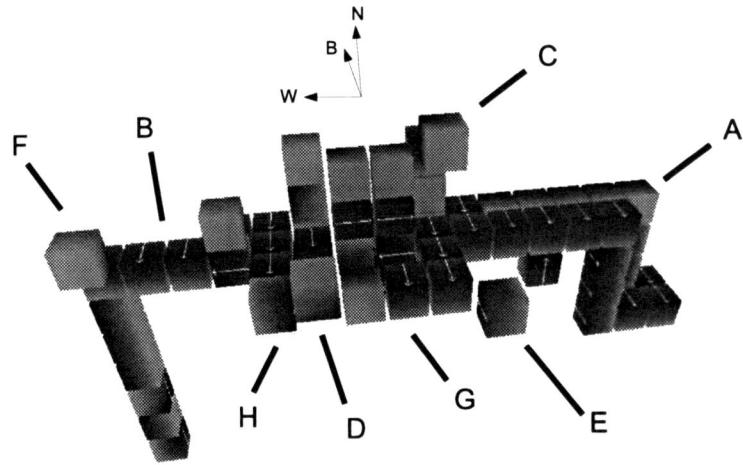

Figure 19. A different view of the CA mechanism

6.6   Unconditional tape moving mechanism UTM

Figure 23 shows the UTM. This mechanism is triggered by the sequencer to move the DT 3 units forwards.

Figure 24 shows the Petri Net for the UTM.

6.7   Sequencer SQ

Figures 25 and 26 show the Sequencer SQ which is responsible for activating several other mechanisms: BR, CA, PM and UTM.

It can be seen that SQ is essentially a track A along which a construct B moves. In several places B encounters other constructs (not shown in figures 25 and 26) and activates the mechanisms to which they belong. Figure 27 shows the Petri Net for SQ.

6.8   Plane moving mechanism PM

Figures 28 and 29 show the plane moving mechanism PM. This mechanism gets activated by SQ after all actions related to reading and acting on information encoded at the current position of the program plane PP have finished. The mechanism tracks paths on the program plane PP that lead from the portion of PP that has just been examined to another state position on PP.

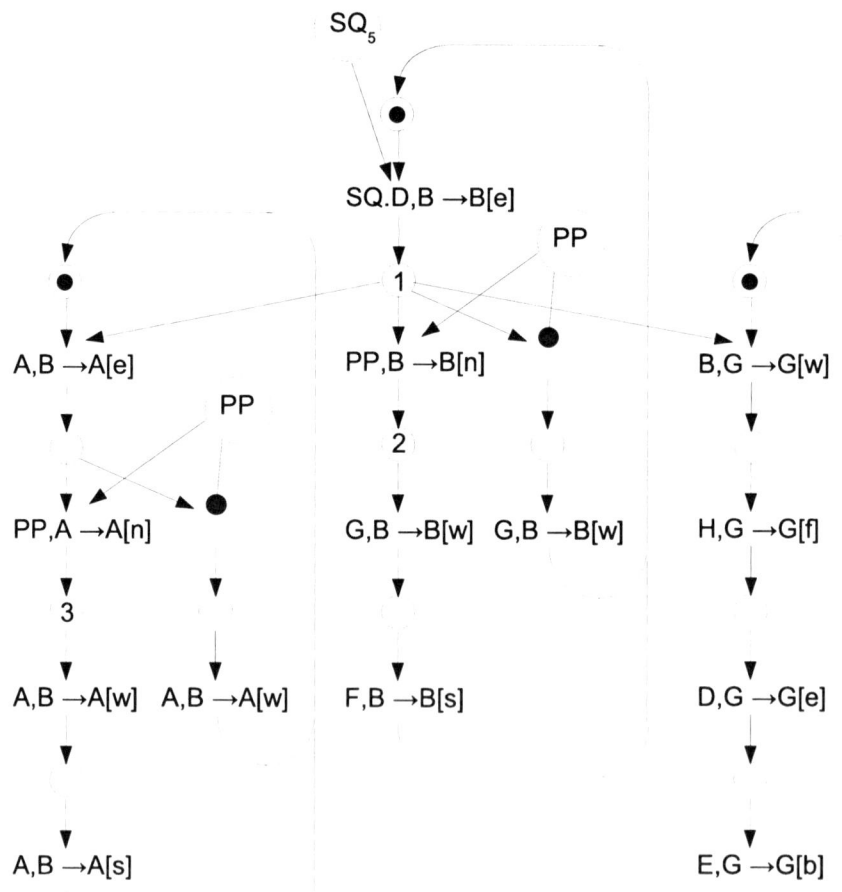

Figure 20. Petri Net for CA.

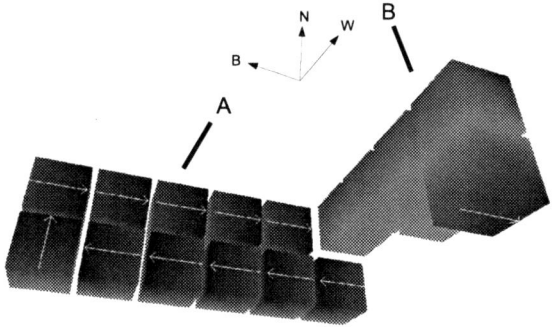

Figure 21. The CTM mechanism

Figure 30 shows the Petri Net for PM. To avoid clutter in this diagram, not all PP states that have arcs leading to transitions are shown.

6.9   Trigger mechanism TR

Figure 31 shows the trigger mechanism TR that starts the sequencer when the program plane PP reaches a state position.

Figure 32 shows the Petri Net for TR.

6.10   Program plane PP

Figure 33 shows a portion of the program plane. This corresponds to a single state in the abstract Turing Machine.

The part labelled A is the part that activates the trigger mechanism TR which in turn activates the sequencer SQ when a state position is reached. The pair of parts labelled B correspond to the two bits of information needed to tell the Turing Machine what to do when it encounters a 0 on the data tape DT. Similarly, the pair of parts labelled C tell the Turing Machine what to do when it encounters a 1 on the data tape. For B and C, a part p with $p.main$ pointing West towards the body of the Turing Machine and $p.secondary$ pointing North represents 1, and a part in any other legal orientation represents 0. In both B and C, the first (i.e. back-most) of the two bits of information tells the machine whether to move DT (1) or not (0). The second of the two bits tells the machine whether to write a 1 or a 0 onto DT. The paths labelled D and E are the paths that the plane moving mechanism PM tracks when leaving this

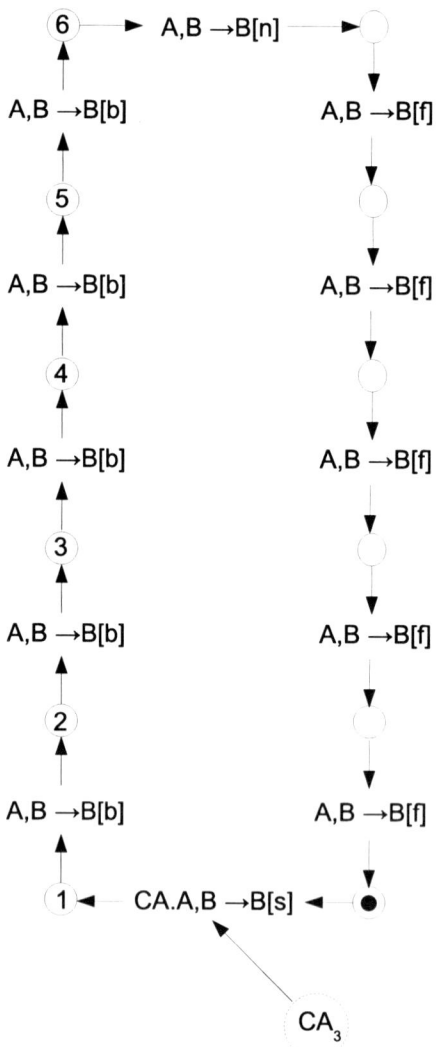

Figure 22. Petri Net for CTM.

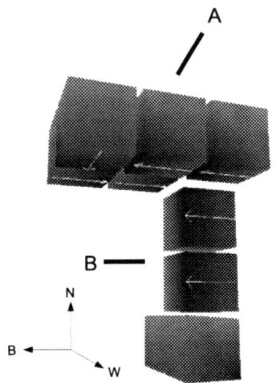

Figure 23. The UTM mechanism.

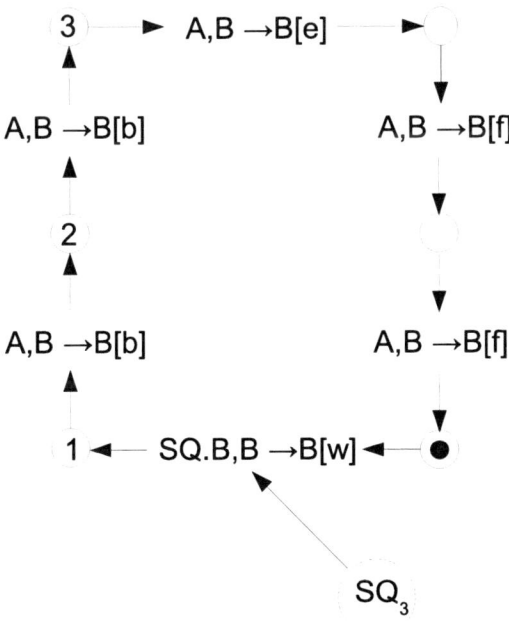

Figure 24. Petri Net for UTM.

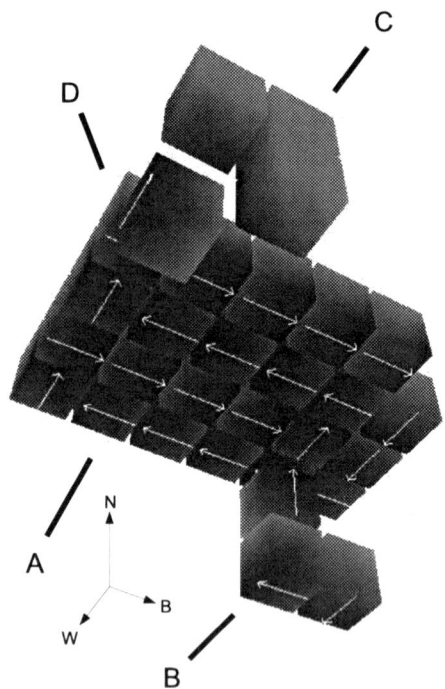

Figure 25. The SQ mechanism.

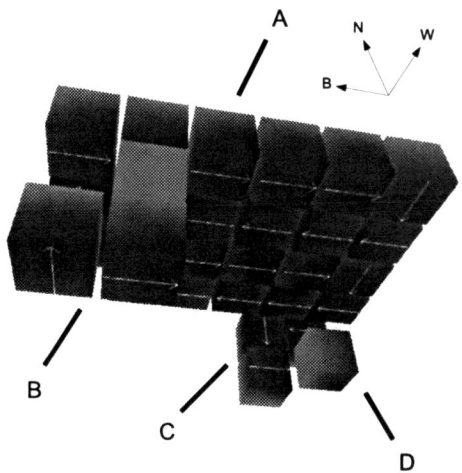

Figure 26. A different view of the SQ mechanism.

state for another state. Which path PM follows depends on whether a 0 or a 1 was encountered on DT.

Figure 34 shows the corner of a track in PP. This figure should be examined in conjunction with the Petri Net for PM in figure 30 to see how PM and PP negotiate a corner.

Figures 35 and 36 show partial Petri Nets for the program plane PP. (The complete Petri Net would be large and not instructive, so is omitted).

7   An example

Supplementary material for this paper can be found at:

http://www.srm.org.uk/downloads/kinturing.zip

This file contains software for simulating the CBlocks3D environment and a file containing an example Turing Machine that contains a program plane for multiplying two unary numbers together, and a data tape containing the numbers 2 and 3. After several minutes of simulation time, the machine will halt with a number 6 on the data tape.

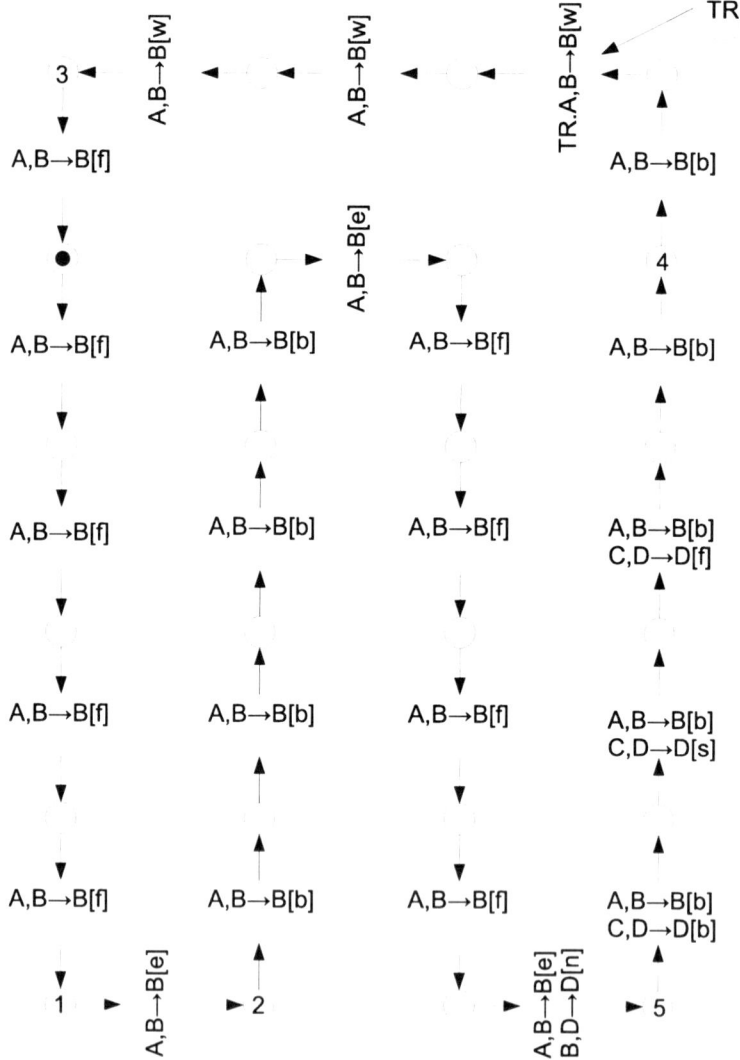

Figure 27. Petri Net for SQ.

8 Discussion

The most notable aspect of the Turing Machine presented here is that it is made from a single type of part and depends upon only one type of sliding kinematic interaction.

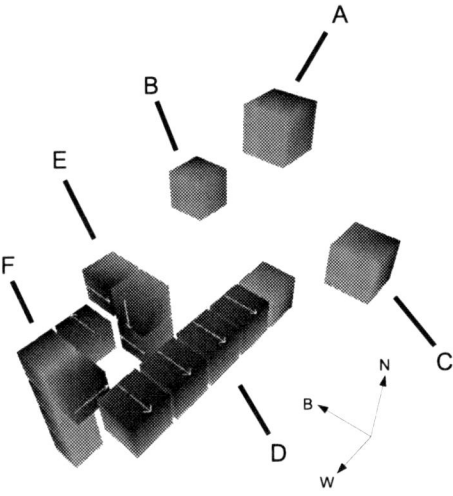

Figure 28. The PM mechanism.

The computational properties of a similar two dimensional kinematic system were investigated in [10]. From the perspective of physical implementation, the CBlocks3D system is much less attractive than the system in [10]. This is partly because CBlocks3D is three dimensional, and it is difficult to engineer 3D substrates with properties different from the ones that nature provides us with. (Unlike in two dimensions, where a surface can be modified two produce a range of properties). Also, in [10] there is an action and reaction between parts, but in CBlocks3D parts are able to slide neighbours without moving themselves.

The CBlocks3D universe as described here does not have construction-closure. i.e. mechanisms can exist in the CBlocks3D universe which cannot be constructed by any mechanism in the universe. The main reason for this is that there is no 'connecting' operation by means of which two constructs may become one.

It may be possible to define 'connecting' and 'disconnecting' operations that occur when two constructs are pushed together or pulled apart by SlideOn parts with opposing secondary axes. Such a system might have construction closure, and it might be possible to devise a programmable-constructor based self-replicating machine in this system.

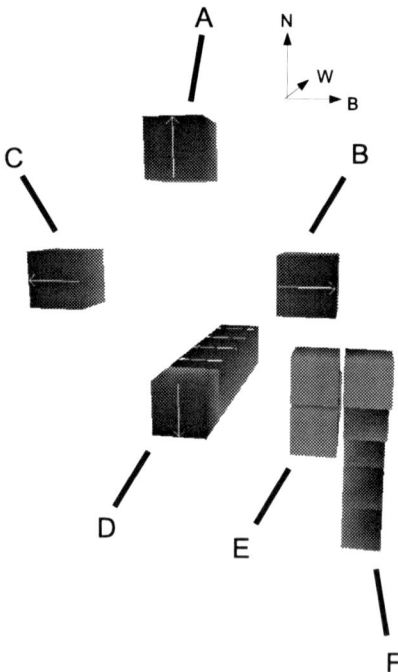

Figure 29. A different view of the PM mechanism.

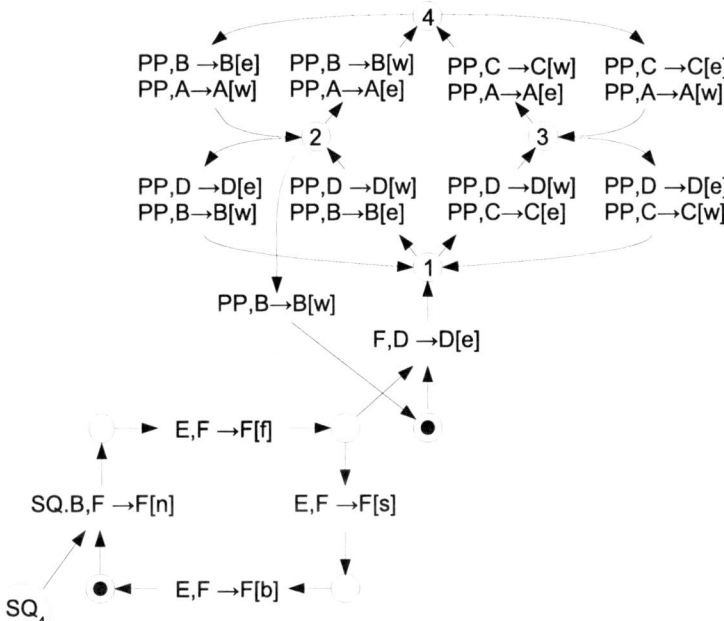

Figure 30. Petri Net for PM.

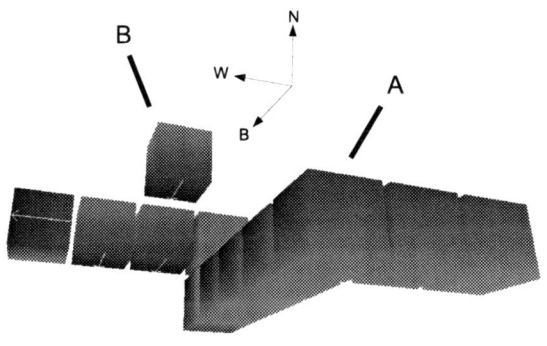

Figure 31. The TR mechanism.

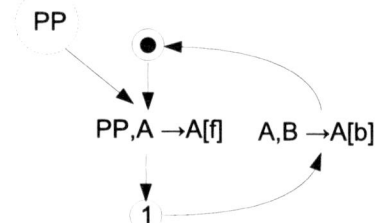

Figure 32. Petri Net for TR.

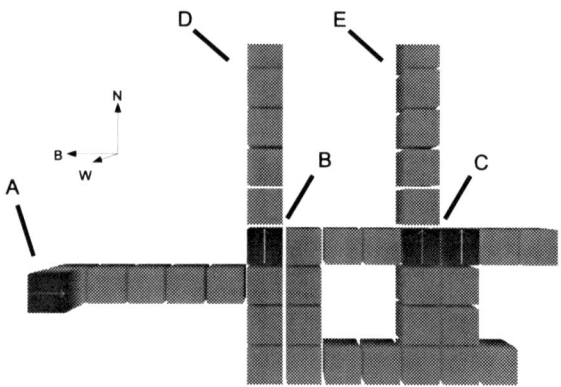

Figure 33. The PP mechanism.

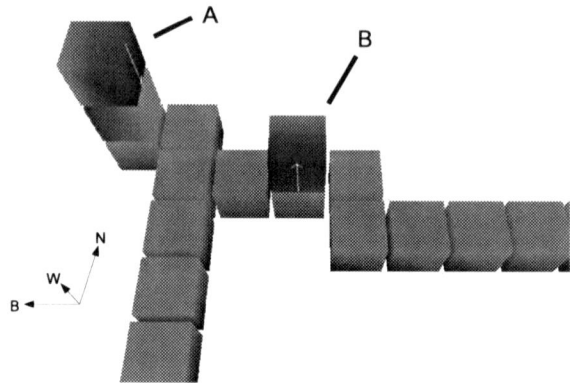

Figure 34. The PP mechanism.

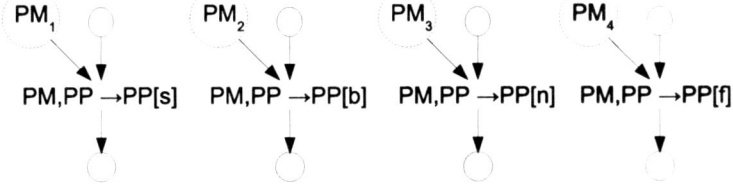

Figure 35. Petri Net for PP. PP being moved by PM.

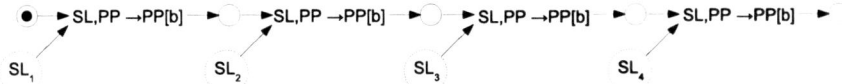

Figure 36. Petri Net for PP showing the action of SL on PP.

References

[1] Arbib, M.A.: Theories of Abstract Automata. Prentice-Hall, Englewood Cliffs, New Jersey (1969) 355–361

[2] Babbage, C.: Passages from the Life of a Philosopher. Chapter 8: Of The Analytical Engine Longman, Green, Longman, Roberts and Green (1864) http://www.fourmilab.ch/babbage/lpae.html Cited on 13 May 2007

[3] Cook, M.: Universality in Elementary Cellular Automata. Complex Systems **15** (2004) 1–40

[4] Fredkin, E., Toffoli, T.: Conservative Logic. J. Theo. Phys. **21(3,4)** (1982) 219–253

[5] Thompson, R.L., Goel, N.S.: Movable Finite Automata (MFA) models for biological systems. I: Bacteriophage assembly and operation. J. Theor. Biol. **131(3)** (1988) 351–385

[6] Laing, R.A.: Some alternative reproductive strategies in artificial molecular machines. J. Theor. Biol. **54** (1975) 63–84.

[7] Merkle, R.C.: Two Types of Mechanical Reversible Logic. Nanotechnology **4** (1994) 114–131 http://www.zyvex.com/nanotech/mechano.html Cited on 13 May 2007

[8] Petri, C.A.: Kommunikation mit Automaten. PhD Thesis, University of Bonn. (1962)

[9] Rendell, P.: Turing Universality of the Game of Life. Collision-Based Computing (2002) 513–539

[10] Stevens, W.M.: Logic circuits in a system of repelling particles. From Utopian to Genuine Unconventional Computers, Luniver Press, Frome, UK (2006) 157–182

[11] Stevens, W.M.: Simulating Self Replicating Machines. J. Intelligent and Robotic Sys. **49(2)** (2007) 135–150

[12] Toth-Fejel, T.: LEGO(TM) to the Stars: Active Mesostructures, Kinematic Cellular Automata, and Parallel Nanomachines for Space Applications. The Assembler **4(3)** (1996)
http://www.islandone.org/MMSG/9609lego.htm
Cited on 13 May 2007

[13] Wikipedia article on Petri Nets.
http://en.wikipedia.org/wiki/Petri_net
Cited on 13 May 2007

[14] Rojas, R.: Konrad Züse's Legacy: The Architecture of the Z1 and Z3. IEEE Annals of the History of Computing. **19(2)** (1997) See also:
http://www.epemag.com/zuse/
Cited on 13 May 2007

Nucleic acid enzymes: The fusion of self-assembly and conformational computing

Effirul I. Ramlan and Klaus-Peter Zauner

School of Electronics and Computer Science
University of Southampton, SO17 1BJ, United Kingdom
{eir05r,kpz}@ecs.soton.ac.uk

Abstract Macromolecules are the predominant physical substrate supporting information processing in organisms. Two key characteristics—conformational dynamics and self-assembly properties—render macromolecules unique in this context. Both characteristics have been investigated for technical applications. In nature's information processors self-assembly and conformational switching commonly appear in combination and are typically realised with proteins. At the current state of biotechnology the best candidates for implementing artifical molecular information processing systems that utilise the combination self-assembly and conformational switching are functional nucleic acids. The increasingly realised prevalence of oligonucleotides in intracellular control points towards potential applications.

The present paper reviews approaches to integrating the self-assembly and the conformational paradigm with allosterically controlled nucleic acid enzymes. It also introduces a new computational workflow to design functional nucleic acids for information processing.

1 Biomolecular computing paradigms

With the feature size of solid-state devices approaching nanometer scale molecules are coming increasingly into focus as an alternative material substrate for the implementation of information processing devices. A wide range of approaches to utilizing molecules in computing are under consideration (Fig. 1). The area of *molecular electronics* investigates possibilities for implementing with organic materials the architectures familiar from silicon-electronics (cf. [1]). Polymer semiconductors and

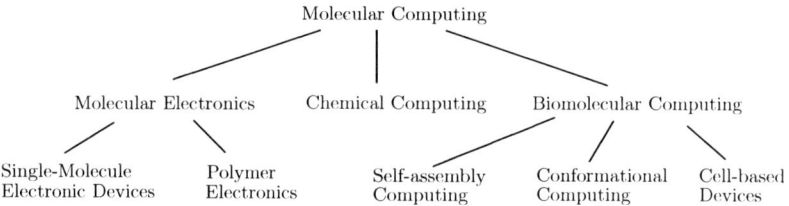

Figure 1. Research directions within the field of molecular computing.

single-molecule transistors are typical research goals. In *chemical computing* excitable chemical reaction systems with diffusive coupling are investigated for their potential as massively parallel processing media (cf. [2]). And *biomolecular computing* is concerned with the use of macromolecules and supramolecular systems and in many cases attempts to exploit mechanisms found in nature. The present paper will concentrate on the latter; a recent review covering also the other directions can be found in [3].

A prominent difference between solid-state materials and macromolecular materials is the large range of properties found in molecules. Biomolecules are mainly composed from only six (C, H, O, N, S, P) out of the 91 naturally occurring chemical elements. The number of possible compounds that could in principle be formed from these six atoms is very large. Even though there are many restrictions on how the atoms can be combined, stable macromolecules comprising hundreds or thousands of atoms can be formed. Macromolecules occurring in organisms are typically formed from a set of building block molecules. These building blocks link through covalent bonds originating at specific atoms, but can be combined in arbitrary order. The twenty commonly occurring amino acids form such a set of building blocks. Linear polymers from up to a few hundred of these amino acids linked in arbitrary sequential order constitute an important class of biomacromolecules, the proteins . Another set of building blocks found in nature are the nucleotides which combine, again in arbitrary order, to long nucleic acid molecules. The exact linear sequence of the building blocks may have a relatively small influence on the properties of the complete macromolecule, as is the case with the deoxyribonucleic acids (DNA), the carriers of genetic information in the cell. But the exact sequence can also be crucial to the properties of the macromolecule, as is typical for proteins. Both cases have practical advantages. The former is ideally suited for representing information, because the physical properties of the macromolecule are largely independent of its information content. The latter case gives rise

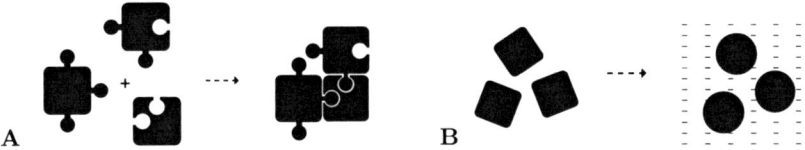

Figure 2. Cartoon of the two basic biomolecular computing paradigms. In *self-assembly computing* (A) information is encoded by molecular shapes. Molecules with complementary shape form supra-molecular clusters through non-covalent binding. Thus shape-encoded input is mapped into features of the cluster as output. In *conformational computing* (B) the physicochemical environment of a macromolecule serves as input signal. Intramolecular dynamics maps this milieu information into a change in conformational state.

to the diverse specificity and large range of material properties that is the basis of the tremendous variety of organisms seen in nature.

Two phenomena are key to the interaction and function of macromolecules: self-assembly and conformational dynamics. Both play also an important role for molecular information processing in nature and each serves as a paradigm for man-made molecular computing schemes. Figure 2 illustrates these paradigms.

Atoms attached through covalent bonds in a molecule can exert weak, short-range attractive forces (van-der-Waals interaction, hydrogen-bonds) on atoms in other molecules. If two macromolecules have complementary surfaces, i.e., surfaces that allow for close proximity of a large number of suitable atom-pairs, then the additive effect of the weak attractive forces results in a stable binding of the complementary molecules. In other words, the potential energy will dominate entropy even at high temperature. Molecules, on the one side, are large enough to have specific shape features. On the other side, they are small enough to be moved around by thermal motion and therefore can explore each others shapes by diffusion.

The self-assembly paradigm (Fig. 2A) effectively converts a symbolic pattern recognition problem into a free-energy minimisation process [4–6]. The self-assembly paradigm can conveniently be implemented with deoxyribonucleic acid (DNA) because it is relatively easy to predict the information among oligonucleotides from thermodynamic data and computer simulations. Over the past decade a number of experimental realisations of self-assembly computing have been reported (e.g., [7–9]). A drawback of self-assembly computing is that the random search of molecules for complementary partners by means of Browninan motion does not scale well to large reaction volumes.

In conformational computing (Fig. 2B) one attempts to exploit the shape changes that large macromolecules can undergo in response to their environment. The freedom of atoms in a molecule to rotate around single covalent bonds equips molecules with considerable flexibility. In proteins in particular have a distinctive agility that is core to their folding from a linear amino acid chain to a compact functional or structural component. This flexibility, however, does not terminate with the folding. The physicochemical milieu in which the macromolecule is embedded modulates the transition probabilities among the molecule's conformational states. Different conformational states commonly result in altered functional activity. A few experimental implementations that make use of the conformational dynamics have been reported (e.g., [10–12]). In the conformational paradigm much of the computation is an intramolecular process and state changes can therfore be fast. However, a problem with this approach is that in practice the conformational effects are at least hard and often impossible to predict. The practical implementations so far rely on molecules occurring in nature or genetically engineered variants of these molecules [10].

In nature's molecular information processing infrastructure both self-assembly and conformational dynamics play an important role. Typically both occur in combination. A protein may undergo a conformational change and as a consequence of this its shape becomes complementary to a region on another macromolecule thus leading to self-assembly. Conversely, a molecule that participates in self-assembly experiences a significant change in its environment as a result of the binding to another molecule and this change can give rise to an altered conformation. In combination self-assembly and conformational switching are a powerful set of primitives on which the entire molecular machinery of cells is build. It would be desirable to combine the self-assembly and the conformational paradigm also for artificial molecular computing schemes. In nature proteins are the key components that integrate self-assembly and conformational switching. Unfortunately, both phenomena are notoriously difficult to predict for proteins. Recently an intriguing alternative has been experimentally demonstrated in form of nucleic acid enzymes [13–15], i.e., DNA or RNA molecules with catalytic activity.

The main goal of the present paper is to provide a self-contained overview of nucleic acid enzymes from an information processing perspective as an introduction to the field that can be read without prior background in molecular biology. The remainder of this paper will first introduce a basic biochemical background on nucleic acids, then review nucleic acid enzymes in general, next describe their application in information processing experiments, and finally turn to software tools supporting the design of nucleic acid enzymes. Moreover, a new work-flow for designing functional RNA is introduced.

2   Properties of nucleic acids

Nucleic acids are macromolecules that play an important role as information carriers in cells. Two types occur, ribonucleic acids (RNA) and deoxyribonucleic acids (DNA), which are named after the structure of a sugar component always present in these molecules (Fig. 3). RNA and DNA are typically long linear polymers that consist of a large number of monomers taken from a set of four different nucleotides. The sequential order in which the nucleotides are interlinked in the nucleic acid molecule can represent information.

The two types of nucleic acids play different roles. RNA has the task of transmitting information within the cell, while DNA transmits information from generation to generation. The genetic information is encoded in a dimer of two complementary nucleotide chains ('single-stranded' DNA) which upon self-assembly assumes the well known double-helical structure ('double-stranded' DNA'). DNA is well suited as carrier of genetic information because of its energy degeneracy with respect to the sequential order of the nucleotides. The properties of RNA molecules are more dependent on the sequence of nucleotides and as a consequence RNA takes on additional roles in the cell aside from representing information.

Each of the monomer units that make up nucleic acids consists of a sugar moiety, a phosphate group, and a base. The sugar component of the monomers in RNA molecules is ribose, hence the name *ribonucleic acid*. Correspondingly, DNA is named *deoxyribonucleic acid* after its sugar component deoxyribose. Figure 3 shows the chemical structures of both sugar components. The hydroxyl group (-OH) at the $2'$-carbon in ribose is not present in deoxyribose [16]. The consequence of this structural difference is twofold. DNA is considerably more stable against hydrolysis and forms more compact double strands, while RNA has more conformational flexibility [17]. The flexibility of macromolecules to change their three-dimensional shape, i.e. their *conformation*, while maintaining the covalent bonds among atoms, i.e. their *configuration* unchanged, is the basis of conformational computing. The chemical stability of DNA and the structural flexibility of RNA are both desirable properties for the molecular computing based on nucleic acid enzymes. Which type of nucleic acid is preferred for the implementation of a particular molecular component will often depend on this tradeoff.

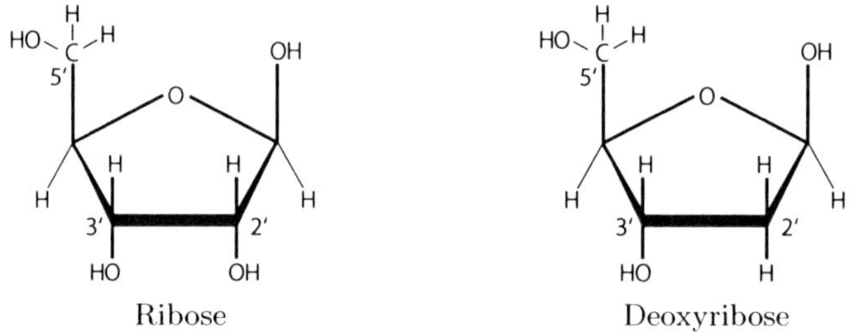

Figure 3. The sugar compounds that form the backbone of RNA (left) and DNA (right). Note the lack of the oxygen at the 2′-carbon in the right panel as compared to the ribose (left). The name *deoxy*nucleic acid refers to the absence of these oxygen atoms in the backbone of DNA.

Within either RNA or DNA the sugar moieties and the phosphate groups of all monomer are identical. The base that forms the third component of each monomer provides the variety requisite for representing information in the sequence of monomers. Each monomer unit carries one of four possible bases. In RNA these are adenine, guanine, cytosine, and uracil, abbreviated as A, G, C, and U. The first three of these bases also occur in DNA, but instead of uracil DNA contains thymine (T). This difference is thought to be of use for DNA repair mechanisms that actively maintain the integrity of a cell's genetic information, but is of no relevance within the context of the present paper.

Of crucial importance for the interaction of nucleic acid molecules is the *complementarity of bases*. The base of a nucleotide can form weak bonds, called hydrogen-bonds, with another nucleotide that carries a complementary base. Hydrogen bonds occur between a hydrogen atom bound to an electronegative atom, and another electronegative atom. They are roughly 20× weaker than a covalent bond.

Among the four possible bases that can occur in a nucleotide, T or U can bind to A with two hydrogen bonds and G can bind to C forming three such bonds. Two nucleotide strands with complementary base sequence will form a dimer that is held together by the additive effect of the hydrogen bonds that can be formed between the complementary bases. This process is called *hybridisation*. The direction of the sequence has to be taken into account if complementarity is considered. The two strands that form a double helix are intertwined running in opposite direction. To indicate the orientation of a single stranded nucleic acid, its ends are named after the unbound carbon atom in the sugar moiety (cf. Fig. 3) as 5′ at one end and 3′ at the other. As a convention, the notation of nucleic acid sequences is written from left to right in 5′ to 3′ direction (i.e., ATTGC always stands for 5′–ATTGC–3′) [16]. In the following figures a diamond symbol (◇) indicates the 3′-end of a strand.

If a nucleic acid has a sequence that is complementary to itself, then it can fold back onto itself and form an intramolecular double-helix [18]. Partial intramolecular hybridisation can result in a complex three-dimensional structures of the molecule. In some instances the three-dimensional structure confers functionality such as a specific catalytic activity.

As mentioned above, RNA is more flexible than DNA and as a consequence it forms more readily intramolecular base-pairs. A single stranded RNA molecule can bind to itself in several regions with the unbound segments present as loops between bound segments or dangling ends. The loops can be grouped into four classes illustrated in Fig. 4. Due to its higher flexibility, in addition to the pairing of complementary bases (A-U/U-A, C-G/G-C), the 'wobble pairing' of G-U (and, reverse oriented, U-G) through two hydrogen-bonds also contributes to the structural variability exhibited by RNA molecules [20]. For a given RNA sequences ('primary structure'), there is often a diverse set of secondary structures it can fold into. Which structure is favoured will depend on the environment of the molecule, for example, the presence or absence of other molecules or ions. Conversely, a diverse set of sequence configurations can yield a particular secondary structure [21, 22]. Subsequently, interactions among secondary structure motifs lead to the formation of a

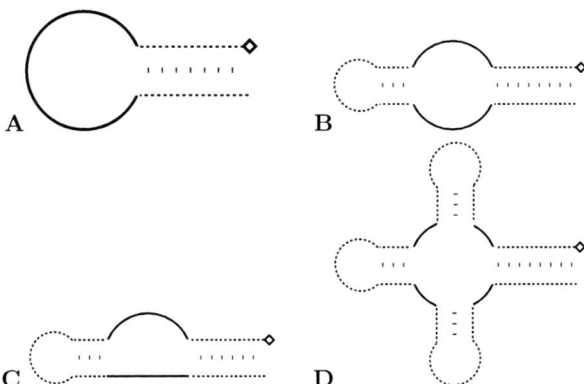

Figure 4. Classification of RNA loop motifs; the named motif is shown with solid lines. Hairpin loop (A), internal loop (B), bulge (C), multi-branch loop (D; a four-way junction is shown). A hairpin loop with the adjacent stem is referred together as stem loop. After [19].

tertiary structure which in some cases entails functionality. Such functional RNA molecules are discussed in the next section. A review of RNA secondary structure and its prediction is offered by [23].

3   Functional nucleic acids

Biological catalysis was thought to be synonymous with catalytically active protein, i.e., enzymes, until RNA molecules with catalytic capability were discovered [24]. These *ribozymes*, as the RNA enzymes are also called, led to the hypothesis that precursors of the cell may have relied on RNA only for the both transmission of genetic information and metabolism, tasks which are in present cells relegated to DNA and protein, respectively. Although it appears unlikely that DNA has catalytic function in nature, it was possible to arrive at DNA enzymes in the laboratory [25].

Ribozymes can be categorised according to size and catalytic activity [26, 27]. The three classes of ribozyme are small catalytic RNAs, group I and II introns, and Ribonuclease P (RNase P). Small catalytic RNAs range in size from 40–160 nt (nucleotides) and are self-cleaving molecules. Group I introns are self-splicing RNA molecules over 700 nt in length while group II introns are more than 1500 nt long. These self-splicing RNAs are found in unicellular organisms. RNase P is over 500 nt long, occurs in all cells, and is required for the production of transfer RNA (tRNA), a key component of the cellular machinery for protein synthesis [24, 26].

Small catalytic RNAs are the most attractive with regard to molecular computing applications. The group of small catalytic RNA comprises hammerhead ribozymes [28], hairpin ribozymes [26], the hepatitis delta virus (HDV) ribozyme [29], and the Neurospora Varkud Satellite ribozyme [30]. Each of these ribozymes has a distinct structure. Nevertheless, all of them catalyse the same reaction. They cleave the phosphodiester bond in RNA, generating a $5'$-product with a $2'$, $3'$-cyclic phosphate terminus and a $3'$-product with a $5'$-hydroxyl terminus. It is thought that the $2'$-hydroxyl group of the ribose moiety of RNA participates in the catalysis [31], however DNA can also act as a catalyst as will be discussed later in this section.

Most of the known natural occurring ribozymes catalyse intramolecular (also called *in-cis*) reactions in which the ribozyme cuts and detaches from part of its own sequence [32]. However, some ribozymes have been successfully modified to split other nucleic acids. To avoid ambiguity, we will use the term *ribozyme core* to refer to the catalytically active RNA molecule in intermolecular (*in-trans*) reactions. Such a reaction is illustrated in Fig. 5. The figure shows the sequence of reaction steps in

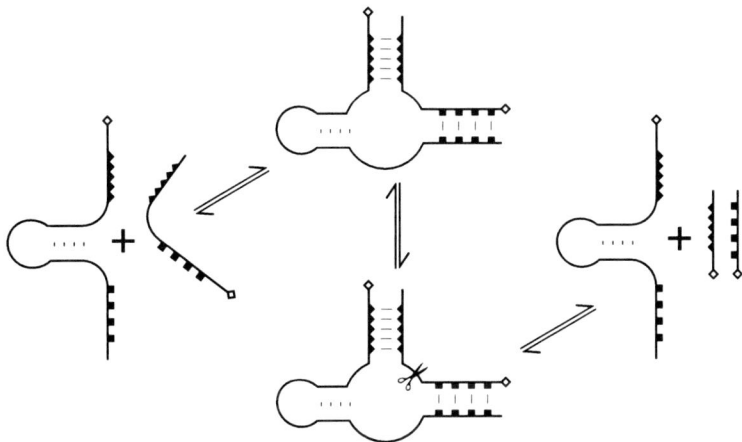

Figure 5. Splitting of an RNA molecule catalyzed by another RNA molecule [33]. The catalytic RNA binds a substrate RNA molecule if it is complementary to the two hybridisation regions indicated by squares and triangles (left). The substrate molecule is cut at the location indicated by scissors (centre). After the two reaction products dissociate from the catalytic RNA, the latter is ready for another reaction cycle (right). The 3′-ends of the RNA sequences are indicated by diamonds (◇).

the catalytic cycle of a hammerhead ribozyme. For brevity the release of both products is shown as a single step, however, the products are likely to dissociate from the ribozyme core one after another [33]. The turnover rate of small ribozymes is typically about 1 cleavage per minute [26, 34].

Among the small catalytic RNAs the Neurospora Varkud Satellite ribozyme has the largest core with a length of with 150 nt. It is followed by HDV ribozyme with a core of at least 90 nt. In both the tertiary structure appears to play an important role for their catalytic function. Even shorter cores have been found in ribozymes of plant viroids and virusiods undergoing site-specific, self-catalyzed cleavage as part of the replication process [32, 35]. Hairpin ribozymes can have cores as short as 70 nt, although in nature they are part of a four-way junction (cf. Fig. 4) [36]. The smallest known natural ribozyme cores are of the hammerhead type and can be as short as 40 nucleotides [32]. Still, smaller ribozyme cores have been engineered. So called minizymes, derived from hammerhead

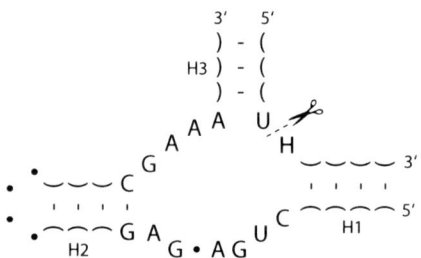

Figure 6. Minimal functional structure of hammerhead ribozyme. Three helical stems (H1, H2, H3) emanate from a junction on the ribozyme core [35,41]. In nature, always either helix H1 or H3 is terminated by a hairpin loop which results in intramolecular catalysis. Hammerhead ribozymes that catalyse the *in-trans* reaction, as depicted in the figure, can be made synthetically [28]. The core region has a specific sequence for all known active structures and is therefore termed 'conserved'. Conserved bases are specified explicitly, with H representing any one of {A, C, U}. A dot (•) stands for any base that will not cause hybridisation in this position; correspondingly two parentheses connected by a dash indicate an arbitrary pair of complementary bases.

ribozymes can be as short as 22 nt, but the reduced size comes at a cost in catalytic efficiency [37].

For some of the ribozyme cores it is feasible to control their catalytic activity. This property is key to the application of ribozymes in molecular computing. In order to understand the mechanisms of controlling the activity of the ribozymes it is useful to consider their secondary structure. The secondary structure that emerges from an RNA sequence is composed from the motifs in Fig. 4 and possibly dangling single-stranded ends. The different types of ribozymes are distinguished by their characteristic combination of loops and helices [38]. The secondary structure of a hammerhead ribozyme is depicted in Fig. 6. Hammerhead ribozymes require the presence of a metal ion (typically Mg^{2+}) to be catalytically active [26]. A ribozyme with a different structure, the so called hairpin ribozyme, is shown in Fig. 7. For convenience the description of these structures is given in the figure captions. A more detailed discussion of their mechanisms can be found in [27,39,40].

It is generally believed that the conformational flexibility of RNA is important for the catalytic process itself [39,44]. The conformational flexibility RNA gives also rise to a large variety of secondary structures. The secondary structure consists of single stranded regions alternating with double stranded regions where stretches of the RNA molecule bind to itself (cf. Figs. 6 and 7). These motifs interact and from the 3-dimensional

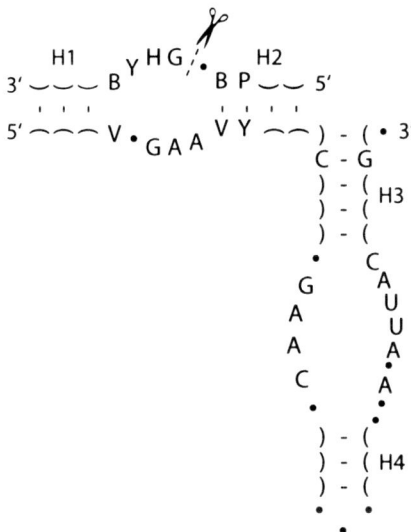

Figure 7. Minimal functional structure of hairpin ribozyme. Two domains are distinguished each of which contains two helices separated by a conserved internal loop. One domain is drawn horizontally and includes the substrate which binds to the core by forming helix H1 and helix H2. The other domain is drawn vertically and contains the helices H3 and H4 [36,42,43]. The following notation is used for the conserved region: $Y \in \{C, U\}$, $V \in \{A, C, G\}$, $B \in \{C, G, U\}$, and $H \in \{A, C, U\}$. See Fig. 6 for explanation of dots and parenthesis.

tertiary structure of the RNA molecule. The conformational flexibility thus supports a diverse set of functional roles [45].

The structural variety of RNA and its concomitant functional diversity make RNA a suitable medium for directed in-vitro evolution [46]. This technique is based on the possibility to copy RNA molecules with aid of protein enzymes. Errors in the copy process yield a population of RNA molecules with slightly varied sequences. Repetitive application of this error-prone replication process will lead to an evolution of the population of molecules. In the absence of other selection pressures, the evolution would favour molecules that are most efficiently reproduced by the participating protein enzymes. However, a selection step can be introduced to assert evolutionary pressure in another direction. The molecules could for example be selected by their binding capabilities towards a particular substrate molecule [47].

A number of ribozymes have been produced through directed evolution [48, 49]. The majority of them possess a ribozymes core that does not resemble any of those found in nature [50]. Directed evolutions provides a technique to enrich the repertoire of RNA structures amenable to molecular computing applications.

Directed evolution can also be applied to DNA and, rather surprising, yields DNA molecules with enzymatic activity (deoxyribozymes) [51–54]. DNA is best known as a memory molecule inscribed with information crucial for the production of macromolecular components in cells. The properties that make DNA suitable for this function are its stability, reliable hybridisation, but also the fact that DNA forms a double-helical structure largely independent of the sequence of bases as long as the two strands that hybridise are complementary. These properties together with the absence of DNA enzymes in nature had let to the view that DNA is not flexible enough to act as a catalyst. It is now, however, well established that DNA does have the structural flexibility to support a range of secondary and tertiary structures [55] and can form a diverse set of tertiary structures with a potential to function as catalysts [25]. Secondary structures of three deoxyribozymes developed through the process of in-vitro selection are depicted in Fig. 8. For the deoxyribozymes shown in panels B and C of Fig. 8 it was found that their catalytic reaction rates are comparable to those of ribozymes [56]. As mentioned above, hammerhead ribozymes require the presence of metal ions to be catalytically active. The deoxyribozymes also require metal ions. The first deoxyribozyme was designed in the presence of Pb^{2+} as cofactor [51] and the deoxyribozymes shown in Fig. 8 all require Mg^{2+}.

From an application perspective the use of DNA has the advantage over RNA that DNA molecules are generally more stable. Furthermore, the DNA-DNA binding is more reliable and results in higher specificity. Given these practical advantages of DNA and the fact that DNA enzymes do not occur in nature it is of particular interest that recently a ribozyme was successfully converted into a deoxyribozyme by means of directed evolution [57]. A DNA sequence that corresponded (apart from the T for U substitution) to a known ribozyme which catalyses a covalent bonding between two RNA oligonucleotides was found to be inactive. However, after acquiring suitable mutations during directed evolution, a deoxyribozyme that also catalyses a covalent bonding between two RNA oligonucleotides—though at a lower efficency and different bond location—was arrived at.

In combination the capability of self-assembly through hybridisation of complementary sequences and the conformational flexibility to form sequence dependent spatial structures with catalytic activity make nucleic acids an attractive material for molecular computing.

4   Nucleic acid enzymes for computing

From the time it became apparent that nucleic acid polymers carry the genetic information in their base sequence, the astounding information density was recognised. Early suggestions for implementing a molecular computer with DNA followed the encoding principle of genetic information (cf. [58]). This would require the formation and cleavage of numerous covalent bonds for their operation and thus require specific sets of enzymes. Major progress in the application of nucleotides for information processing came about two decades later with Adleman's insight that random oligonucleotides could be the basic tokens for information

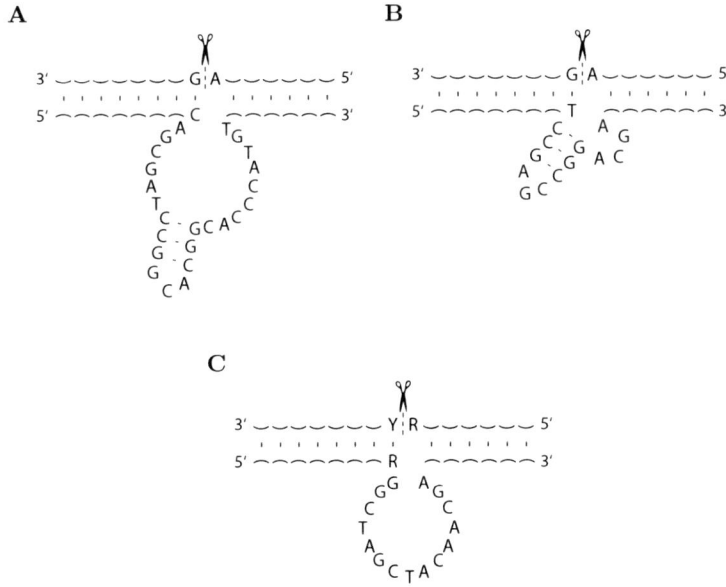

Figure 8. Secondary structures of three deoxyribozymes. Panel A shows a deoxyribozyme that resembles the general structure of the hammerhead ribozymes (cf. Fig. 6). It is characterised by a specific ('conserved') region of 15 nt connected to a stem-loop and is capable of cleaving substrates that contain a G-A-joint. A deoxyribozyme with a different structure, but also applicable only to substrates sequences with a G-A-joint is shown in panel B. The deoxyribozymes in panels A and B both have been applied for information processing [13]. Panel C shows a deoxyribozyme that is less constrained in the substrate junction it will cleave [53]. The notation for the binding region is: Y∈{C, T} and R∈{A, G}.

processing [7]. His method employed enzymes only to stabilise (through covalent bonds) the products of a self-assembly process (hybridisation of partially complementary oligonucleotides) but not in the information processing itself, and accordingly did not require enzymes with sequence specificity.

More recently these two lines of thought have come together with the use of nucleic acid enzymes [13–15, 59, 60]. Key to this approach is the possibility to control the activity of a ribozyme or deoxyribozyme with oligonucleotides. Such allosterically controlled nucleic acid enzymes have been investigated as sequence specific biosensors, where they have the advantage over molecular beacons that they catalytically amplify the recognition event [61]. The concept of allosteric ribozymes is illustrated in Fig. 9. Within certain constraints, the base sequence for the binding site of the control oligonucleotide (labeled OBS in the figures) can be chosen of the sequence on which the nucleic acid enzyme will act. It is therefore possible to have an oligonucleotide sequence start (or stop) the production of another, largely independent, oligonucleotide sequence. Moreover, it is possible to engineer nucleic acid enzymes to be controlled by more than one oligonucleotide. For instance, the molecule shown in Fig. 10 was designed by adding an allosteric control to thy deoxyribozyme shown in Fig. 8B [13]. It is inactive unless two effector

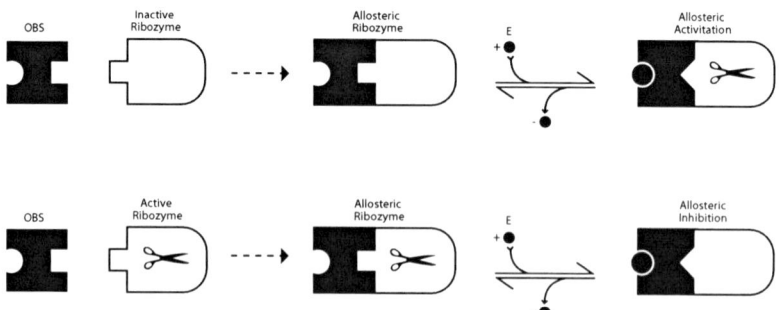

Figure 9. Allosterically activated ribozyme (top) and allosterically inhibited ribozyme (bottom) [62, 63]. The allosteric ribozyme is composed of two components (left of the dashed arrow), a oligonucleotide binding site (OBS) and a ribozyme part. The two components are covalently bound and from a single nucleic acid molecule (centre). Upon binding an effector oligonucleotide (E) the conformation of the binding site changes and affects the conformation of the ribozyme component. The latter conformational change will activate (top) or inhibit (bottom) the catalytic activity of the ribozyme part. The same scheme can also be realised with deoxyribozymes.

molecules with specific base sequences are present. The behaviour of the molecule can be interpreted as an AND logic gate. Note, however, that the possibility to catalyse the production of oligonucleotides as output signal with a base sequence independent of the sequences that serve as input signals (effector molecules) allows for applications other than logic AND operations.

A hammerhead ribozyme requiring the presence of two specific oligonucleotides for it to become active is shown in Fig. 11. While the deoxyribozyme gate in Fig. 10 is inactivated by blocking the substrate binding site, the ribozyme in Fig. 11 is controlled by a different mechanism. In the absence of effector oligonucleotides the molecule will self-hybridise to form a structure that is not a ribozyme. Hybridisation with the effector molecules overcomes the self-hybridisation of the inactive conformation and the molecule changes into a structure with a hammerhead ribozyme component. A comparison of the the multi-branch loop on the far right of Fig. 11 with the structural requirements of a hammerhead ribozyme depicted in Fig. 6 reveals how the straitening of the oligonucleotide bind-

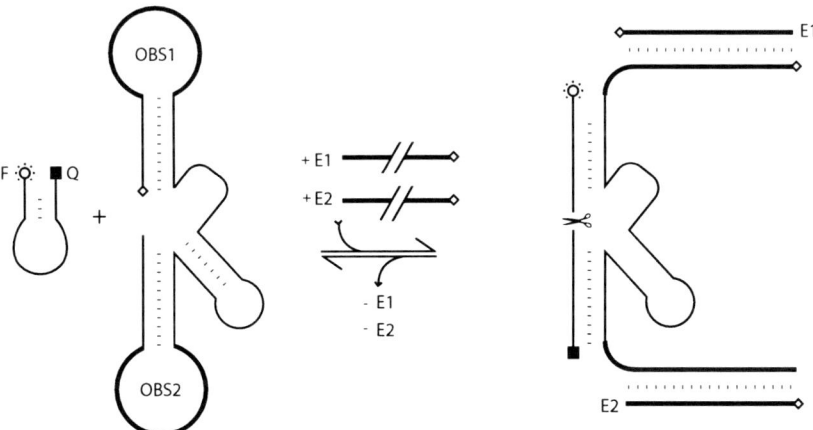

Figure 10. Deoxyribozyme acting as a logic AND gate after [13]. The molecule is designed in such a way that it can self-hybridise to block its own substrate binding site. This self-hybridisation is weaker than the binding of the effector molecules (E1, E2) to their oligonucleotide binding sites (OBS1, OBS2). Only in the presence of both effector molecules is the substrate binding site accessible and accordingly the deoxyribozyme catalytically active. By supplying a molecular beacon (far left) as substrate the output of the gate can be determined optically. If the deoxyribozyme is catalytically active it will cleave the beacon molecule, thus separate the quencher (Q) from the fluorophore (F), and consequently give rise to a fluorescence signal.

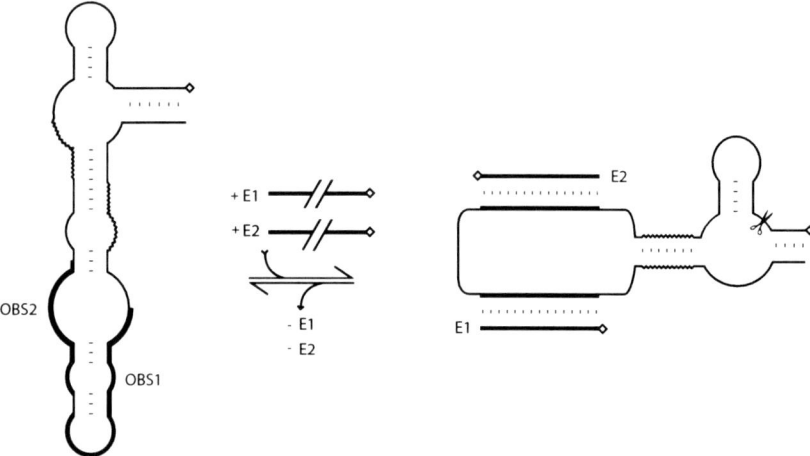

Figure 11. Two-input molecular switch based on allosterically controlled hammerhead ribozyme after [15]. In the absence of effector molecules (E1, E2) the inactive conformation (left) is more stable. Upon binding of the effector molecules to their corresponding oligonucleotide binding sites (OBS1, OBS2) the ribozyme changes into a catalytically active conformation (right). Crucial for the formation of the hammerhead conformation is the correct self-hybridisation in the helix II region shown by crinkled lines in both conformations. The oligonucleotide binding sites are indicated by bold lines in both conformations.

ing sites upon hybridisation with two DNA effector molecules induces catalytic activity.

 The conformational dynamics of RNA molecules allows for a relatively straight forward design of allosteric control structures into known ribozymes along the line of the concept represented in Fig. 9. Accordingly hammerhead ribozymes have been engineered with a wide variety of effector molecules [62]. One strategy is to add effector binding sites at the crucially important helix II of the hammerhead structure (cf. Fig. 6). Due to the conformational flexibility of RNA it is then likely that an effector molecule binding to the ribozyme will affect the helix II conformation and thus disrupt the catalytic function.

 For the application of ribozymes as signal processing components RNA structures that can be controlled with nucleic acid oligonucleotides as effector molecules are of particular interest. This is the case because the control oligonucleotide may conceivably be the product of a reaction catalyzed by another ribozyme and therefore enable the implementation of small molecular control networks. Different approaches to controlling a hammerhead ribozyme by means of oligonucleotide effectors are illustrated in Fig. 12. All four have been demonstrated in experiments [64, 66–68]. The first three (A–C) follow a common design philosophy. Starting from the basic hammerhead ribozyme structure shown in Fig. 6 an RNA sequence is engineered that does not fulfil the requirements for a hammerhead ribozyme, but can overcome this deficiency by hybridising with an effector oligonucleotide.

The earliest implementation of an engineered allosteric control mechanism in a ribozyme [64] is based on an RNA molecule that can form a hammerhead ribozyme, but has a preferred secondary structure that does not resemble the hammerhead motif and shows no catalytic activity Fig. 6A. The self-hybridisation that stabilises the preferred conformation (left side in Fig. 6A.) can be overcome by a suitable effector molecule, the binding of which is energetically more favourable than the self-hybridisation. Upon binding the effector molecule the RNA sequence folds into an active hammerhead conformation. This control strategy is the one that has been used in the molecular switch shown in Fig. 11 [15].

The hammerhead motif of the ribozyme in Fig. 12B is inactivated by self-hybridisation between the 3'-end of the ribozyme and its conserved junction region [66]. Between the region of the ribozyme participating in helix III and the region near the 3'-end that is complementary to part of the conserved core is an effector binding site. The binding of an oligonucleotide effector to the binding region is energetically favoured over the self-hybridisation in the core region. Accordingly the binding of the effector releases the hybridisation of the core and activates the hammerhead structure. As mentioned earlier, the helix II is a necessary part of the hammerhead motif and its stability is important for the enzymatic activity of hammerhead ribozymes [28]. Fig. 6C shows a control strategy based on a RNA sequence that contains the essential components of a hammerhead motif short of the complementary regions that could form the helix II. Binding of the effector induce a pseudo-half-knot structure that together with helix formed between the effector strand and the ribozyme apparently forms a pseudo-stem capable of activating the ribozyme [67].

In contrast to the three allosteric ribozymes just described, the ribozyme shown in Fig. 6D is always in a catalytically active state and can cleave a sequence that will bind fully to form helix I and helix III (cf. Fig. 6 for helix positions). However, the catalytic activity with regard to substrate sequences that bind only partially in the helix III region can be controlled by an effector molecule [68]. The effector binds to the dangling 3'-end of the ribozyme and the dangling 5'-end of a suitable substrate. It thus facilitates the binding between the ribozyme and a substrate that would not be cleaved without the effector. Note, that

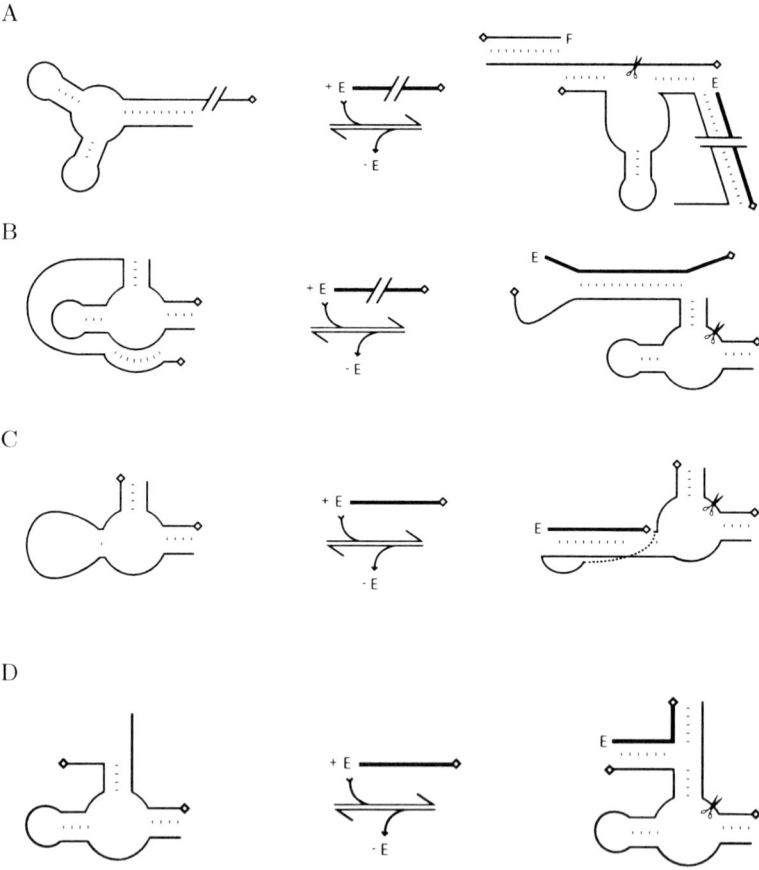

Figure 12. Four different strategies to control a hammerhead ribozyme. In all cases the ribozyme is active only in the presence of an oligonucleotide effector (E). Panel A: Formation of hammerhead structure upon binding of effector [64]; a DNA facilitator strand (F) enhances the binding of substrate to ribozyme [65]. Panel B: Effector releases conserved core junction from hybridisation [66]. Panel C: Effector enables formation of helix II [67]. Panel D: Effector supports binding of substrate [68].

the effector sequence in this case will influence on which substrates the ribozyme acts.

The combination of molecular motifs found in nature, molecules developed through directed evolution, and rational design decisions have led to a set of allosterically controlled functional nucleic acids suitable as components for simple molecular information processors. From the diverse family of catalytically active nucleic acids that have been found [47, 50, 69] it appears likely that the set of available components will grow.

5   Design of functional RNA for computation

Nucleic acids cannot compete with proteins when it comes to conformational dynamics or self-assembly . But from the practical perspective of implementation, using nucleic acids has the important advantage that both their conformations and their self-assembly can be predicted far more readily than the behaviour of proteins.

Computational tools are available for predicting the likely secondary structure of RNA molecules and for suggesting sequences that can be expected to fold to a desired structure. In particular the prediction of RNA secondary structure has received much attention in the literature and mature software is available for this task. *MFOLD* [70] and *RNAfold* in the *Vienna Package* [71] are commonly applied. They differ in the model used to calculate the free energy of folded RNA structures with the former using nearest neighbour rules to assign free energies to loops, while the latter employs a partition function [72] to estimate the probabilities of folding alternatives. Both programs can also be employed for predicting DNA conformation by substituting the appropriate thermodynamic parameters [18].

With a view to computing, predicting the folding of a single nucleotide strand in isolation, however, is not sufficient. As exemplified by the allosteric regulation mechanisms discussed in the previous section, the secondary structure which will be assumed by a nucleic acid strand may depend on the presence or absence of other oligonucleotides. Predicting the interaction among multiple nucleic acid molecules is therefore of particular interest. The *MFOLD* [73] program and the *Vienna Package* [74,75] offer the prediction of RNA-RNA and DNA-DNA co-folding, but lack the thermodynamic parameters for predicting RNA-DNA interactions. The *RNAsoft* package [76] allows for predicting the interaction of multiple RNA strands.

For designing RNA molecules with specific properties it is often useful if a secondary structure can be specified and a suitable nucleotide sequence that will fold into the specified structure is generated from the specification. This, so called, inverse folding problem is addressed by *RNAinverse* [71], *RNAdesigner* [76], and *INFORNA* [77]. Generally there is a large number of possible sequences that will fold into a given secondary structure; the aforementioned programs typically provide one arbitrary sequence from this pool.

By combining these tools Penchovsky and Breaker devised a workflow to engineer information processing units from nucleic acid sequences [15]. They used this workflow to design the biomolecular AND gate illustrated in Fig. 9. In the following we will describe the approach of [15] in more detail and subsequently discuss a modification to the workflow that places less restrictions on the inactive conformation. The design of a simple gate will serve to exemplify the approach and to investigate the design space. The simplest logic gates have only one bit input. Two such gates exit. The NOT gate that inverts the input and the PASS gate (sometimes also called "identity" or YES gate) that forwards the input signal. From a purely logic viewpoint PASS gates serve no purpose, in practice they can reform a degraded signal or adjust signal delay [3]. The molecular pass gates considered below are more powerful than one-bit logic gates and an essentially arbitrary input sequence of limited length can be recoded into a different output sequence.

The workflow employed by Penchovsky and Breaker to engineer an RNA PASS gate [15] starts with a design for an approximately 80 nucleotides long RNA molecule which contains a highly sensitive hammerhead ribozyme in its sequence. The bases for the sequence of the molecule are fixed, except a region of about 10–20 nucleotides long which, however, is crucial for maintaining the active hammerhead conformation. If this region can participate in internal hybridisation the molecule will undergo a large change in conformation to a catalytically inactive state. The binding of an effector oligonucleotide to this region will prevent internal hybridisation and thus stabilise the catalytically active conformation of the hammerhead ribozyme. Accordingly, the region acts as an oligonucleotide binding site (OBS, cf. section 4) that exerts allosteric control over the catalytic activity of the ribozyme.

The aim is to design the sequence for this OBS such that it is likely to allow the switching between the active and inactive state in a real RNA molecule. This is achieved by first selecting a candidate sequence for the OBS and inserting it into the fixed sequence of the sensitive hammerhead ribozyme. The sequence is generated by randomly assigning bases to the positions in the sequence while obeying the constraint in the first row of Tab. 1. To judge the plausibility for this RNA sequence design to be practicable and likely to be operative if implemented as a real

Table 1. The constraints imposed on potential candidate sequences following [15].

Stage Filter	Condition to satisfy
1 Identical nucleotides	No more than three identical consecutive nucleotides in the oligonucleotide binding site(s)
2 Active state conformation	Active hammerhead conformation in the presence of effector
3 Base-pairing percentage	In the absence of effector(s) 30%–70% of the oligonucleotide binding region is hybridised
4 Energy gap	Energy gap between the inactive and active state is within -6 kcal/mol to -10 kcal/mol
5 Temperature tolerance	Structure is preserved over a temperature range of $20°$–$40°$C
6 Ensemble diversity	For neither active nor inactive state the ensemble diversity (cf. [78]) exceeds 9 units
7 Folding efficiency	The RNA molecule must fold, in the absence of the effector, to the inactive conformation within 480 units in *Kinfold* [79]

RNA molecule Penchovsky and Breaker introduced a filter cascade, the steps of which are summarised as rows two to six in Tab. 1. If a generated sequences passes these five filter steps it is taken as a model design for the secondary structure of the OBS region in the desired gate. This model is specified by the complete secondary structure and a partial sequence which commits to all bases except those located in the the OBS region. By repeatedly running *RNAinverse* with this specification one obtains a set of complete sequences for the logic gate which differ from each other only in the OBS region. Note that the secondary structure of sequences generated by *RNAinverse* may not strictly conform to the specified conformation, but does not differ by more than two base pairs. Only sequences that have a thermodynamic stability comparable to the model design are maintained. The folding efficiency of these sequences is then verified (last row of Tab. 1). In [15] a second stage of processing which derives from the sequence designs that have successfully passed the filter chain alternative sequences with similar folding and thermodynamic stability is suggested. In our tests the yield of this second stage was only about 3% better than that of the first stage. The value of the second stage presumably lies in providing sequence alternatives to designs that are already favoured, e.g., because they have been verified experimentally. In contrast to the highly constraint design procedure of Penchovsky and Breaker [15] outlined above, we introduce now a protocol for designing RNA gates. This protocol relaxes the constraints on the secondary structure of the inactive conformation a gate can assume. Yet, in general, the length of functional nucleotide sequences composed of four bases in arbitrary order gives rise to a combinatorially large design space in which a random search without appropriate constraints would not be efficient in generating useful designs. However, the design space spanned by the RNA gate designs found in the literature can be used to narrow the search. In Tab. 2 the properties of existing implementations of RNA logic gates are summarised. The parameter ranges derived from the nucleic acid gates described in the literature are shown in Tab. 3; these values were used in the design of the PASS gates described below.

Our protocol for designing an allosterically controlled ribozyme comprises three generating steps, outlined in Tab. 4, to arrive at a sequence design. The generating of the sequence is followed by a series of validation

Table 2. Structural elements of biomolecular logic gates designed from deoxyribozymes and ribozymes. For OBS, helices, hairpin loops, bulges, and internal loops the entries in the table give the size of the structural element in nucleotides.

Gate	Length	OBS	Number of Junctions	Helices	Hairpin loops	Bulges	Internal loops	Ref.
PASS$_1$	60	15	2	3, 7	3, 15	-	-	[13]
PASS$_2$	63	15	2	5, 6	4, 15	-	-	[60]
PASS$_3$	80	22	3	5, 8, 16	4, 7	1, 1	3	[15]
PASS$_3$	80	22	3	5, 8, 16	4, 7	1	4	[15]
NOT$_1$	52	15	1	5	15	-	-	[13]
NOT$_2$	92	22	3	4, 10, 13	6, 6	-	2, 4, 8	[15]
AND$_1$	78	15, 15	3	3, 8, 9	3, 15, 15	5	-	[13]
AND$_2$	85	15, 15	3	5, 6, 6	4, 15, 15	-	-	[60]
AND$_3$	73	15, 15	2	8, 9	15, 15	-	-	[59]
AND$_4$	112	16, 16	3	5, 8, 21	4, 7	1	2, 5, 10	[15]
OR	103	20, 20	3	5, 8, 23	4, 7	1, 1	2, 4, 6	[15]
a AND $\neg b$	76	15, 15	2	5, 7	15, 15	-	-	[13]

Table 3. Parameters for structure generation derived from Tab. 2.

Type	Probability	Maximum no.	Length Range
Helix	0.50	-	4–15
Hairpin Loop	-	0–3	4–15
Internal Loop	0.45	0–3	2–8
Bulge	0.05	0–1	1–8
Junction	-	0–3	4–8
OBS	-	1	15–22
Linker	0.55	2	0–5

steps. In the first generating step the conformation for the catalytically active ribozyme is determined by specifying the secondary structure of an extension to the hammerhead core composed of helix II, two linkers and an OBS region. To this end for each position in the sequence its participation in internal hybridisation is selected by generating a dot-bracket representation for the secondary structure of the molecule. In generating the secondary structure constraints derived from Tab. 2 and detailed in Tab. 3 are invoked by a generation algorithm that follows [80].

The second generating step of the protocol assigns nucleotides to the positions in the sequence, except the OBS region. This assignment of the nucleotides adheres to the secondary structure generated in the first step. It repeatedly uses a simplified version of the initialisation routine of *RNAdesigner*. Rather than applying the subsequent refinement of the nucleotide assignment offered by *RNAdesigner* [80], we reject sequence assignments that do not fold to the specified secondary structure in *RNAfold* and repeat the initialisation. For the folding tests of the nucleotide assignment, the unassigned OBS region is set to a repetition of a hypothetical non-binding base as suggested by Penchovsky and Breaker [15].

The hammerhead ribozyme can reliably be deactivated by binding to its conserved core-region and thus distorting its secondary structure [62,81–83]. Therefore, in the third generating step of then protocol, first a sequence complementary to the conserved CUGAUGAG-region of the hammerhead core is inserted at a random location within the two linkers and the OBS, i.e., in the hairpin loop attached to helix II. Afterwards the remaining unassigned positions in the sequence are filled by drawing randomly from the four possible nucleotides {A, U, C, G}. The resulting sequence is likely to be inactive due to internal hybridisation of the hammerhead core with a section of the OBS, parts from a linker, or

Table 4. Proposed computational procedure for designing allosterically controlled hammerhead ribozyme gates. In contrast to the method described in [15] the procedure starts with the conformation of the active ribozyme.

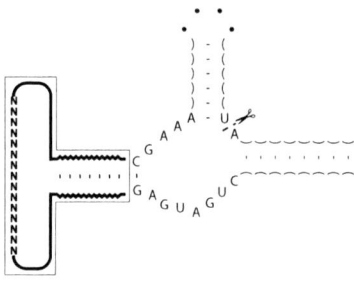

Design of the secondary structure of a potential gate based on a hammerhead core. First the structure (but not sequence) of the helix II and associated allosteric control domain is generated. This extension of the ribozyme is comprised of four parts: a helix that attaches to the ribozyme core, an effector binding region (NN···NN), and two linker sequences connecting binding site and helix. The lengths of the extension (23–62 nt) and the helix (4–15 nt) are chosen randomly. The remaining part of the structure (binding region and linkers) is filled with the constraints listed in Tab. 3.

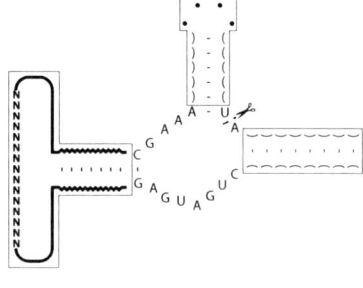

Assign the conserved nucleotides of the hammerhead core. The remaining sequence positions except the binding site are assigned by searching for a base sequence that will fold into the structure designed in the previous step. This can be achieved with *RNAinverse from Vienna* [71], *RNAdesigner from RNAsoft* [76, 80], or *INFORNA* [77]. To arrive at the active conformation of the gate, a non-binding pseudo-base (N) is assigned to all positions in the binding region during the search process.

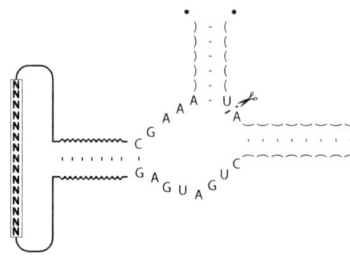

Replace the pseudo-bases in the binding region with real bases. This is done such that the structural elements initially generated for the linker and binding parts of the extension become manifest. This can be verified with *MFOLD* [70] or *RNAfold from Vienna* [71].

from helix II. In the folding predictions a hybrid molecule composed of ribozyme and substrate is considered and an interference of the substrate with the OBS is unlikely.

The subsequent validation of the generated sequence designs involves all steps of the filter cascade in Tab. 1. This screening process prunes out 99% of the generated sequence candidates as depicted in Fig. 13. Starting with a pool of 58 423 candidates as input to the first filter stage, 586 designs passed the entire filter chain. A manual inspection of the dot-plot graph [71] for all 586 designs confirmed in every case that the conserved sequence region of the hammerhead core is blocked by hybridisation in the inactive conformation and free in the active conformation. To further evaluate the plausibility of the remaining computational designs we calculate the equilibrium constants for the three possible dimers that can form when ribozyme molecules and effector molecules interact. The calculation is based on the free-energy values provided by *RNAcofold* [71, 74]. and the assumption of a fixed, equal concentration for the monomeric ribozyme (R) and monomeric effector (E) [75]. Any point in the area of the triangle depicted in Fig. 14 corresponds to a calculated (cf. Eq. 13

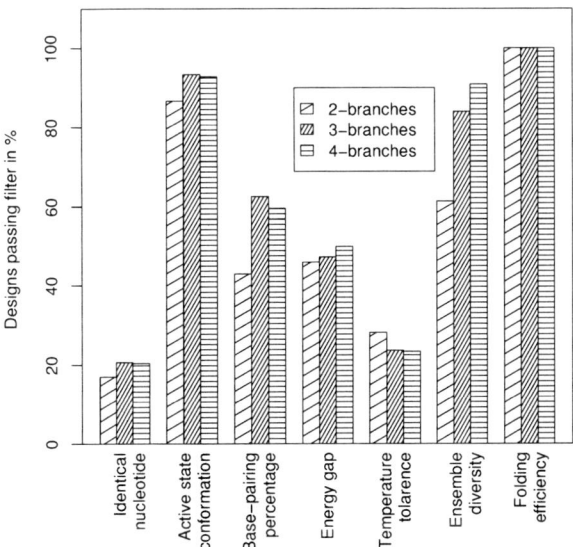

Figure 13. Permissiveness of each step in the filter chain of Tab. 1 when applied to generated candidate sequences for PASS gates. The filter stages are applied consecutively from lefty to right. For each step the sequences adhering to the filter condition is shown as percentage of input sequences supplied to this filter stage.

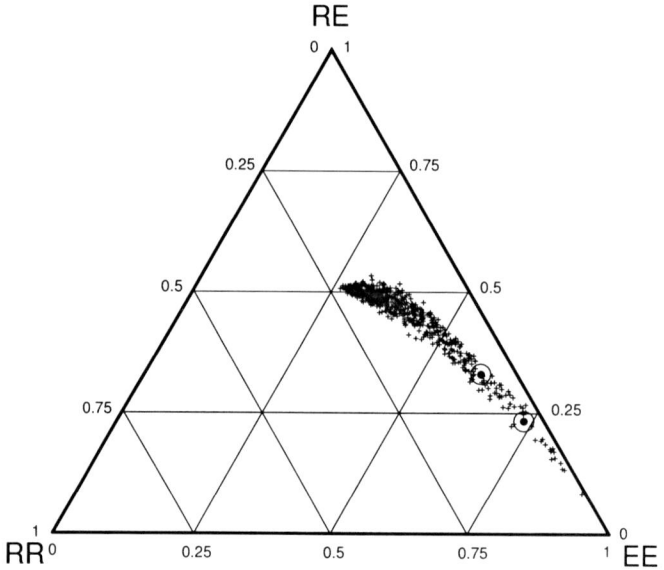

Figure 14. Estimated binding between ribozyme (R) and effector (E) for different PASS gate designs. The plot indicates the values obtained for PASS gates that have been designed with the method of Tab. 4 and evaluated with the filter chain in Tab. 1 labeled as +, and the two experimentally validated designs from [15] labeled ⊙.

of [84]) combination of the relative concentrations of the three possible dimers that can form (RR, EE, RE) . From this analysis it appears that the enlarged degrees of freedom in the design procedure outlined in Tab. 4 can yield PASS-gate designs with good ribozyme-effector binding (RE). Note, however, that due to the lack of tools for simulating RNA-DNA hybridisation, the values shown for the designs from Penchovsky and Breaker have been calculated for and RNA effector, while in [15] an experimentally more convenient DNA effector molecule was applied.

Samples of structures that were derived with the procedure outlined in table 4 and have passed the screening with the filter chain in Tab. 1 are shown in Fig. 15. The structural features of these inactive state conformations are determined in the generation process outlined in the first line of Tab. 4. The three structures in panel A, B, and C are representative for the classes of molecules that have inactive conformations with 2-branches, 3-branches, and 4-branches, respectively. In each structure the oligonucleotide binding region for the effector molecule is indicated by a bold line section. The structure in panel A bears some resemblance to a design proposed by Porta and Lizardi [64] (cf. Fig. 12A), while the structure in panel B is similar to the designs by Penchosky and Breaker [15].

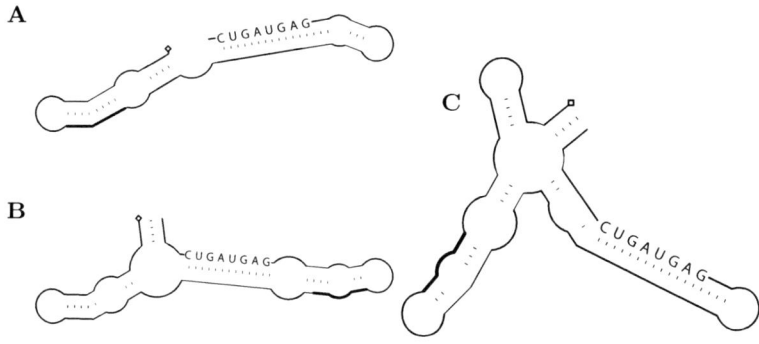

Figure 15. Inactive conformation of allosterically controlled hammerhead ribozymes designed to act as PASS gates.

6 Conclusion

Ribonucleic acids are an attractive computing substrate for three reasons. Firstly, they support both basic paradigms of molecular computing: self-assembly and conformational switching. Secondly, the intramolecular and intermolecular hybridisation of nucleic acids can reasonably well be predicted with existing computational tools. Thirdly, nucleic acids play a very important role for memory and control in every living cell. Ribonucleic acids are also challenging to work with in the laboratory, however. Yet, these challenges are at present more manageable than the computational challenge one would face if one would attempt to design information processing components with nature's main information processing substrate, namely proteins.

The application of allosterically controlled nucleic acids in bioimmersive computation has the potential to open up extraordinary possibilities. Smart drugs that can sense the internal state of cell and intervene in the intracellular regulatory mechanisms may come within reach [85] and engineered molecular control mechanisms that can be integrated into cells would be a powerful tool for life-science research [86].

Before the potential of these long-term aims can be realised obstacles in the laboratory need to be tackled and much better computational design procedures are required. A crucial issue will be the prediction of the interactions within complex mixtures of molecules. At present secondary structure prediction for multiple RNA strands is at its infancy and simulation tools capable of predicting DNA-RNA interactions do not exist. It will be necessary to develop a general methodology and supporting computational tools to create purpose-designed sets of interacting allosterically controlled nucleic acids.

Acknowledgements

E.R. is grateful to the Ministry of Higher Education Malaysia and the University Malaya for financial support.

References

[1] Petty, M.C., Bryce, M.R., Bloor, D., eds.: An Introduction to Molecular Electronics. Oxford University Press, Oxford (1995)

[2] Adamatzky, A., De Lacy Costello, B., Asai, T.: Reaction-Diffusion Computers. Elsevier Science, Amsterdam (2005)

[3] Zauner, K.P.: Molecular information technology. Critical Reviews in Solid State and Material Sciences **30** (2005) 33–69

[4] Conrad, M.: Self-assembly as a mechanism of molecular computing. In: Proceedings of the 11th Annual International IEEE-EMBS Conference, Piscataway, NJ, IEEE (1989) 1354–1355

[5] Conrad, M.: Molecular computing: The lock-key paradigm. Computer (IEEE) **25** (1992) 11–20

[6] Conrad, M.: Emergent computation through self-assembly. Nanobiology **2** (1993) 5–30

[7] Adleman, L.M.: Molecular computation of solutions to combinatorial problems. Science **266** (1994) 1021–1024

[8] Mao, C., LaBean, T.H., Reif, J.H., Seeman, N.C.: Logical computation using algorithmic self-assembly of DNA triple-crossover molecules. Nature **407** (2000) 493–496

[9] Winfree, E.: Algorithmic self-assembly of DNA: Theoretical motivations and 2d assembly experiments. J. of Biomolecular Structure & Dynamics **11** (2000) 263–270

[10] Hampp, N.: Bacteriorhodopsin as a photochromic retinal protein for optical memories. Chem. Rev. **100** (2000) 1755–1776

[11] Zauner, K.P., Conrad, M.: Enzymatic computing. Biotechnology Progress **17** (2001) 553–559

[12] Baron, R., Lioubashevski, O., Katz, E., Niazov, T., Willner, I.: Elementary arithmetic operations by enzymes : A model for metabolic pathway based computing. Angew. Chem. Int. Ed. **45** (2006) 1572–1576

[13] Stojanovic, M.N., Mitchell, T.E., Stefanovic, D.: Deoxyribozyme-based logic gates. Journal of the American Chemical Society **124** (2002) 3555–3561

[14] Stojanovic, M.N., Semova, S., Kolpashchikov, D., Macdonald, J., Morgan, C., Stefanovic, D.: Deoxyribozyme-based ligase logic gates and their initial circuits. Journal of the American Chemical Society **127** (2005) 6914–6915

[15] Penchovsky, R., Breaker, R.R.: Computational design and experimental validation of oligonucleotide-sensing allosteric ribozymes. Nature Biotechnology **23** (2005) 1424–1433

[16] Berg, J.M., Tymoczko, J.L., Stryer, L.: Biochemistry. Fifth edn. W. H. Freeman and Company, New York (2003)

[17] Bloomfield, V.A., Crothers, D.M., Tinaco, Jr., I.: Nucleic Acids: Structures, Properties and Functions. First edn. University Science Books, California (2000)

[18] SantaLucia, Jr., J., Hicks, D.: The thermodynamics of DNA structural motifs. Annual Review of Biomolecular Structure **33** (2004) 415–440

[19] Tinoco, Jr., I., Bustamante, C.: How RNA folds. Journal of Molecular Biology **293** (1999) 271–281

[20] Varani, G., McClain, W.H.: The G·U wobble base pair. EMBO reports **1** (2000) 18–23

[21] Draper, D.E.: Strategies for RNA folding. Trends in Biochemistry Sciences **21** (1996) 145–149

[22] Zuker, M.: On finding all suboptimal foldings of an RNA molecule. Science **244** (1989) 48–52

[23] Higgs, P.G.: RNA secondary structure: physical and computational aspects. Quarterly Reviews of Biophysics **33** (2000) 199–253

[24] Altman, S.: Enzymatic cleavage of RNA by RNA (Nobel lecture). Angewandte Chemie International Edition **29** (1990) 749–758

[25] Breaker, R.R.: DNA enzymes. Nature Biotechnology **15** (1997) 427–431

[26] Doudna, J.A., Cech, T.R.: The chemical repertoire of natural ribozymes. Nature **418** (2002) 222–228

[27] Symons, R.H.: Small catalytic RNAs. Annual Review of Biochemistry **61** (1992) 641–671

[28] Birikh, K.R., Heaton, P.A., Eckstein, F.: The structure, function and application of the hammerhead ribozyme. Eur. J. Biochemistry **245** (1997) 1–16

[29] Ferré-D'Amaré, A.R., Zhou, K., Doudna, J.A.: Crystal structure of a hepatitis delta virus ribozyme. Nature **395** (1998) 567–574

[30] Lafontaine, D.A., Norman, D.G., Lilley, D.M.J.: Structure folding and activity of the VS ribozyme: Importance of the 2-3-6 helical junction. EMBO Journal **20** (2001) 1415–1424

[31] Fedor, M.J., Williamson, J.R.: The catalytic diversity of RNAs. Molecular Cell Biology **6** (2005) 399–412

[32] Forster, A.C., Symons, R.H.: Self-cleavage of plus and minus RNAs of a virusoid and a structural model for the active sites. Cell **49** (1987) 211–220

[33] Long, D.M., Uhlenbeck, O.C.: Self-cleaving catalytic RNA. FASEB Journal **7** (1993) 25–30

[34] Wedekind, J.E., McKay, D.B.: Crystal structure of lead-dependent ribozyme revealing metal binding sites relvant to catalysis. Nature Structural Biology **6** (1999) 261–268

[35] Symons, R.H.: Plant pathogenic RNAs and RNA catalysis. Nucleic Acids Research **25** (1997) 2683–2689

[36] Walter, N.G., Burke, J.M.: The hairpin ribozyme: Structure, assembly and catalysis. Current Opinion in Chemical Biology **2** (1998) 24–30

[37] McCall, M.J., Hendry, P., Jennings, P.A.: Minimal sequence requirements for ribozyme activity. In: Proceedings of the National Academy of Sciences, USA. Volume 89. (1992) 5710–5714

[38] Lilley, D.M.J.: Structure, folding and catalysis of the small nucleolytic ribozymes. Current Opinion in Structural Biology **9** (1999) 330–338

[39] Doherty, E.A., Doudna, J.A.: Ribozyme structures and mechanisms. Annual Review of Biochemistry **69** (2000) 597–615

[40] Tanner, N.K.: Ribozymes: The characteristics and properties of catalytic RNAs. FEMS Microbiology Reviews **23** (1999) 257–275

[41] Hertel, K.J., Herschlag, D., Uhlenbeck, O.C.: A kinetic and thermodynamic framework for the hammerhead ribozyme reaction. Biochemistry **33** (1994) 3374–3385

[42] Fedor, M.J.: Structure and function of the hairpin ribozyme. Journal of Molecular Biology **297** (2000) 269–291

[43] Porschke, D., Burke, J.M., Walter, N.G.: Global structure and flexibility of hairpin ribozymes with extended terminal helices. Journal of Molecular Biology **289** (1999) 799–813

[44] Hohng, S., Wilson, T.J., Tan, E., Clegg, R.M., Lilley, D.M.J., Ha, T.: Conformational flexibility of four-way junctions in RNA. Journal of Molecular Biology **336** (2004) 69–79

[45] Nagai, K., Mattaj, L.W.: RNA-Protein Interactions. First edn. Oxford University Press, New York (1994)

[46] Joyce, G.F.: Directed molecular evolution. Scientific American **267** (1992) 48–55

[47] Famulok, M., Szostak, J.W.: In vitro selection of specific ligand-binding nucleic acids. Angewandte Chemie International Edition **31** (1992) 979–988

[48] Breaker, R.R., Joyce, G.F.: Inventing and improving ribozyme function: Rational design versus iterative selection methods. Trends in Biotechnology **12** (1994) 268–275

[49] Chapman, K.B., Szostak, J.W.: In vitro selection of catalytic RNAs. Current Opinion in Structural Biology **4** (1994) 618–622

[50] Tang, J., Breaker, R.R.: Structural diversity of self-cleaving ribozymes. In: Proceedings of the National Academy of Sciences, USA. Volume 97. (2000) 5784–5789

[51] Breaker, R.R., Joyce, G.F.: A DNA enzyme that cleaves RNA. Chemistry & Biology **1** (1994) 223–229

[52] Breaker, R.R., Joyce, G.F.: A DNA enzyme with Mg^{2+}-dependent RNA phosphoesterase activity. Chemistry & Biology **2** (1995) 655–660

[53] Santoro, S.W., Joyce, G.F.: A general purpose RNA-cleaving DNA enzyme. In: Proceedings of the National Academy of Sciences, USA. Volume 94. (1997) 4262–4266

[54] Joyce, G.F.: Directed evolution of nucleic acid enzymes. Annual Review of Biochemistry **73** (2004) 791–836

[55] Seeman, N.C.: DNA in a material world. Nature (2003) 427–431

[56] Emilsson, G.M., Breaker, R.R.: Deoxyribozymes: New activities and new applications. Cellular and Molecular Life Sciences **59** (2002) 596–607

[57] Paul, N., Springsteen, G., Joyce, F.: Conversion of a ribozyme to a deoxyribozyme through in vitro evolution. Chemistry & Biology **13** (2006) 329–338

[58] Liberman, E.A.: Analog-digital molecular cell computer. BioSystems **11** (1979) 111–124

[59] Stojanovic, M.N., Stefanovic, D.: Deoxyribozyme-based half-adder. Journal of the American Chemical Society **125** (2003) 6673–6676

[60] Stojanovic, M.N., Stefanovic, D.: A deoxyribozyme-based molecular automaton. Nature Biotechnology **21** (2003) 1069–1074

[61] Kuwabara, T., Warashina, M., Taira, K.: Allosterically controllable ribozymes with biosensor functions. Current Opinion in Chemical Biology **4** (2000) 669–677

[62] Soukup, G.A., Breaker, R.R.: Nucleic acid molecular switches. Trends in Biotechnology **17** (1999) 469–476

[63] Silverman, S.K.: Rune goldberg goes (ribo)nuclear? molecular switches and sensors made from RNA. RNA **9** (2003) 377–383

[64] Porta, H., Lizardi, P.M.: An allosteric hammherhead ribozyme. Bio/Technology **13** (1995) 161–164

[65] Goodchild, J.: Enhancement of ribozyme catalytic activity by a contiguous oligodeoxynucleotide (facilitator) and by 2'-O-methylation. Nucleic Acids Research **20** (1992) 4607–4612

[66] Burke, D.H., Ozerova, N.D.S., Nilsen-Hamilton, M.: Allosteric hammerhead ribozyme TRAPs. Biochemistry **41** (2002) 6588–6594

[67] Komatsu, Y., Yamashita, S., Kazama, N., Nobuoka, K., Ohtsuka, E.: Construction of new ribozymes requiring short regulator oligonucleotides as a cofactor. Journal of Molecular Biology **229** (2000) 1231–1243

[68] Wang, D.Y., Lai, H.Y., Feldman, A.R., Sen, D.: A general approach for the use of oligonucleotide effectors to regulate the catalysis of RNA-cleaving ribozymes and DNAzymes. Nucleic Acids Research **30** (2002) 1735–1742

[69] Ellington, A.D., Szostak, J.W.: In vitro selection of RNA molecule that bind specific ligands. Nature **346** (1990) 818–822

[70] Zuker, M., Mathews, D.H., Turner, D.H.: Algorithm and thermodynamics for RNA secondary structure prediction: A practical guide. In: Proceedings of a joint FEBS Advanced Course and NATO Advanced Study Insitute. (1998) 11–17 Cotober 1998, Poznań, Poland.

[71] Hofacker, I.L., Fontana, W., Stadler, P.F., Bonhoeffer, L.S., Tacker, M., Schuster, P.: Fast folding and comparison of RNA secondary structures. Chemical Monthly **125** (1994) 167–188

[72] McCaskill, J.S.: The equilibrium partition function and base pair binding probabilities for RNA secondary structure. Biopolymers **29** (1990) 1105–1119

[73] Markham, N.R.: Hybrid: a software system for nucleic acid folding, hybridizing and melting prediction. Master's thesis, Rensselaer Polytechnic Institute (2003)

[74] Mückstein, U., Tafer, H., Hackermüller, J., Bernhard, S., Stadler, P.F., Hofacker, I.L.: Thermodynamics of RNA-RNA binding. Bioinformatics (2006)

[75] Bernhart, S.H., Tafer, H., Mückstein, U., Flamm, C., Stadler, P.F., Hofacker, I.L.: Partition function and base pairing probabilities of RNA heterodimers. Algorithms for Molecular Biology **1** (2006) doi: 10.1186/1748-7188-1-3.

[76] Andronescu, M., Anguirre-Hernández, R., Condon, A., Hoos, H.H.: RNAsoft: A suite of RNA secondary structure prediction and design software tools. Nucleic Acids Research **31** (2003) 3461–3422

[77] Busch, A., Backofen, R.: INFO-RNA - a fast approach to inverse RNA folding. Bioinformatics **22** (2006) 1823–1831

[78] Penchovsky, R., Ackermann, J.: DNA library design for molecular computation. Journal of Computational Biology **10** (2003) 215–229

[79] Flamm, C., Fontana, W., Hofacker, I.L., Schuster, P.: RNA folding at elementary step resolution. RNA **6** (2000) 325–338

[80] Andronescu, M., Fejes, A.P., Hutter, F., Condon, A., Hoos, H.H.: A new algorithm for RNA secondary structure design. Journal of Molecular Biology **336** (2004) 607–624

[81] Tang, J., Breaker, R.R.: Rational design of allosteric ribozymes. Chemistry & Biology **4** (1997) 453–459

[82] Koizumi, M., Soukup, G.A., Kerr, J.N.Q., Breaker, R.R.: Allosteric selection of ribozymes that respond to the second messengers cGMP and cAMP. Nature Structural Biology **6** (1999) 1062–1071

[83] Breaker, R.R.: Engineered allosteric ribozymes as biosensor components. Current Opinion in Biotechnology **13** (2002) 31–39

[84] Schuster, P.: Prediction of RNA seconday structures: from theory to models and real molecules. Reports on Progress in Physics **69** (2006) 1419–1477

[85] Benenson, Y., Gil, B., Ben-Dor, U., Adar, R., Shapiro, E.: An autonomous molecular computer for logical control of gene expression. Nature **429** (2004) 423–429

[86] Simpson, M.L.: Rewiring the cell: synthetic biology moves towards higher functional complexity. Trends in Biotechnology **22** (2004) 555–557

On the computational complexity of physical computing systems

Ed Blakey

The Queen's College, Oxford, OX1 4AW
edward.blakey@queens.ox.ac.uk

Abstract By far the most studied model of computation is the Turing machine model. It is the suggestion of Church's Thesis that, as far as *computability* is concerned, consideration of this model alone is sufficient; but what of *complexity*?

The traditional notion of computational resource, in terms of which computational complexity may be defined (since, in essence, complexity is nothing more than required resource viewed as a function of input size), caters almost exclusively for algorithmic models of computation such as the Turing machine model. Accordingly, the most commonly encountered computational resource is run-time of a Turing machine or algorithm or equivalent.

We here extend the definition of resource, notably so as to include the physical notion of precision with which we can make measurements; this allows characterization in a more insightful way of the complexity of computations performed by analogue, DNA, quantum and other physical computers.

The primary intent of this (ongoing) work is not that it be of practical use by, for example, offering physical solution methods that improve upon the speed, efficiency or similar offered by existing—especially digital—computers. Rather, the work is a theoretical study: the wider, more physical framework of computation is presented as a context in which to generalize the closely related notions of resource and complexity. The ultimate aim, from this theoretical viewpoint, is to shed light on whether complexity is inherent in *problems* as opposed to *solution methods*, whether the latter be physical or algorithmic.

1 Introduction

Computational complexity is a measure of the resource needed to perform a computation. More specifically, if we suppose that the computation is the evaluation of a function (or the evaluation of any of several values of a multifunction) f at a given point (which supposition, for our purposes, loses no generality), then the complexity evaluated at n can be thought of as the greatest resource, as x ranges through values of size n, required to evaluate $f(x)$.

Perhaps because a vast majority of practical computation conforms to an algorithmic model (specifically, real-world computations are typically performed by digital computers running programs that implement algorithms), and perhaps because of an overestimation of the ambit of Church's Thesis,[1] resource is, traditionally, virtually always taken to be a property—usually run-time, but sometimes space or another measure satisfying Blum's axioms[2]—of an algorithm, or, equivalently, a Turing machine or similar.

Some methods of computing, however, are such that these algorithm-focused measures fail to capture the true complexity of the system. For example, an analogue computer that evaluates function f at point x may have run-time, memory usage, etc. that are independent of x, but require that a physical measurement be made with precision that increases with the size of x. The relevant resource, then—a resource which is overlooked by the traditional notion—is precision of measurement; the constant time and space complexities of the system offer less insight than its 'precision complexity,' with which we quantify the required precision of measurement as a function of the size of x.

Many physical systems, such as the analogue computer of the previous paragraph, do not, therefore, fit meaningfully into the traditional hierarchy of complexity classes, simply because these classes are defined in terms of measures—notably run-time—that do not accurately reflect the systems' complexity.

We propose, instead, a notion of complexity that considers the *dominant* resource, be it time, space, precision of measurement, etc. (or, equivalently,[3] the *total* resource) required to perform a computation as a function of the size of the input.[4] We define corresponding complexity

---

[1] Church's Thesis is introduced in [7] and discussed in [5] and [12], amongst many others.

[2] Blum's axioms, introduced in [3], ensure that (a) a measure of resource is defined precisely at inputs at which the computation being measured is defined, and (b) it is a decidable problem to confirm that a given value is indeed the measure of resource corresponding to a given input.

[3] The equivalence here holds because each computational system uses only a finite number of resource types, of which, by definition of dominance (see Definition 3.11), the dominant and total share common $\mathcal{O}$-notation behaviour.

[4] In some contexts, however, we retain the distinction between time, space, precision and other complexities.

classes. By construction, this notion of complexity includes as a special case the existing one, since, if the most relevant resource of a system is an algorithmic property, then the two definitions coincide for this system.

Remark 1.1. Note that, in a great majority of cases, the direct deduction of a *problem's* complexity is impossible; instead, we may place upper bounds on its complexity by considering that of specific *solution methods*. By being able to measure in a comparable and consistent way the complexity not only of Turing machines and equivalent, but also of physical/analogue computing systems, the set of considered solution methods for a problem becomes more complete, and the bound on the complexity of the problem more accurate.

Before we extend formally the notions of resource and complexity, motivated in part by Remark 1.1, consider the following examples of physical computing systems, on the precision complexity of which we informally comment.

2 Examples of physical systems

Apart from the provision of a means of input and output, an algorithm, Turing machine or similar is a closed, self-contained system, within which all calculation is performed. A physical (e.g. analogue) computer, on the other hand, may tacitly exploit the fact that it sits in a universe that itself 'calculates' according to physical laws: the physical system can defer some processing to its environment.

For example, a ball rolling from rest down a slope of constant incline travels a distance proportional to the square of the time for which it rolls; such a physical system may, therefore, be used to calculate squares. Whereas a Turing machine that squares its input neither knows nor cares about the presence of gravity, then, a physical computer may exploit such.

With this distinction between self-contained algorithms and environment-exploiting analogue systems in mind, consider the following examples of physical computation.

2.1 Addition

Suppose that we wish to find the sum of two given natural numbers n_1 and n_2. This addition can be performed by (naïvely simplistic) physical means by cutting two pieces p_1 and p_2 of some material with respective lengths n_1 and n_2 units, placing the pieces end-to-end, and measuring their combined length.

We consider now the accuracy with which we need to cut p_1 and p_2 and measure the combined length in order to guarantee arrival at the correct result $N := n_1 + n_2$. To this end, we model the various errors arising in the system and define the way in which to interpret non-integer combined lengths. We consider first an additive, then a multiplicative, model of error; which is the more appropriate is determined by the exact details of implementation of the physical process.

Additive error It may be the case that the error introduced during the cutting of p_1 and p_2 is best modelled additively, in that (for $j \in \{1, 2\}$) addend n_j is represented by p_j's length l_j, which we know only to lie in the interval $[n_j - \epsilon_{i_1}, n_j + \epsilon_{i_1}]$ for some non-negative, real *input error* term ϵ_{i_1}. This occurs, for example, if the ruler with which we measure p_j can slip by up to ϵ_{i_1} units before we cut p_j.

The combined length $l_1 + l_2$ (which addition the physical system performs accurately), then, lies in the interval $[N - 2\epsilon_{i_1}, N + 2\epsilon_{i_1}]$, where, recall, $N = n_1 + n_2$. If we measure this combined length with additive *output error* $\epsilon_{o_1} \geq 0$ (that is, length l is measured as a member of $[l - \epsilon_{o_1}, l + \epsilon_{o_1}]$)—for example to allow for slipping of the ruler as before, though this time during measurement of the combined length—then we shall obtain a result R in the interval $[N - \epsilon_b, N + \epsilon_b]$, where $\epsilon_b = 2\epsilon_{i_1} + \epsilon_{o_1}$.

In order to interpret R, we assume—reasonably—that our physical process is such that values of ϵ_b closer[5] to zero occur with greater probability than those further from zero; we note also that the sought total N is a natural number. Accordingly, we interpret R as the natural number requiring the least ϵ_b, which, in this case, is the nearest natural number to R.[6] Hence, if $R < \lfloor R \rfloor + \frac{1}{2}$, then we interpret R as $\lfloor R \rfloor$; else, we interpret R as $\lceil R \rceil$.

We are guaranteed the correct answer, in that our interpretation of R is N, then, if and only if the interval $[N - \epsilon_b, N + \epsilon_b]$ in which we know R to lie is a subset of the interval $[N - \frac{1}{2}, N + \frac{1}{2})$ of which elements are interpreted as N. This is the case if and only if both $N - \epsilon_b \geq N - \frac{1}{2}$ and $N + \epsilon_b < N + \frac{1}{2}$, which is the case if and only if $\epsilon_b < \frac{1}{2}$.

Since this constraint on ϵ_b, and hence the constraint on the precision of measurement in the system, does not depend on N (or, in any other way, on the input values), our intuition should be that the precision complexity of the system is constant. Formally, the region

---

[5] In this paragraph, 'closer', 'further' and 'nearest' are defined in terms of the distance $|a - b|$ between real numbers a and b.

[6] If there are two such natural numbers, we arbitrarily choose the greater of the two.

$\{(\epsilon_{i_1}, \epsilon_{o_1}) : \epsilon_{i_1}, \epsilon_{o_1} \geq 0 \wedge 2\epsilon_{i_1} + \epsilon_{o_1} < \frac{1}{2}\}$ in the plane with axes ϵ_{i_1} and ϵ_{o_1}—that is, the region within which the system is guaranteed to perform correctly the addition of n_1 and n_2—has area $\frac{1}{16}$, regardless of N. The precision complexity, then, which one would expect intuitively to depend solely and inversely on this area—and which in Definition 3.7 we accordingly define to be one divided by this area—is 16, again regardless of N.

Note that the system also has an intuitive notion of space complexity that is linear in N, since p_1 and p_2, the respective lengths of which total N units, are physically realized. Similarly, drawing a measure along the combined length of p_1 and p_2 takes an amount of time proportional to N; the system has time complexity linear in N.

The overall complexity of the system, to which precision complexity does not contribute, is therefore linear in N.

Multiplicative error Suppose instead that the errors in our physical adding process are better modelled multiplicatively, so that (for $j \in \{1, 2\}$) addend n_j is represented by p_j's length l_j, which is in the interval $\left[\frac{n_j}{\epsilon_{i_1}}, \epsilon_{i_1} n_j\right]$ for some real input error term ϵ_{i_1} satisfying $\epsilon_{i_1} \geq 1$. This occurs, for example, if we measure p_1 and p_2 with a unit-length ruler, of which each use can suffer a slip of up to ϵ_{i_1} units.

The combined length $l_1 + l_2$, then, is in the interval $\left[\frac{N}{\epsilon_{i_1}}, \epsilon_{i_1} N\right]$. If we measure this with multiplicative output error $\epsilon_{o_1} \geq 1$ (so that our measurement of a length l lies in $\left[\frac{l}{\epsilon_{o_1}}, \epsilon_{o_1} l\right]$), then we shall obtain a result R in the interval $\left[\frac{N}{\epsilon_b}, \epsilon_b N\right]$, where $\epsilon_b = \epsilon_{i_1}\epsilon_{o_1}$.

In order to interpret R, we assume, analogously to the additive case above, that ϵ_b is likely to be small, and note that N is a natural number. We accordingly interpret R as the natural number requiring the least ϵ_b (making an arbitrary choice—here, the larger—in the event of non-uniqueness): if $R < \sqrt{\lfloor R \rfloor (\lfloor R \rfloor + 1)}$, then we interpret R as $\lfloor R \rfloor$; else, we interpret R as $\lceil R \rceil$.

We are guaranteed the correct answer, in that our interpretation of R is N, then, if and only if the interval $\left[\frac{N}{\epsilon_b}, \epsilon_b N\right]$ in which we know that R lies is a subset of the interval $\left[\sqrt{N(N-1)}, \sqrt{N(N+1)}\right)$ of which elements are interpreted as N. This is the case if and only if $\frac{N}{\epsilon_b} \geq \sqrt{N(N-1)}$ and $\epsilon_b N < \sqrt{N(N+1)}$, which is the case if and only if $\epsilon_b < \sqrt{\frac{N+1}{N}}$.

In contrast to the additive case above, this constraint on ϵ_b depends on N. Specifically, the region $\left\{ (\epsilon_{i_1}, \epsilon_{o_1}) : \epsilon_{i_1}, \epsilon_{o_1} \geq 1 \wedge \epsilon_{i_1} \epsilon_{o_1} < \sqrt{\frac{N+1}{N}} \right\}$— within which the system is guaranteed to perform correctly the addition of n_1 and n_2—has area $\int_1^r \frac{r}{\epsilon_{i_1}} - 1 d\epsilon_{i_1} = r \ln r - (r-1)$, where $r = \sqrt{\frac{N+1}{N}}$. The precision complexity, then, is $(r \ln r - (r-1))^{-1}$, which is bounded above by $\frac{8}{3} N (3N + 2)$ and below by N^2, and is hence in $\mathcal{O}(N^2)$.

Further, the system has space and time complexities in $\mathcal{O}(N)$, the former for the same reason as the additive-error system above, and the latter because of the repeated use of the unit-length ruler. Hence, precision is the most relevant resource when considering this system's overall complexity, which is seen to be quadratic in N.

2.2   Greatest common divisor

Consider now the problem of finding the greatest common divisor of two given natural numbers. Recall that the traditional method, Euclid's algorithm, has time and space complexities logarithmic in the input values[7] (and constant precision complexity, since there is no notion of input/output error in the algorithmic model of which it is an instance—see Theorem 3.1).

The following physical method of solution exploits the behaviour (in particular, the interference) of transverse waves.

Suppose that we wish to find the greatest common divisor of positive natural numbers n_1 and n_2. Suppose further that our apparatus allows us to instantiate two transverse standing waves with wavelengths of our choosing, and that these waves are superposed so as to interfere. In particular, we instantiate in the interval $[0, 2\pi)$ two waves with respective wavelengths $\frac{4\pi}{n_1}$ and $\frac{4\pi}{n_2}$, travelling from 2π to zero, and reflect the waves at zero so as to produce standing waves w_1 and w_2. Then, for $i \in \{1, 2\}$, w_i has set M_{n_i} of maxima, where, for $j \in \mathbb{N}^+$,
$$M_j := \left\{ x \in [0, 2\pi) : \frac{jx}{2\pi} \in \mathbb{Z} \right\} = \left\{ 0, \frac{2\pi}{j}, \frac{4\pi}{j}, \ldots, \frac{2(j-1)\pi}{j} \right\}; \ w_i, \text{ therefore,}$$
has $|M_{n_i}| = n_i$ maxima in the interval $[0, 2\pi)$.

Recall that w_1 and w_2 are allowed to interfere. Since $M := \bigcap_{i=1}^2 M_{n_i}$ is non-empty (zero, for example, is in this intersection), the resultant superposition wave w has maxima, with a value twice that of the maxima of the individual waves, at points at which both w_1 and w_2 have maxima; that is, precisely at elements of M.

---

[7] Lemma 11.7 of [11] establishes the former; the latter follows from the observation that space complexity cannot '$\mathcal{O}$-exceed' time complexity, simply because moving a Turing machine tape head to a new cell takes time.

Proposition 2.1. $M = M_{\gcd(n_1, n_2)}$.

Proof. Suppose that $x \in M_{\gcd(n_1, n_2)}$; then $x \in [0, 2\pi)$, and

$$x = \frac{2\pi m}{\gcd(n_1, n_2)} \tag{1}$$

for some $m \in \mathbb{Z}$. Since $\gcd(n_1, n_2)$ divides both n_1 and n_2, $\gcd(n_1, n_2) = \frac{n_1}{k_1} = \frac{n_2}{k_2}$ for some $k_1, k_2 \in \mathbb{Z}$. By (1), then, $x = \frac{2\pi m k_1}{n_1}$, which is in M_{n_1} since $x \in [0, 2\pi)$ and $m k_1 \in \mathbb{Z}$; similarly, $x = \frac{2\pi m k_2}{n_2} \in M_{n_2}$. Hence, $x \in M$.

Conversely, suppose that $x \in M = \bigcap_{i=1}^{2} M_{n_i}$; then $x \in [0, 2\pi)$, and $x = \frac{2\pi m_1}{n_1} = \frac{2\pi m_2}{n_2}$ for some $m_1, m_2 \in \mathbb{Z}$. So $\frac{2\pi}{x}$ is a common divisor of n_1 and n_2, and hence divides their *greatest* common divisor $\gcd(n_1, n_2)$; say $\gcd(n_1, n_2) = \frac{2\pi m}{x}$ with $m \in \mathbb{Z}$. Then $x = \frac{2\pi m}{\gcd(n_1, n_2)} \in M_{\gcd(n_1, n_2)}$. □

By measuring the frequency of the maxima of w—in effect, by counting $|M| = |M_{\gcd(n_1, n_2)}| = \gcd(n_1, n_2)$—then, we find the greatest common divisor of n_1 and n_2. Further, this calculation is performed with constant time and space complexities (the process is virtually instantaneous, and, crucially, takes no longer as n_1 and n_2 increase; further, larger n_1 and n_2 require no larger apparatus); what of the system's precision complexity?

For this example, the constraints on input error are independent of those on output error; we may evaluate each individually. Consider first input error.

Recall that, as input to the system, the natural number values n_1 and n_2 are encoded as wavelengths $\frac{4\pi}{n_1}$ and $\frac{4\pi}{n_2}$; suppose that the actual, implemented wavelengths $\frac{4\pi}{l_1}$ and $\frac{4\pi}{l_2}$ suffer an additive error: for $j \in \{1, 2\}$, $\frac{4\pi}{l_j} \in \left[\frac{4\pi}{n_j} - \epsilon_{i_1}, \frac{4\pi}{n_j} + \epsilon_{i_1} \right]$, and so $l_j \in \left[\frac{4\pi n_j}{4\pi + \epsilon_{i_1} n_j}, \frac{4\pi n_j}{4\pi - \epsilon_{i_1} n_j} \right]$. Suppose further that the system corrects non-integer values of l_j by rounding to the nearest integer l_j' (which, arbitrarily, we take to be the greater of two equally near integers), for example by continuously adjusting the wavelength in the range $\left[\frac{4\pi}{l_j + \frac{1}{2}}, \frac{4\pi}{l_j - \frac{1}{2}} \right)$ until the corresponding wave has a maximum at 2π. Now, a small additive change in n_1 and n_2 rarely leads to a small additive change in $\gcd(n_1, n_2)$;[8] consequently, it is necessary (and sufficient)—in order that the correct computation be performed—that the nearest integer l_j' to l_j coincide with n_j. This occurs precisely

---

[8] This is because of the subtle interaction between the additive and multiplicative structures of the natural numbers, and, specifically, the seeming randomness of the primes.

when, for each $j \in \{1,2\}$, the interval $\left[\frac{4\pi n_j}{4\pi+\epsilon_{i_1}n_j}, \frac{4\pi n_j}{4\pi-\epsilon_{i_1}n_j}\right]$ in which l_j lies is a subset of the interval $\left[n_j - \frac{1}{2}, n_j + \frac{1}{2}\right)$ of which elements are rounded to n_j, which is the case if and only if $\frac{4\pi n_j}{4\pi+\epsilon_{i_1}n_j} \geq n_j - \frac{1}{2}$ and $\frac{4\pi n_j}{4\pi-\epsilon_{i_1}n_j} < n_j + \frac{1}{2}$, which is the case if and only if $\epsilon_{i_1} < \frac{4\pi}{n(2n+1)}$, where $n = \max\{n_1, n_2\}$.

Now consider output error. Suppose that the system can distinguish two maxima if and only if a distance of ϵ_{o_1} or more separates them. Then the measured output is accurate if and only if the distance $\frac{2\pi}{\gcd(l_1', l_2')}$ between consecutive maxima of the resultant wave w is no less than ϵ_{o_1}.

Hence, we have that the region (in the space spanned by axes ϵ_{i_1} and ϵ_{o_1}) in which the system correctly evaluates the greatest common divisor of n_1 and n_2 is $\left\{ (\epsilon_{i_1}, \epsilon_{o_1}) : 0 \leq \epsilon_{i_1} < \frac{4\pi}{n(2n+1)} \wedge 0 \leq \epsilon_{o_1} \leq \frac{2\pi}{\gcd(l_1', l_2')} \right\}$. One divided by the area of this region is $\frac{n(2n+1)\gcd(l_1', l_2')}{8\pi^2}$, which is cubic in $n = \max\{n_1, n_2\}$.

2.3   Other examples

There are many other existing examples of physical computing systems (which we acknowledge and bear in mind while developing our framework of computation, whilst not describing in detail):

- famously, soap bubbles formed between parallel plates can be used, since they tend naturally to states of minimal surface area, to solve the problem of finding a minimum-length spanning network connecting given points (see [9]);
- mechanical means can be used to solve differential and integral equations (see [6] for a description of the Differential Analyzer, for example);
- DNA computing techniques can tackle the directed Hamiltonian path problem (see [1]), amongst other graph-theoretic/combinatoric problems;
- and so on.

3   Definitions

3.1   Physical solution methods

Perhaps surprisingly, the notion of a physical computation system per se does not, for our purposes, require formal definition, since there are no set conditions that an object need satisfy in order to constitute such a system. Instead, it is how we view the system as a computing device, with particular regard to input and output, that is important here. Of interest is the system (or process) together with our choice from its physical attributes of what we consider to be the inputs and outputs of the computation. Objects under consideration, then, are physical processes, along with two sets—inputs and outputs—of distinguished physical parameters, and a rule for interpreting outputs.

Definition 3.1. *A* physical solution method *(or* PSM*) is a tuple* $\Phi :=$ (I, π, O, ι), *where:*

- *I is a tuple of positively and finitely many pairs* (i, V_i), *where i is an input of* Φ *and* V_i *a set of values;*
- π *is the* process *of* Φ;
- *O is a tuple of positively and finitely many pairs* (o, V_o), *where o is an* output *of* Φ *and* V_o *a set of values; and*
- ι *is the* interpretation function *of* Φ, *which maps each output value of* Φ *(see Definition 3.2) to an output value.*[9]

Remark 3.1. The intuition behind Definition 3.1 is as follows.

A physical computation system has parameters that admit alteration so as to supply input to the system; for coordinate (i, V_i) of I, i acts as an input parameter, and V_i as the set of values to which i can be set.

After setting the input parameters, a process—typically presented as a description of the physical set-up[10] along with an 'algorithm' consisting of physical instructions—is carried out;[11] π models this process.

Output is then taken from the system in the form of a measurement of certain parameters; each such is modelled by a coordinate (o, V_o) of O: o models the measurable parameter, and V_o the set of values that o may take.

Finally, the output value tuple measured by the user is converted by ι into an 'interpreted' output value tuple. This conversion can be thought of as a form of error correction: the interpretation function takes *arbitrary* output and interprets it as *corrected* output, typically based upon a priori knowledge about the form of the expected output.[12]

---

[9] In practice, ι is used to map *measured* output values to *interpreted* output values—see Definition 3.5.

[10] We do not assume that the same physical set-up can universally accommodate all input values; larger inputs may require larger apparatus with more physical storage, for example.

[11] The system may execute the process automatically, or user intervention, uniquely defined by the description of the process, may be required.

[12] For example, if the computation is such that the expected answer is an integer, then the measured output value, a real number, say, may by way of interpretation be rounded off to the nearest integer.

Note that 'physical instruction', 'parameter' and other terms used in this remark are neither formally defined nor used in Definition 3.1. This renders the class of PSMs sufficiently wide to include instances of many models of computation.

Definition 3.2. *Given a PSM* $\Phi = (I, \pi, O, \iota)$ *with* $I = \left((i_1, V_{i_1}), \ldots, (i_p, V_{i_p}) \right)$, *an* input value *of* Φ *is an assignment of 'allowed' values to the inputs of* Φ*; that is, a tuple* $\left(v_{i_1}, \ldots, v_{i_p} \right) \in V_{i_1} \times \ldots \times V_{i_p}$ *of values, where* i_j *is considered to be assigned* v_{i_j}.

Similarly, supposing that $O = \left((o_1, V_{o_1}), \ldots, (o_q, V_{o_q}) \right)$, *an* output value *of* Φ *is an assignment of 'allowed values' to the outputs of* Φ*; that is, a tuple* $\left(v_{o_1}, \ldots, v_{o_q} \right) \in V_{o_1} \times \ldots \times V_{o_q}$ *of values, where* o_j *is considered to be assigned* v_{o_j}.

Definition 3.3. *For a PSM* $\Phi = (I, \pi, O, \iota)$, *define the relation* C_Φ *to be the set of pairs* (x, y) *such that, on assigning to* Φ *the input value* x *and executing process* π *of* Φ, *(uninterpreted) output value* y *can result. As is normal with relations, write '$xC_\Phi y$' for '$(x, y) \in C_\Phi$'.*

Definition 3.4. *A PSM* Φ *is said to be* deterministic *(a D PSM) if, for every input value* x *of* Φ, *there exists a unique output value* y *such that* $xC_\Phi y$*; then* C_Φ *may be viewed as a function, and we denote this* y *by* $C_\Phi(x)$. *A PSM is said to be* non-deterministic *(an N PSM) if it is not deterministic.*

3.2   Error; Precision complexity

Definition 3.5. *Recall from Remark 3.1 the process by which a PSM performs a computation. We now elaborate on this process, and include in our model the fact that, in practice, errors can be introduced into the system during both the application of input values and the measurement of output values.*

There are four stages in performing a computation using a PSM Φ[13] *(say* $\Phi = (I, \pi, O, \iota)$ *with* $I = \left((i_1, V_{i_1}), \ldots, (i_p, V_{i_p}) \right)$ *and* $O = \left((o_1, V_{o_1}), \ldots, (o_q, V_{o_q}) \right)$*):*

1. INPUT. *The user attempts to apply to* Φ *his intended input value— the input value* $x := \left(v_{i_1}, \ldots, v_{i_p} \right)$ *at which he wishes to evaluate* C_Φ *(or, if* Φ *is non-deterministic, for which he wishes to find some* y *such that* $xC_\Phi y$*)—by adjusting the inputs of* Φ*. Due to lack of precision in this adjustment, however, the intended input value* x *may*

---

[13] Usually, Φ is deterministic, and the computation is evaluation of C_Φ at some input value.

not coincide with the implemented input value $x' := \left(v'_{i_1}, \ldots, v'_{i_p} \right)$ —
the actual input value received by the PSM. There is (assuming that the PSM is of practical use), however, a relationship between each v_{i_j} and v'_{i_j} in terms of some real error term ϵ_{i_j}—the specific input error *for i_j.*[14] *Define the* generic input error *(or input error) of Φ to be the tuple $\epsilon_I := \left(\epsilon_{i_1}, \ldots, \epsilon_{i_p} \right) \in \mathbb{R}^p$. Given an intended input value, then, the corresponding implemented input value is not necessarily uniquely determined, but is at least bounded in some sense by the input error. The error introduced whilst inputting a value may, therefore, be modelled by the* input error relation R_{ϵ_I}, *a relation, parameterized by ϵ_I, between intended and implemented input values.*

2. EXECUTION. *The user then executes the process π of Φ.*

3. MEASUREMENT. *Next, the user attempts to measure the output value from Φ. Again due to lack of precision, this time during measurement, the* true output value[15] $y' := \left(v'_{o_1}, \ldots, v'_{o_q} \right)$ —*that supplied by Φ—and the* measured output value $y'' := \left(v''_{o_1}, \ldots, v''_{o_q} \right)$ —*that measured by the user—may not coincide. Again, however, there is a relationship between each v'_{o_j} and v''_{o_j} in terms of some real error term ϵ_{o_j}—the* specific output error *for o_j. Let the* generic output error *(or output error) of Φ be the tuple $\epsilon_O := \left(\epsilon_{o_1}, \ldots, \epsilon_{o_q} \right) \in \mathbb{R}^q$, and let the* error[16] *of Φ be the pair $\epsilon := (\epsilon_I, \epsilon_O) \in \mathbb{R}^p \times \mathbb{R}^q$.*[17] *Given a true output value, then, the corresponding measured output value is not necessarily uniquely determined, but is at least bounded in some sense by the output error. The error introduced during measurement of an output value, as during input, may, therefore, be modelled by a relation parameterized by ϵ_O, the* output error relation R_{ϵ_O} *between true and measured output values.*

---

[14] For example, it may be the case that v_{i_j} and v'_{i_j} differ by no more than specific input error $\epsilon_{i_j} \geq 0$; cf. n_j and l_j in Sect. 2.1.

[15] Note that this value y' satisfies $x' C_\Phi y'$. The physical system computes with perfect accuracy, even if the user's ability to input values and measure results is imperfect.

[16] For our purposes, the error of a PSM is an independent variable—we are free to alter the error, and wish to consider those values of the error for which the system performs the correct computation (at least given input of a certain size).

[17] We sometimes view error ϵ as an element of $\mathbb{R}^{p+q}$, using the obvious map $((a_1, \ldots, a_p), (b_1, \ldots, b_q)) \mapsto (a_1, \ldots, a_p, b_1, \ldots, b_q)$ from $\mathbb{R}^p \times \mathbb{R}^q$ to $\mathbb{R}^{p+q}$. In such contexts, then, ϵ is the tuple concatenation, rather than the pair, of ϵ_I and ϵ_O.

4. INTERPRETATION. *Finally, the user applies the interpretation func-tion of Φ to the measured output value y'' to find the interpreted output value $y := \iota\,(y'')$.*

Say that intended input value x (Φ, ϵ)-yields interpreted output y if and only if there exist x', y' and y'' such that $x R_{\epsilon_I} x'$, $x' C_\Phi y'$, $y' R_{\epsilon_O} y''$ and $\iota\,(y'') = y$ (where $\epsilon = (\epsilon_I, \epsilon_O)$);[18] let $Y_{\Phi,\epsilon} = \{(x,y) : x\ (\Phi, \epsilon)\text{-yields } y\}$, and write '$x Y_{\Phi,\epsilon} y$' for '$(x,y) \in Y_{\Phi,\epsilon}$'.

Intuitively, if the specific input and output errors of a PSM Φ constrain sufficiently tightly that intended and implemented input, and also true and measured output, values be close together,[19] then the error in the system is sufficiently small that some intended input and corresponding yielded interpreted output values will be related by C_Φ—these small errors introduced during input and measurement are successfully corrected during interpretation. We are, in particular, interested in the set of errors ϵ that guarantee that an input value will be related by C_Φ to any corresponding (Φ, ϵ)-yielded interpreted output value; the behaviour (usually, the dwindling) of this set of 'good' errors as the input size increases is the basis of our notion of precision complexity. These ideas are now formalized.

Definition 3.6. *A size function for PSM Φ is a map from the set of input values of Φ to the set $\mathbb{R}^+$ of non-negative, real numbers.*

Definition 3.7. *Let Φ be PSM (I, π, O, ι) with $I = \big((i_1, V_{i_1}), \dots, (i_p, V_{i_p})\big)$ and $O = \big((o_1, V_{o_1}), \dots, (o_q, V_{o_q})\big)$, let σ be a size function for Φ, and let x be an input value for Φ.*

- *An error $\epsilon \in \mathbb{R}^{p+q}$ is said to be precise for x if, for all y such that $x Y_{\Phi,\epsilon} y$, it is the case that $x C_\Phi y$.[20]*
- *Let $\mathcal{E}_\Phi\,(x)$ denote the set $\{\epsilon \in \mathbb{R}^{p+q} : \epsilon$ is precise for $x\}$ of errors precise for x.*
- *Let $\mathcal{V}_\Phi\,(x)$ denote the Euclidean, $(p+q)$-dimensional volume[21] of $\mathcal{E}_\Phi\,(x)$. Thus, $\mathcal{V}_\Phi\,(x) \in [0, \infty]$.*

---

[18] Intuitively, an attempt by the user to input x, execute the process, and measure and interpret the output can result in the output y if and only if x (Φ, ϵ)-yields y.

[19] This can happen by the specific errors being close to zero if the error modelled is additive, one if multiplicative, and so on.

[20] Intuitively, then, an error is precise if and only if it is sufficiently 'small' that, at least on input x, it is corrected during interpretation.

[21] For our purposes, the choice of measure in terms of which multidimensional volume is defined is unimportant; we arbitrarily choose Lebesgue measure.

- Let the precision required by Φ given x be $P_\Phi(x) := \frac{1}{V_\Phi(x)} \in [0,\infty]$.[22]
- Define Φ's precision complexity relative to σ to be the function $PC_{\Phi,\sigma} : \mathbb{N} \to [0,\infty]$ with

$$PC_{\Phi,\sigma}(n) = \sup \{P_\Phi(x) : \sigma(x) = n\} \ .$$

When size function σ is understood, 'relative to σ' and the subscript 'σ' are sometimes omitted.

Remark 3.2. In many practical cases, we have the following. If input values x_1 and x_2 are such that $\sigma(x_1) \le \sigma(x_2)$, then, for a 'practically useful' PSM $\Phi = (I, \pi, O, \iota)$ and 'sensible' size function, we have that $\mathbb{R}^{|I|+|O|} \supseteq \mathcal{E}_\Phi(x_1) \supseteq \mathcal{E}_\Phi(x_2) \supseteq \emptyset$, whence $\infty \ge V_\Phi(x_1) \ge V_\Phi(x_2) \ge 0$ and $0 \le P_\Phi(x_1) \le P_\Phi(x_2) \le \infty$. PC_Φ is therefore a non-decreasing function of the size of an input value.

It can be seen that the conclusions of Sect. 2—notably that the additive- and multiplicative-error systems of addition have precision complexities respectively constant and quadratic in $n_1 + n_2$, and that the greatest common divisor system has precision complexity cubic in $\max\{n_1, n_2\}$[23] — remain valid under the formal definition of precision complexity.

Theorem 3.1. An instance (e.g. a Turing machine) of a traditional, algorithmic model of computation can be expressed as a PSM Φ, in that the relation between inputs and outputs of the instance is $Y_{\Phi,\epsilon}$ (where choice of ϵ is unimportant). Further, $PC_{\Phi,\sigma}(n) = 0$ for all n and σ.

Proof. Note that, in an algorithmic model of computation, there is assumed to be no error during input and measurement (and hence no need for interpretation): the intended input value is faithfully passed to the algorithm, which supplies an output that can be accurately read.

This situation is expressible in the more general, error-accommodating framework of PSMs defined above by letting the input/output error relations and the interpretation function be identity maps (and by letting the process of the PSM, $\Phi = (I, \pi, O, \iota)$, say, be a physical implementation— a digital computer with sufficient physical storage, for example—of the algorithmic instance itself).

Since, in particular, the input/output error relations R_{ϵ_I} and R_{ϵ_O} are chosen to be identity maps for any generic errors ϵ_I and ϵ_O, every error

---

[22] For this purpose, we use the standard extensions $\frac{1}{0} = \infty$ and $\frac{1}{\infty} = 0$ of real number division.

[23] Note that these precision complexities are essentially viewed as functions of input values rather than sizes of input values; we have, therefore, implicitly used an identity size function.

is precise for all input values. Hence, for all x, $\mathcal{E}_\Phi(x) = \mathbb{R}^{|I|+|O|}$, which has infinite volume, and so the precision complexity of an algorithm, Turing machine or similar is constantly $\sup\left\{\frac{1}{\infty}\right\} = 0$.   □

Hereafter, we assume that instances of algorithmic models of computation are presented as Turing machines. Church's Thesis suggests that, as far as computability is concerned, this is adequate; further, complexity seems to be affected only polynomially by change of algorithmic model (see Remark 3.3).

Definition 3.8. *In light of Theorem 3.1, define an* algorithmic solution method *(or ASM) to be a PSM with identity input/output error relations and interpretation function.*

3.3   Time complexity; Space complexity

To exploit fully the fact (proven in Theorem 3.1) that algorithms, Turing machines, etc. (viewed as ASMs) are a special case of PSMs, we define for PSMs measures of time and space complexity; naturally, when the solution method in question is an ASM, we wish these measures to agree with the existing algorithmic definitions. This extension of the class of solution methods of which we can ascertain such complexity measures allows fair comparison of ASMs with other PSMs, as well as meaningful comparison of different measures (time, space, precision, etc.) of complexity.

Definition 3.9. *Let Φ be a PSM, σ a size function, and x an input value.*

- *Define the* run-time *$T_\Phi(x) \in [0, \infty]$ of Φ given x to be the (possibly infinite) time taken from (a) supply to Φ of x until (b) receipt from Φ of an (interpreted) output value.*[24]
- *Define Φ's* time complexity *relative to σ to be the function $TC_{\Phi,\sigma} : \mathbb{N} \to [0, \infty]$ with*

$$TC_{\Phi,\sigma}(n) = \sup\{T_\Phi(x) : \sigma(x) = n\}\,[25]\;.$$

Remark 3.3. Recall, for example from Definition 2.5 of [11], the traditional notion of time complexity: run-time of a Turing machine given

---

[24] The units in which this time is measured are of no importance, since we primarily consider $\mathcal{O}$-notation behaviour. For the sake of run-time being well-defined, however, let us take seconds as our unit.

[25] As with Definition 3.7, 'relative to σ' and the subscript 'σ' are sometimes omitted when σ is understood.

some input is the (possibly infinite) number of time steps in the computation of the machine on the input, and the machine's time complexity evaluated at n is, as here, the supremum over inputs of size n of run-time.

Other algorithmic models—the random access machines of Sect. 2.6 of [11], for example—have similar notions of time complexity; further, for each random access machine, there is an equivalent[26] Turing machine with time complexity only polynomially different from that of the random access machine, and vice versa.

Theorem 3.2. *For all functions f, $\mathcal{O}(f)$ contains the time complexity (according to Definition 3.9) of an ASM if and only if it contains the traditionally defined time complexity (according to Definition 2.5 of [11], say) of the corresponding[27] Turing machine.*

Proof. Fix a function f.

Suppose that ASM Φ is such that $TC_\Phi \in \mathcal{O}(f)$. Since the process of Φ is a direct implementation of the corresponding Turing machine T, every operation—of which there are only finitely many—achievable by Φ in one second can be performed by T in a finite number of steps. Let s be the maximum of these finitely many finite numbers of steps. Then the run-time of T given some input is no more than s times the run-time (in seconds) of Φ given the same input; hence, T has time complexity in $\mathcal{O}(sf) = \mathcal{O}(f)$.

Similarly but conversely, suppose that a Turing machine T has time complexity in $\mathcal{O}(f)$. Let Φ be the corresponding ASM, which is an implementation (on a digital computer, for example) of T; each 'atomic operation'—of which there are only finitely many—that T can perform in one time step is implementable by the process of Φ in finite time. Let t be the maximum of these finitely many finite times. Then the run-time of Φ given some input is no more than t times the run-time of T given the same input; hence, $TC_\Phi \in \mathcal{O}(tf) = \mathcal{O}(f)$. $\square$

Definition 3.10. *Let Φ be a PSM, σ a size function, and x an input value.*

- *Define the space $S_\Phi(x)$ needed by Φ given x to be the physical, three-dimensional volume required by Φ to perform its computation with input x.[28]*

---

[26] Two algorithms or similar are *equivalent* if they compute the same function/relation.

[27] The correspondence here is in the sense of Theorem 3.1

[28] Similarly to Definition 3.9, the units in which this volume is measured are not important, since we primarily consider $\mathcal{O}$-notation behaviour. So that space is well-defined, however, let us measure in cubic metres.

– *Define Φ's space complexity relative to σ to be the function $SC_{\Phi,\sigma}$: $\mathbb{N} \to [0, \infty]$ with*

$$SC_{\Phi,\sigma}(n) = \sup\{S_\Phi(x) : \sigma(x) = n\}\,^{29}\;.$$

Remark 3.4. Compare with this definition the traditional notion of space complexity (given, for example, in Definition 2.6 of [11]): the space required by a machine on an input is essentially the number of tape cells used by the machine in processing the input. Space complexity evaluated at n is, as here, the supremum over inputs of size n of required space.

Theorem 3.3. *For all functions f, $\mathcal{O}(f)$ contains the space complexity (according to Definition 3.10) of an ASM if and only if it contains the traditionally defined space complexity (according to Definition 2.6 of [11], say) of the corresponding Turing machine.*

Proof. Note that there exists a multiplicative constant that bounds the size of each of a Turing machine cell and a unit of physical data storage in terms of the other[30]; then the proof mirrors that of Theorem 3.2. □

Recall from Sect. 2.2 that the problem of finding the greatest common divisor of two given natural numbers has:

– an algorithmic solution (Euclid's algorithm), which has time and space complexities logarithmic in the input values, and precision complexity constant in same; and
– a physical solution (presented in Sect. 2.2), which has time and space complexities constant in the input values, and precision complexity cubic in same.

The onus, it seems, can be moved from one resource type to another by selecting different solution methods.

When classifying problems according to their complexity, it is intuitively more insightful to describe, for example, Sect. 2.2's physical solution to the greatest common divisor problem as a *cubic-precision*, rather than *constant-time*, solution, since the former focuses on the more relevant resource. This notion of 'most relevant resource' is now formalized, and corresponding complexity classes are defined.

---

[29] 'Relative to σ' and the subscript '$_\sigma$' are sometimes omitted when σ is understood.

[30] This follows from the fact that there is not only a practical, but also theoretical, limit on the density of data storage (that is, the number of bits that can be stored per cubic meter, say).

Definition 3.11. – *A dominant resource type of a PSM Φ is a type A of resource (time, space, precision, etc.) such that, for any resource type B, the B complexity BC_Φ is in $\mathcal{O}(AC_\Phi)$.*[31]
- *For a resource type A and a function f, let $\mathbf{C}(f, A)$ be the class of problems for which there exists a DPSM Φ with dominant resource type A such that $AC_\Phi \in \mathcal{O}(f)$.*
- *For a resource type A and a function f, let $\mathbf{NC}(f, A)$ be the class of problems for which there exists an NPSM Φ with dominant resource type A such that $AC_\Phi \in \mathcal{O}(f)$.*
- *Let $\mathbf{C}(f)$ and $\mathbf{NC}(f)$ be the unions over all resource types of $\mathbf{C}(f, r)$ and $\mathbf{NC}(f, r)$ respectively.*

4   Quantum computing systems

We suggest that the framework of computation presented here is a context in which the true complexity of quantum computing systems may be captured.

An arbitrary algorithm (or Turing machine or similar) can, by definition of complete, be expressed as a conversion of input to output via operations exclusively taken from some complete set of what are deemed to be 'atomic' operations. For a given input value, the number of such operations performed during this conversion is an accurate measure (or, depending on viewpoint, a definition) of run-time.[32]

Similarly, an arbitrary quantum computing system can be expressed as the preparation of several quantum bits, followed by a sequence of applications to subsets of these quantum bits of 'atomic' unitary operations taken from a complete set, followed by a measurement of the system. As in the classical case, an enumeration of the invocations of these atomic operations gives a measure of the system's complexity; indeed, this is the basis of an existing definition of complexity of quantum computing devices (see pages 191–194 of [10]). Also as in the classical case, however, the result is essentially a measure of run-time, which is not, we suggest, particularly relevant—in the sense of the preamble to Definition 3.11—to quantum systems.[33]

---

[31] The slight abuse of notation here and in the remainder of this definition is such that, if resource type A is time, then 'AC_Φ' stands for 'TC_Φ', and so on.

[32] The more accurately the complete set reflects the environment (e.g. the chipset; specifically the atomic instructions thereof) in which the algorithm is implemented, the more accurate the measure.

[33] The advantage of quantum computers over their classical counterparts arises from the use of entangled states and the effective parallelism that such use

The broadened notions of resource and complexity presented here, particularly those defined in terms of precision, seem to better encapsulate the nature of the true complexity of a quantum system; this complexity, after all, arises because of our limited ability to take precise measurements from the system. We believe that introduction of a type of resource similar to required precision (Definition 3.7), and hence introduction of a corresponding measure of complexity (see Sect. 5.2, Further Resource Types) will offer for quantum computing systems a notion of complexity more insightful than run-time.

5   Conclusion

5.1   Summary

We note in Remark 1.1 that our reasoning about *problems'* complexity is done largely by considering specific *solution methods*, but that the methods traditionally analysed to this end—Turing machines and instances of other algorithmic models of computation—impose a restricted view of complexity. It is, therefore, desired to introduce a framework of computation that allows a broader analysis (making use, specifically, of new complexity measures) of more solution methods (including physical systems such as analogue and quantum computers, etc.).

We present in Sect. 2 examples of physical computing systems, with which the new framework deals particularly well; this offers an intuitive introduction to the chief ideas (notably error and precision complexity) of this paper.

In Sect. 3, we formalize the ideas of *physical solution methods*, the extended class of solution methods about the complexity of which we can reason (Definition 3.1); *input/output error* (Definition 3.5); and *precision complexity* (Definition 3.7). We embed the traditional class of solution methods (algorithms, etc.) in the extended class (Theorem 3.1) and endow the extended class with notions of *time* and *space complexities* (Definitions 3.9 and 3.10) that extend those already defined for the embedded subclass (Theorems 3.2 and 3.3). We also define the notion of *dominant resource*, and introduce *complexity classes* corresponding to the measures of complexity presented (Definition 3.11).

---

allows; a disadvantage is the strictly constrained way in which information can be read from the quantum system. The run-time of such a system, then, is a reflection of neither the 'amount of computation' being performed (due to the parallelism) nor the 'difficulty' in using the system (which chiefly arises during measurement).

5.2   Future work

Hierarchy of complexity classes. The obvious next step for this project (which, we reiterate, is ongoing) is to investigate the hierarchy in which the complexity classes of Definition 3.11 lie, and to establish the relationship between the new and existing hierarchies. Many results and open problems in the field of computational complexity concern the equality, containment, etc. of (traditional) complexity classes; by establishing the correspondence between the two hierarchies, these results can be transferred and open problems restated (with insight hopefully being offered by such restatement).

The nature of computation. As well as results concerning the hierarchies, there are fundamental questions about the nature of computation:

– Is complexity inherent in *problems*, or is it an artefact of our choice of *solution method*?
– What is the source of the 'true complexity' of quantum computers? of DNA computers? of analogue computers? . . .
– Given an arbitrary problem, how free are we to engineer the dominant resource type by selecting different solution methods?
– etc.

that may succumb to analysis in the framework presented here.

Further resource types Note that, regardless of type of resource, the corresponding measure of complexity and complexity classes are defined in terms of the resource type in the same way.[34] The motivation is that this allows fair comparison of different measures of complexity; a byproduct is that the framework is easily extensible: the introduction of a new resource type leads automatically to a new measure of complexity and corresponding complexity classes.

Another direction in which the work can be extended, then, is by considering further types of resource.[35]

---

[34] Specifically, the complexity (evaluated at n) is the supremum over inputs x of size n of the resource required by x, and the classes are as specified in Definition 3.11.

[35] Here, something like Blum's axioms is needed in order to constrain what is allowed to be a resource; since these axioms admit, for example, measures of complexity exponentially generous and exponentially stringent compared to standard time complexity, further criteria should also be stipulated.

Probabilistic computation Recall Definition 3.4: we allow computation to be non-deterministic, but have not explicitly considered the probabilities that accordingly arise. This offers a way in which the work can be extended; such extension would be especially useful when pursuing the ideas of Sect. 4.[36]

5.3   Other work

Existing work in the vein of this paper includes:

- presentation ([2]) of a Turing-machine-like framework to accommodate real-number computations, rather than traditional, $\{0, 1\}$-based (without loss of generality) computations, though without focus on precision;
- a study ([4]) of computability with (though not complexity of) a specific model of analogue computation; and
- a study ([13]) of analogue complexity, in which precision of measurement is seen as a factor constraining the set of problems that can be solved, rather than—as here—a dependent-variable property of problem instances and hence a resource type.

The differences (of which some are mentioned above) between these and the current work render the framework of computation and complexity presented here sufficiently novel to warrant study.

Acknowledgements We thank Bob Coecke and Joël Ouaknine for their support, supervision and suggestions.

References

[1] Adleman, L. M.: *Molecular Computation of Solutions to Combinatorial Problems.* Science **266** (1994) pp. 1021–1024
[2] Blum, L.; Cucker, F.; Shub, M.; Smale, S.: *Complexity and Real Computation.* Springer (1997)
[3] Blum, M.: *A Machine-Independent Theory of the Complexity of Recursive Functions.* J. of the Assoc. for Computing Machinery **14** no. 2 (1967) pp. 322–336
[4] Bournez, O.; Campagnolo, M.; Graça, D.; Hainry, E.: *The General Purpose Analog Computer and Computable Analysis are Two Equivalent Raradigms of Analog Computation.* Theory and Applications of Models of Computation (TAMC'06) **3959** LNCS (2006) pp. 631–643

---

[36] Such pursuit may also benefit from the ideas of [8] and similar.

[5] Bovet, D. P.; Crescenzi, P.: *Introduction to the Theory of Complexity.* Prentice Hall (1994)

[6] Bush, V.: *The Differential Analyzer: a New Machine for Solving Differential Equations.* J. of the Franklin Inst. **212** (1931) pp. 447–488

[7] Church, A.: *An Unsolvable Problem of Elementary Number Theory.* American J. of Math. **58** (1936) pp. 345–363

[8] Dowling, M. R.; Nielsen, M. A.: *The Geometry of Quantum Computation.* arXiv:quant-ph/0701004v1 (2006)

[9] Miehle, W.: *Link-Length Minimization in Networks.* Operations Research **6** no. 2 (1958) pp. 232–243.

[10] Nielsen, M. A.; Chuang, L.: *Quantum Computation and Quantum Information.* Cambridge University Press (2000)

[11] Papadimitriou, C.: *Computational Complexity.* Addison-Wesley (1995)

[12] Sipser, M.: *Introduction to the Theory of Computation.* PWS (1997)

[13] Vergis, A.; Steiglitz, K.; Dickinson, B.: *The Complexity of Analog Computation.* Mathematics and Computers in Simulation **28** no. 2 (1986) pp. 91–113

116

Mapping virtual self-assembly rules to physical systems

Navneet Bhalla[1], Peter J. Bentley[2], and Christian Jacob[1,3]

[1] Dept. of Computer Science, Faculty of Science, University of Calgary, 2500
University Drive N.W., Calgary, Alberta, Canada T2N 1N4
bhalla@cpsc.ucalgary.ca
[2] Dept. of Computer Science, University College London, Malet Place,
London, United Kingdom, WC1E 6BT
P.Bentley@cs.ucl.ac.uk
[3] Dept. of Biochemistry & Molecular Biology, Faculty of Medicine,
University of Calgary, 2500 University Drive N.W., Calgary, Alberta, Canada
T2N 1N4 cjacob@ucalgary.ca

Abstract Throughout nature, decentralized components emerge into complex forms. It is through their interaction that components, governed by simple rules, self-assemble to create specific entities. The programs constituting these entities are based on the rules present in a given system and are executed on the physically and chemically encoded information comprising the components and their environment. A three-level approach is presented here which encompasses specifying a set of rules, modeling these rules to determine the outcome of a specific system in software, and translating to a physical system based on the set of rules present. The benefit of this approach is that no knowledge of the end result is required to create the physical system, mirroring the bottom-up process in nature. Five experiments, based on an example implementation of this approach, show that the translated physical systems self-assemble into the desired entities achieved by the simulations. These successful results demonstrate how this three-level approach is used for mapping virtual self-assembly rules to physical systems.

1 Introduction

Entities in nature self-assemble, self-repair, self-reproduce, and in some instance are capable of morphogenesis. They are products of parallel construction processes. Their designs are autonomous, robust, and adaptive.

Only recently has our technology displayed, with limited capabilities, such characteristics from nature.

The creation of new technologies that mirror nature is applicable to further developing many fields including developing new non-classical computing paradigms, such as embodied computation. Expressing computation not purely mathematically, but also physically bears the potential for computing devices that are autonomous, robust, and adaptable to unexpected changes. Being able to create self-assembling systems is one of the keys to realizing such computing paradigms.

However, designing and creating self-assembling systems continues to be extremely challenging. One aspect that remains an open problem is how to design a set of components and their environmental conditions, such that the set of components self-assemble to create a desired entity. Techniques have been developed for a few specific applications, but a general approach has not been achieved [1] [2] [3] [4].

Natural self-assembly is primarily dictated by the morphology of the components and their environmental conditions, as well as the physical and chemical properties of the components and the environment within a given system [5] [6]. One could view entities in nature constituting programs that are based on the rules present in a given system [7]. The programs are executed on the physically and chemically encoded information, comprising the components and their environment.

A three-level approach for designing and creating self-assembling systems based on this view is presented here. This approach encompasses specifying a set of rules, modeling these rules to determine the outcome of a specific system in software, and translating to a physical system based on the set of rules present. This is consistent with the definition of self-assembly, refined here, as being a reversible process that can be controlled through appropriately designed components and their environment [8].

This approach is distinguished from the ones described above in two ways. (1) This approach translates a set of rules to create a physical system. (2) More importantly, this approach does not use global information derived from the target entity or the final configuration of components. This bottom-up aspect is critical in terms of embodied computation. For example, to perform a particular computation, a system may need to go through many states (i.e. physical structures) to complete. Consequently, an approach that needs to evaluate its end state in order to create a set of components will not scale effectively.

An example system is presented that demonstrates an implementation of this three-level approach together with five experiments. The example system and the experiments are used to demonstrate how this three-level approach is capable of mapping virtual self-assembly rules to physical systems under various constraints.

2 Background

L.S. Penrose and R. Penrose were the first to show a mechanical analogue to natural self-assembly, specifically self-reproduction in the form of templated self-assembly [9]. They created two component types, labeled A and B, that connected in either an AB or BA configuration (Figs. 1 and 2). Multiples of these A and B components were confined to a track in a random ordering, that when shaken, allowed components to move horizontally and interact with one another. By placing either an AB or a BA seed complex on the track, it would cause neighbouring A and B or B and A components to self-assemble into AB and BA complexes respectively.

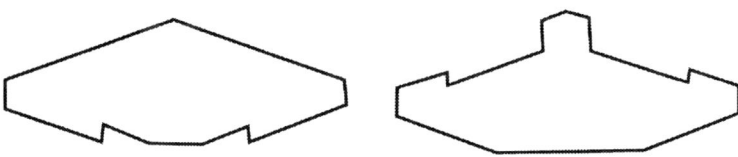

Figure 1. Component type A (left) and component type B (right) in their initial neutral positions [9]

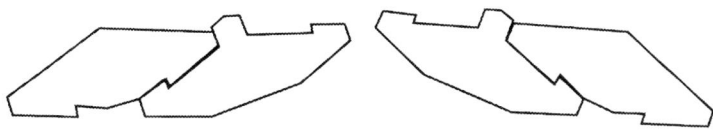

Figure 2. Complex AB (left) and complex BA (right) [9]

Although Penrose and Penrose did not discuss this in particular, they were able to achieve artificial self-reproduction through the morphology of each component type and the design of the environment. There are ten pieces of information, physically encoded through shapes (hooks, latches, and neutral sites), which dictate the self-assembly process (Fig. 3).

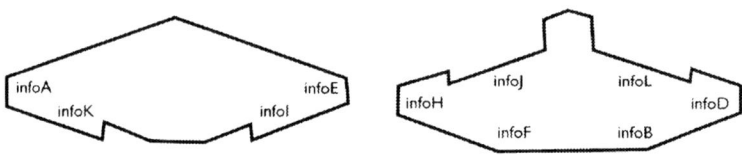

Figure 3. Component type A (left) and component type B (right) identifying regions of physically encoded information using shapes

There are six rules present in this system. The information associated with each rule is physically encoded by shape. Two rules directly dictate the self-assembly of the two components by hooking-and-latching together (Fig. 4),

$$infoI \text{ fits } infoJ \tag{1}$$
$$infoK \text{ fits } infoL \tag{2}$$

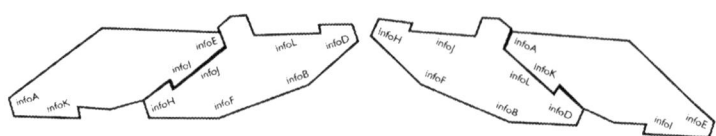

Figure 4. AB assembly (left) using rule (1) and BA assembly (right) using rule (2)

and four rules influence the motion of neighboring components (Fig. 5) and (Fig. 6),

$$infoB \text{ rotates } infoA \tag{3}$$
$$infoA \text{ rotates } infoD \tag{4}$$
$$infoF \text{ rotates } infoE \tag{5}$$
$$infoE \text{ rotates } infoH. \tag{6}$$

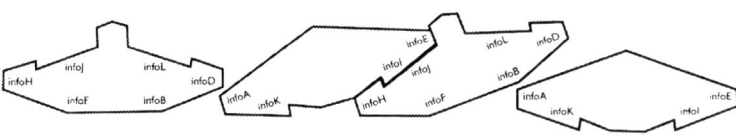

Figure 5. Rotation rules (3) and (4) promote the creation of new AB complexes

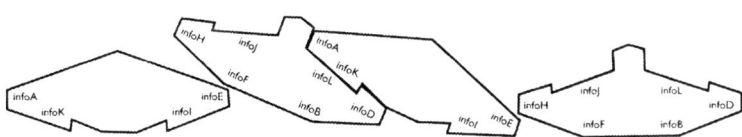

Figure 6. Rotation rules (5) and (6) promote the creation of new BA complexes

The constraints of the system include the environment and the initial positioning of the components. These constraints, along with initially placing either an AB or BA seed complex, ensures which information comes into contact. Thus, the system is able to replicate AB and BA complexes accordingly.

By showing that it is possible to physically mirror self-reproduction in nature artificially, their pioneering work has influenced many fields, including computation. Adleman first showed that computing using self-assembling molecules was possible [10]. An algorithm was devised to solve a seven node directed Hamiltonian path problem. Single strands of DNA were created to encode partial sample solutions of the problem (representing the edges and vertices of the directed graph). These partial sample solutions were then allowed to interact and self-assemble based on Watson-Crick complementarity. The linear double-stranded self-assembled structures representing potential solutions were then filtered using biochemistry and molecular biology techniques to identify the true self-assembled solution.

With the success of Adleman's experiment research has focused on scaling DNA Computing to solve larger problems. This requires a shift from the original brute-force approach. Developing both new DNA structures, as well as new algorithmic techniques, has shown tremendous promise in furthering this method of embodied computing.

Wang tiles, square tiles with coloured edges, are computationally equivalent to a Turing machine [11]. This is achieved primarily on the motion of the tiles (cannot be rotated or reflected) and adjacent tiles must have the same colour. The rules of this system allow for the self-assembly of a two-dimensional structure that can tile a plane. The resulting tilings can be viewed as representing the state of a Turing machine. The physical realization of Wang tiles has been achieved through the creation of DNA tiles [12]. These tiles use interwoven strands of DNA to create the square body of the tile (double-stranded) with single strands (representing colours) extending from the edges of the tiles. These tiles follow the same assembly rules based on Watson-Crick complementarity. This has lead to the Tile Assembly Model (an extension of Wang tiles using DNA tiles) [13]. It has also lead to new algorithmic proposals for using DNA tiles to perform mathematical operations. These include multiplication and cyclic convolution product [14]. Multiplication is based on the self-assembly of a two-dimensional structure. Cyclic convolution product requires new DNA tiles that allow for the self-assembly of a three-dimensional structure.

Understanding how structures emerge through self-assembly is vital to continue to expand embodied computing, as experienced in DNA Computing. It is also important in other disciplines, such as modular robotics and material science. Techniques have been developed to create a set of components that self-assemble into a single desired structure. These techniques all use information from the desired entity's global structure to create the set of components. Algorithms evaluating the geometry of a global structure have been used in modular robotics [1] [2] [3]. Algorithms using the global form to act as a template that dictate the mechanisms in which components can recreate an image of the template have been used to create DNA origami [4].

These techniques have all been successful in creating a set of components that self-assemble into a desired structure. However, these techniques will not scale effectively to problems requiring the self-assembly process to create multiple structures. Such a case is likely in embodied computation where a computation may require multiple states (structures) to complete. Inspirations from nature can be taken to develop bottom-up design approaches to create self-assembling emergent systems.

3   Three-level approach

Emergent systems are present throughout nature. Based on the rules present, decentralized systems self-assemble to create various structures. These rules are in relation to the physically and chemically encoded information comprising a set of components and their environment. DNA serves as the rule set for living organisms. Through transcription to RNA and translation to proteins, these rules are mapped to physical shapes. Using this analogy, an approach that maps a set of rules directly to physical shapes to create self-assembling systems is independent of the desired entity. This serves as the basis for the approach proposed here, which is described in three levels (Fig. 7).

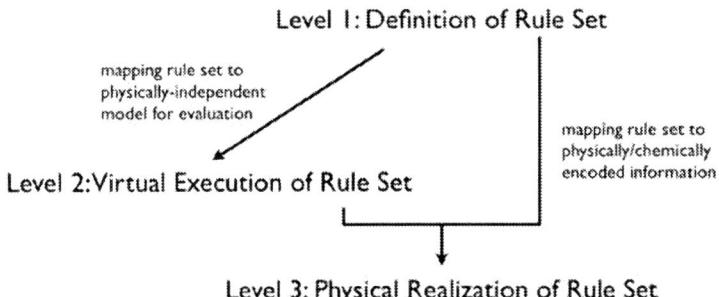

Figure 7. Three-level approach with transitions, showing that translation can be done directly from level 1 to level 3

Level one: Definition of rule set The highest level, level one, is concerned with being able to describe any rule set. The rules could be of several types. For example, rules represented mathematically could be of the string re-writing form, e.g. L-Systems [15, 16]. Rules expressed abstractly could mirror those seen with proteins, e.g. A fits B, C+D fits E, F breaks G+H (where the letters represent physically and chemically encoded information). As well, rules in which physically and/or chemically encoded information is effected by natural phenomena should also be addressed. Such rules could include temperatureT breaks I+J. For example, the physical equivalent to this rule represents that DNA is denatured at a high enough temperature (i.e. double strand breaks apart into two single strands).

Level two: Virtual execution of rule set Execution of a set of level-one rules is accomplished at the mid level, level two. Information comprising the components and their environment, as well as the rule set present, are expressed using modeling. This could be implemented through a simulation, for example. In order to evaluate an emergent system, it has to be allowed to run [7]. This is the primary purpose of this level. The outcome of the simulation can determine if, in order for a desired self-assembled structure to emerge, the level-one rules or the information related to the different component types and their environment need to be modified. Representations used for the information and the rules should be independent of any physical medium. This has two benefits. The first is that it allows for the simulation to remain simple and computationally inexpensive to run (modeling the interactions of complex forms is a difficult task). The second is that it then allows for the integration of various materials and self-assembly mechanisms, which may be required depending on the desired structures.

Level three: Physical realization of rule set At level three, information comprising the components and/or their environment, together with the level-one rules present in the system, are mapped from their virtual representations to a physical system. Since this mapping is done directly, it is independent of the results observed from the level-two modeling. The consequence of this is that mapping can occur before, during, or after a simulation run. Components and their environment can be mapped in two ways. The first way is if a database of physically and/or chemically encoded rules is specified. An algorithm can then choose entries in the database and assign them to the information comprising the components and/or environment. The second method would be to use optimization techniques to create translated designs of physically and/or chemically encoded information comprising the components and/or environment. Designs in a database and those created by optimization techniques can always be exchanged, for cases when partial knowledge exists for a particular system. The use of a database and optimization techniques within the three-level approach is illustrated in Fig. 7.

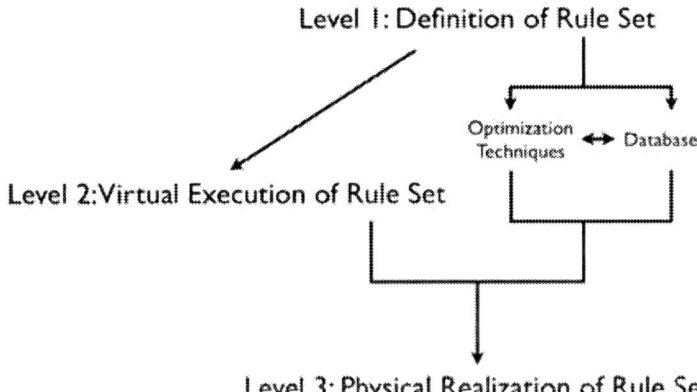

Figure 8. The use of a database of physical encodings and optimization techniques to map a rule set to physical systems

The objective of this approach is to provide a method in which components and their environment can be designed in a bottom-up manner to create scalable self-assembling systems. This is achieved by being able to directly map a set of virtual self-assembly rules to physical systems.

4   System design

To demonstrate how this three-level approach can be used, the following example system was constructed. Its purpose is to show how to create a set of physical two-dimensional components that self-assemble into a single desired entity. In general this is a multi-component implementation. Information is located along a components surface. The location of this information and its arrangement defines a components type. The interaction between the components is dictated by the rules present in the system, acting on the information associated between components. A breakdown of this system is provided in terms of the three levels of the proposed approach.

4.1   Level one: Definition of rule set

To allow for the assembly of components, one rule type is specified in this example system. Abstractly, if two pieces of complementary information come into contact (i.e., they fit together), it will cause them to assemble. The rule type is defined as

$$A \text{ fits } B \rightarrow A + B \tag{7}$$

where the letters A and B are in reference to physically and/or chemically encoded information. The predicate 'fits()' is the condition that A and B must satisfy. If the condition is met then A and B assemble together, $+$. For the purpose of this example system this rule type is commutative, that is

$$B \text{ fits } A \rightarrow B + A = A + B. \tag{8}$$

In certain cases it may be beneficial for this rule type to not be commutative, based on the mechanisms in which two components self-assemble. As well, since this system is in two-dimensions, components only need to line up when this rule type is satisfied. In three-dimensions, due to the extra degree of freedom, the orientation of the components must be addressed when this rule type is satisfied.

To comply with the definition of self-assembly being a reversible process, a second rule type is required. Abstractly, if two assembled pieces of information experience a force above a certain threshold their assembly is broken. This rule type is defined as,

$$force X \text{ breaks } A + B \rightarrow A; B \tag{9}$$

where $forceX$ is in reference to a physical force, the term 'breaks()' is the condition that must be satisfied, and a semicolon denotes two separated pieces of information. This rule type is not commutative. How the first rule type (and its commutative property) as well as how this second rule type are utilized is described in the implementations of level two and three.

4.2   Level two: Virtual execution of rule set

As described previously, the purpose of level two is to evaluate a set of level one rules through simulation. The simulation should be disconnected from any physical medium. However, the simulation cannot be disconnected from the general principles of self-assembly. In [17], a framework for designing self-assembling systems is detailed. The framework consists of seven parts:

1. Components,
2. Environment,
3. Energy,
4. Assembly Protocol,
5. Spatial Relationship,
6. Localized Communication, and
7. Rule Set.

Components are defined by their properties, such as shape, mass, and material properties. The physical and chemical properties of the environment will influence the manner in which components self-assembly. Features in the environment can serve as templates. Or, the environment can simply provide a boundary to which components are confined to. In order for self-assembly to occur, components require energy. This energy could be available internally or transfered to the components by the environment. An assembly protocol defines the method in which components can self-assemble. These methods are highly dependent on the scale of the system, as well as the physical and chemical properties of the components and their environment. The spatial relationship between components and/or elements in their environment defines the underlying pattern formations capable by a system. Pattern formation greatly influences the range of achievable formations by a system. Localized communication dictates how components interact with one another and their environment. Communication can be viewed as physically or chemically encoded information. The rule set can be viewed as being external to components, acting on their physically and/or chemically encoded information (e.g. the growth of crystals). Or, the rule set can be internalized within components (e.g. DNA within cells). The simulation, implemented using Breve [18], is described in reference to this framework.

In the simulation, components are represented as spheres of unit radius. Points on the surface of a component serve as information locations. These surface points can occur anywhere, but a four-point system is used in this example. The location of this information and its arrangement defines a component's type. Information is represented by a capital letter. If no information is associated with a point, the - symbol is used (Fig. 9).

The primary purpose of the environment is to provide a boundary to which a given set of components are confined. Components are assigned a velocity. This energy allows a component to move around the

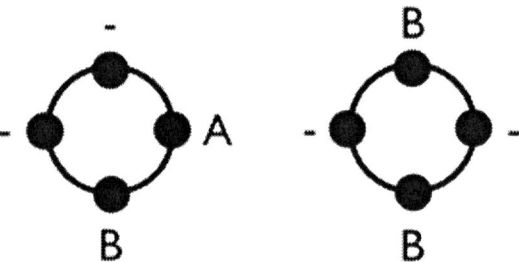

Figure 9. Two example component types with their information arrangements

environment and rotate in two-dimensions, as well as interact with other components.

The assembly protocol used in this system is based on stickiness. If two components collide with complementary information (associated with a fits rule) at their points of collision, they stick together. If the information is not complementary, the two components reflect off each other, emulating rigid-body dynamics and inelastic collisions.

The arrangement of information locations, of each component type, defines the spatial relationship and underlying pattern formations achievable by a given system. As a decentralized system, components communicate locally. In this case, communication occurs when two components collide. The information associated with the points of collision is the data exchanged between two components. If the information satisfies the rule set present in the system, the two components perform the associated action, otherwise they reflect off each other. A summary of this virtual system is provided in Table 1.

4.3   Level three: Physical realization of rule set

The conceptual representation of this simple virtual self-assembling system, serves as the basis for the translation process, at level three. Information abstractly represented in the virtual system is mapped to physical systems. For demonstration purposes, the physical system constitutes a set of mechanical components that are placed on a tray. The tray is shaken parallel to its surface. This energy is transferred to the components in the form of vibrations, allowing them to interact. Component interaction is based on physically encoded information. This information

Table 1. Parameter settings for the virtual simulation

Framework	Virtual System
Components	spheres of unit radius; type defined by information arrangement
Environment	spherical boundary to contain components
Energy	components have a velocity
Assembly Protocol	stickiness
Spatial Relationship	arrangement of information locations
Localized Communication	information is exchanged at points of collision
Rule Set	fits rule type and breaks rule type

is represented through form and an assembly protocol, and is the basis of the design space (the set of feasible designs).

Component form is based on a key-lock-neutral design concept. The base shape of a component is a square. The shape of a key (convex polygon) and a lock (concave polygon) are designed such that these shapes fit together. Neutral sites represent regions where components cannot assemble together. This form space is described in Figure 10. The physical form of a component is constructed out of a nonmagnetic material.

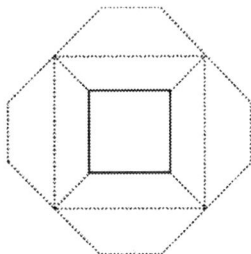

Figure 10. Form space of the components, where each region associated with each edge of the minimum shape (square, shown with a solid line) can be a convex (key), concave (lock), or neutral (flat edge) shape

In order to allow components to assemble together, an assembly protocol based on three-bit magnetic encodings is used here. Three magnetic discs, with magnetic north and south arbitrarily assigned to represent zero and one for a binary representation, are used to create a particular encoding.

The sets of three magnetic discs encodings are placed along the edges of key and lock regions. The magnetic discs are arranged such that the flat surface of the magnetic discs is parallel to the edge's surface to which they are placed in. In this system, physical information is represented as a combination of shape and a magnetic encoding.

To help prevent keys and locks from interacting amongst themselves, two measures are taken. First, the magnetic encodings are divided based on their pairings into two sets: keys and locks. This prevents encodings within the same set from assembling. Secondly, the magnetic encodings 011 and 100 are reversed from their incremented binary representation. This is done to maintain consistency with the first measure. Table 2 lists the set of physical encodings possible in this system, taking these two measures into account.

Table 2. Physical encodings

Physical Encoding #	Shape	Magnetic Encoding
0	Neutral	(none)
1	Lock	000
2	Lock	001
3	Lock	010
4	Lock	100
5	Key	011
6	Key	101
7	Key	110
8	Key	111

Another unintended effect can occur when magnetic encodings are exposed at the edges. In this scenario it is possible for them to assemble (Fig. 11). This is due to the fact that there are no physical barriers from preventing two flat edges from sliding along each other. For example, two physical encodings such as seven and eight can assemble together if only the last bit of encoding seven and the first bit of encoding eight come into contact. The measure taken to prevent this error is pushing the magnets into the edges of the component, and covering them with small pieces of nonmagnetic material. This allows for information associated with components to align selectively with one another. Since the magnets cannot come into direct contact, the mixed fields generated from the encodings must be strong enough to overcome the force of friction of the components, as well as the force generated from other component collisions and collisions with the side walls of the tray. This also helps prevent a third unintended effect form occurring, namely a key and a lock from assembling when only a subset of each encoding matches.

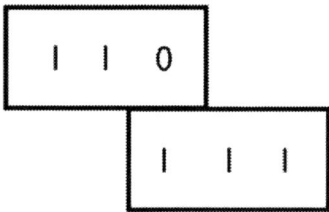

Figure 11. Two physical encodings, 7 and 8, assembling caused by being exposed at the surface of two flat edges

The rules from the level two modeling are mapped to physical systems, based on these design principles. Since the fits rule type (Eq. 7) used here is commutative, it allows for a pair-wise representation of the information associated with this rule. The database of physical encodings (one to eight) is provided to the system. The mapping procedure goes through each fits rule present in the system. For each fits rule, a two-step process occurs. First, a physical encoding is selected at random for the first piece of information, and its associated rule is assigned to the second piece of information. Second, these two encodings are removed from the available database. To illustrate this procedure, the components in Figure 12 are translated to their physical designs (Fig. 13) based on the rules,

$$C \text{ fits } D \rightarrow C + D, \text{and} \tag{10}$$
$$E \text{ fits } F \rightarrow E + F. \tag{11}$$

The breaks rule type (Eq. 9) here determines the location of the magnetic encodings within a component. For each piece of physical information associated with a breaks rule, their assigned magnetic encoding is placed in the interior of the nonmagnetic material. Otherwise, the assigned magnetic encoding is placed at the surface.

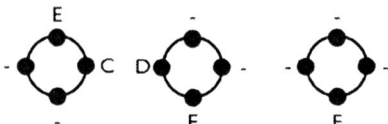

Figure 12. Example component types with information arrangement (right, top, left, bottom), for component type 1 - left (C,E, -, -), component type 2 - centre (-, -, D, E), and component type 3 - right (-, -, -, F)

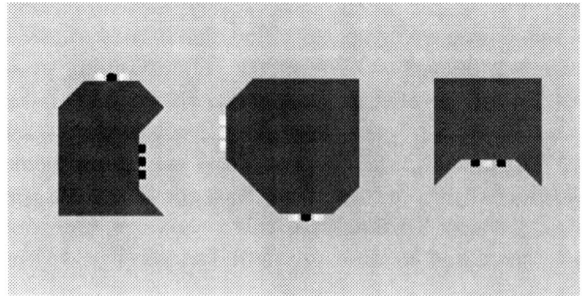

Figure 13. Translated components from Figure 12 with component type 1 - left (1, 6, 0, 0), component type 2 - centre (0, 0, 3, 1), and component type 3 - right (0, 0, 0, 8) with physical encodings from Table 2 referred to as (right, top, left, bottom)

5 Experiments and results

The objective here is to determine if it is possible to create self-assembled desired entities by using a bottom-up rule-based approach to map virtual self-assembly rules to physical systems. The three-level approach, in the context of the previously described example system, was used to see if it was capable of achieving this objective.

Five entities were designed as the target structures as the result of self-assembly. Figure 14 shows the five desired entities: line, T-shape, L-shape, open square, and Y-shape. Attempting to create each of the desired entities was done in three phases, each corresponding to the three-level approach.

Figure 14. Five desired entities (from left to right): line, T-shape, L-shape, open square, and Y-shape

5.1 Level one: Experiments' rule set

These five structures were chosen because they offer various degrees of complexity in terms of the number of components required and the symmetry or asymmetry of the desired entities structures. As a consequence, each of the five desired entities cannot be created through pattern formation exclusively.

Therefore, a level one rule set is required for each system in order to create the five desired entities. Both rule types, fits and breaks, are used in each system. Leveraging each rule set, in terms of the information required and the resulting component types, is the objective of determining whether or not each rule set is able to create its respective desired entity. A summary of the rule set used for each of the five experiments is given in Table 3.

5.2 Level two: Simulation experiments

Level two virtual simulations were used to test the ability of each rule set being able to create its respective desired entity. The component types used for each experiment are also provided in Table 3.

Table 3. System design for the five experiments

	System Design		Desired Entity	
Experiments	Component Types (right, top, left, bottom)	Rules	Number of Components	Symmetric vs. Asymmetric
Line	Type 1: (A, -, A, -) Type 2: (B, -, -, -)	A fits B forceX breaks A+B	3	symmetric
T-shape	Type1: (A, -, A, C) Type2: (-, B, -, -) Type3: (-, D, -, A)	A fits B C fits D forceX breaks A+B forceX breaks C+D	5	symmetric and asymmetric
L-shape	Type1: (A, C, -, -) Type2: (-, -, B, -) Type3: (-, E, -, D) Type4: (-, -, -, F)	A fits B C fits D E fits F forceX breaks A+B forceX breaks C+D forceX breaks E+F	4	asymmetric
Open Square	Type 1: (A, C, -, -) Type 2: (H, -, B, -) Type 3: (-, -, B, -) Type 4: (G, -, -, H)	A fits B C fits D E fits F G fits H forceX breaks A+B forceX breaks C+D forceX breaks E+F forceX breaks G+H	8	symmetric
Y-shape	Type 1: (A, -, E, C) Type 2: (-, D, -, G) Type 3: (B, -, -, G) Type 4: (-, -, F, G) Type 5: (-, -, -, H)	A fits B C fits D E fits F G fits H forceX breaks A+B forceX breaks C+D forceX breaks E+F forceX breaks G+H	7	symmetric and asymmetric

Initially, components are randomly placed and oriented in the environment of each simulation. This is to ensure that no bias is entered into the self-assembly process. The number of component types for each simulation is equal, and is set to ten. However, in certain experiments the ratio of components needed to create the desired entity is not equal. In experiment one, the ratio is 1:2. Since the focus of these simulation experiments is being able to create a single desired entity and not on efficiency, it is felt that this does not affect the objective of the experiments negatively.

In the simulations, components (of unit radius and unit mass) are assigned a velocity to move around their environment, emulating rigid-body dynamics and inelastic collisions. Components were assigned a velocity, randomly generated in the range of 1-5 units/second. Periodically, components velocity was increased or decreased fractionally to emulate changes in energy within the system. The purpose of this was to help serve the breaks rule type. When information associated with components experienced collisions with each other or with the environment (circular, with radius 30 units) and their velocity was greater than 4 units/second, the breaks rule type was applied.

5.3   Level two: Simulation results

Using these parameter settings, a level two simulation was performed for each system. Ten simulation runs were conducted for each system. A subset of the components was able to self-assemble into the desired entity of each system, during each run. Due to the number of components present in each system being larger than the number of components required to create each desired entity, multiple desired entities were created in some instances. This was done with the intent to speed up the simulations, and was effective as seen in the performance of the simulations. As well, component types one through five were assigned (when present in a system) colours black, white, red, blue, and green respectively. The results of the simulations are shown in Figure 15.

The set of components for each simulation was able to self-assemble into its respective desired entity. Systems with a lower number of components were able to self-assemble faster than systems with a larger number of components. Entities with symmetric properties were also able to self-assemble faster than the ones with asymmetric properties. These characteristics were observed qualitatively.

For each desired entity, there was no particular self-assembly sequence followed by the components. Various substructures would occur during the self-assembly process, and is reflected in Figure 15.

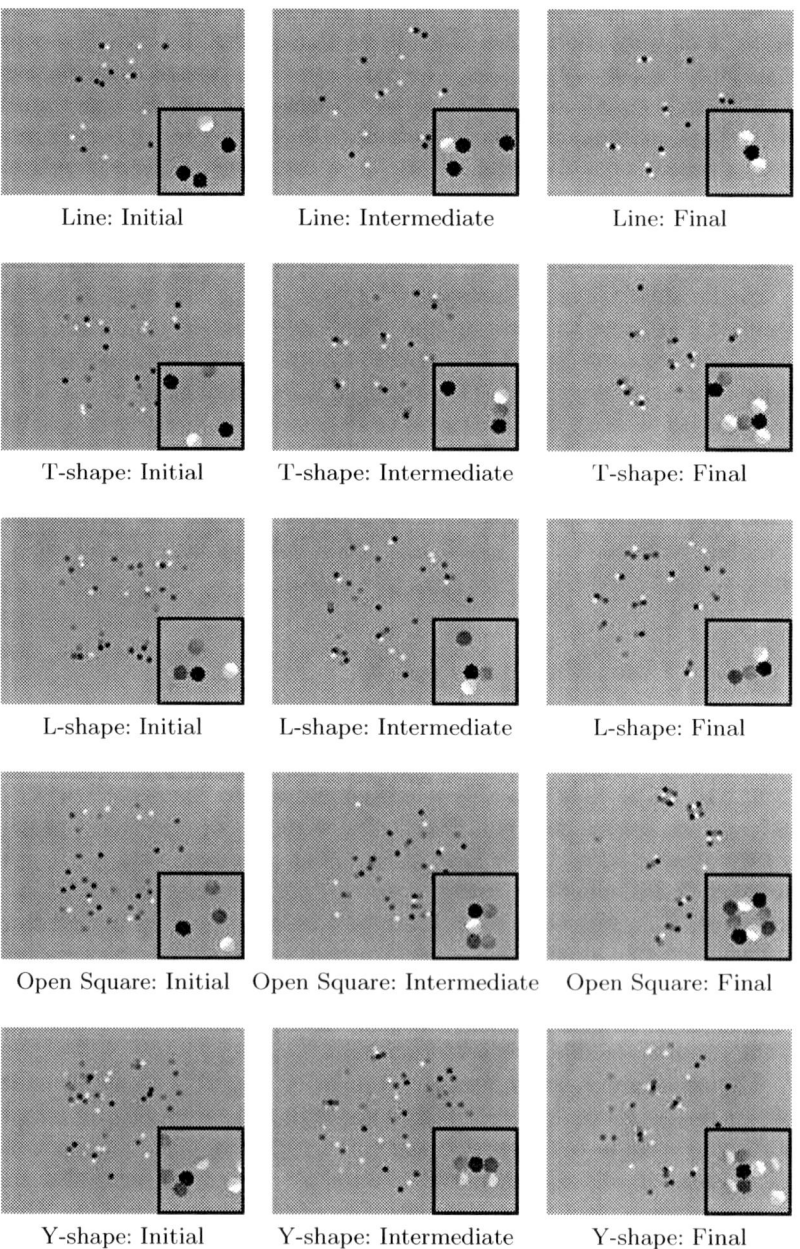

Figure 15. Results of the five simulations

As well, components in the simulations were able to disassemble, Figure 16. This was due to the breaks rule type. Although the components were able to break apart from collisions with each other and with the environment, with time corresponding information associated with components were able to self-assemble again. The virtual simulations are reversible, and therefore are in keeping with the definition of self-assembly.

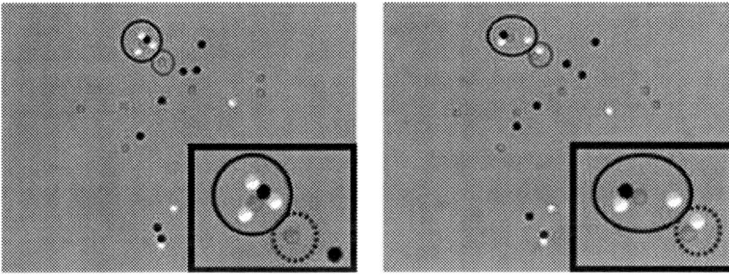

Figure 16. Before configuration (left) and after configuration (right), where the structure within the solid oval is broken apart by the component within the dashed oval, and a further collision with the environment's side wall (not shown).

5.4   Level three: Physical experiments

With the success of the virtual simulations, a level three translation was performed on each system. This was done to test if the translated component types of each system could self-assemble into its respective desired entity, based on the rule set of each system. The virtual self-assembly rules mapped to physical systems are shown in Figure 17 to Figure 21 (magnets shown for clarity) for each system.

The component translations of each system were physically built by hand, to determine if the virtual simulations could be replicated physically. Components were randomly placed on the surface of a tray. The tray was shaken by hand, parallel to its surface in a jarring motion. This energy was transferred to the components in the form of vibration, allowing them to interact. The jarring motion resulted in components coming to rest, giving information associated with colliding components to establish an assembly (when a fits rule applied). As well, fluctuations in energy allowed assembled information to disassemble (when a breaks

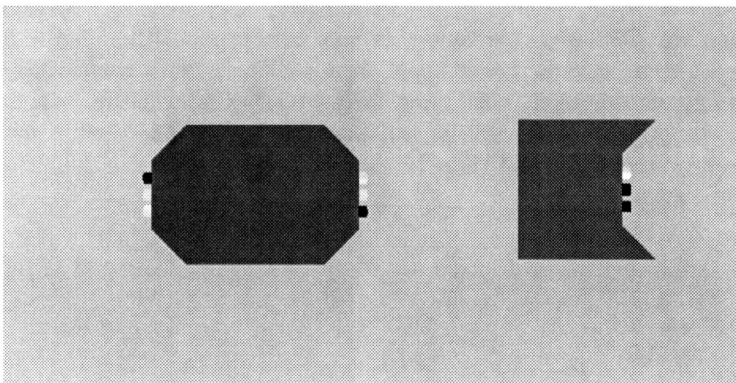

Figure 17. Line: translated component types with physical encodings from left to right: type 1 (7, 0, 7, 0), and type 2 (2, 0, 0, 0). The physical encodings (from Table 2) are listed in the order (right, top, left, bottom).

Figure 18. T-shape: translated component types with physical encodings, starting from the top, then left to right: type 1 (3, 0, 3, 4), type 2 (0, 6, 0, 0), and type 3 (0, 5, 0, 3). The physical encodings (from Table 2 are listed in the order (right, top, left, bottom).

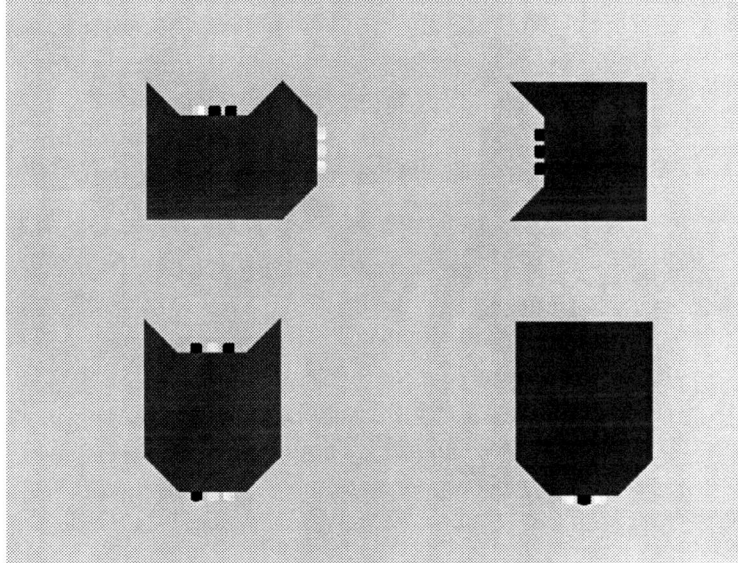

Figure 19. L-shape: translated component types with physical encodings, starting from the top, then left to right: type 1 (8, 2, 0, 0), type 2 (0, 0, 1, 0), type 3 (0, 3, 0, 7), and type 4 (0, 0, 0, 6). The physical encodings (from Table 2) are listed in the order (right, top, left, bottom).

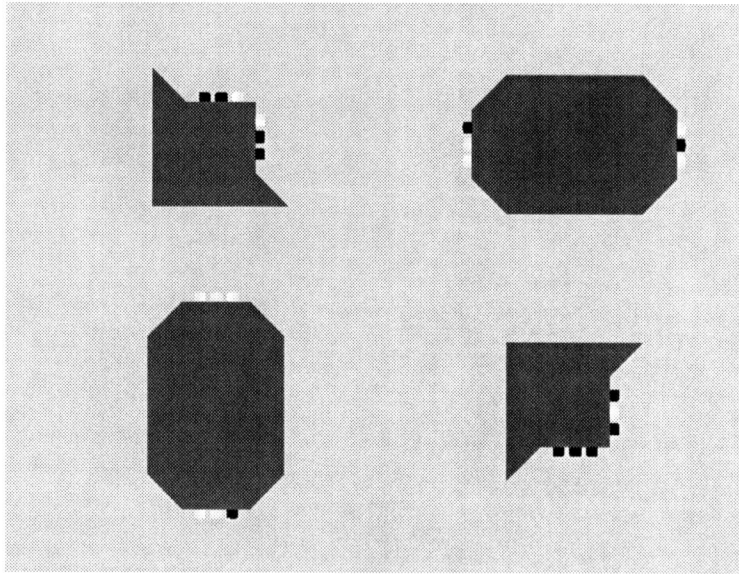

Figure 20. Open Square: translated component types with physical encodings, starting from the top, then left to right: type 1 (2, 4, 0, 0), type 2 (6, 0, 7, 0), type 3 (0, 8, 0, 5), and type 4 (3, 0, 0, 1). The physical encodings (from Table 2) are listed in the order (right, top, left, bottom).

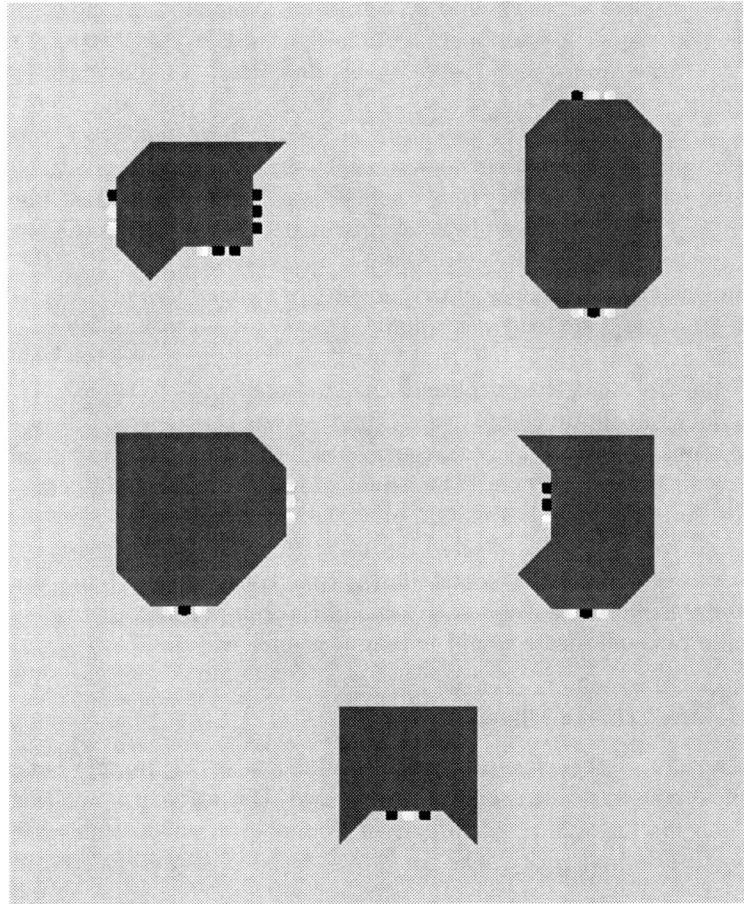

Figure 21. Y-shape: translated component types with physical encodings, starting from the top, then left to right: type 1 (1, 0, 7, 4), type 2 (0, 5, 0, 6), type 3 (8, 0, 0, 6), type 4 (0, 0, 2, 6), and type 5 (0, 0, 0, 3). The physical encodings (from Table 2) are listed in the order (right, top, left, bottom).

rule applied) during component collisions with other components and with the sidewalls of the tray.

Components were constructed out of foam board, neodymium magnetic discs (NdFeB, 3mm diameter and 1.5 mm thick, with N50 rating), general-purpose adhesive, and scotch tape. Depending on the resulting translated shape, a component's dimensions would vary between 2 cm by 2cm (base square shape representing lock sites), 4 cm by 4 cm (outer square representing neutral sites) and 2 cm short edge and 4 cm long edge (irregular octagon representing key sites), in reference to Figure 10. The height of a component measured 0.5 cm. Magnets were placed approximately 5 mm apart (from their centers). When a breaks rule was translated, magnets were placed approximately 3 mm within the interior of the component's edge.

In the simulations, the environment was circular. Its physical equivalent (the tray's surface) was square. This was done to simplify the construction process. It is felt that this doesn't compromise the physical experiments, since the purpose of the environment is to serve only as a boundary to contain components. The tray was constructed out of foam board, push pins, and general-purpose adhesive. The dimensions of the tray were 50 cm by 50 cm. The height of the sidewalls was 10 cm.

In the physical experiments, only the number of components types required to create each desired entity was present in the environment. This was done for two reasons. It was time consuming to build components by hand. Also, due to the scale of the components, increasing the number of components would require a greater tray size.

5.5   Level three: Physical results

The results of the physical system are shown in Figure 22. For each physical system, three runs were performed. During each run, the set of components for each physical system were able to self-assemble into its respective desired entity. The same qualitative observations seen in the simulations were also seen in the physical systems. Systems with a lower number of components were able to self-assemble faster than systems with a larger number of components. Entities with symmetric properties were also able to self-assemble faster than the ones with asymmetric properties.

Also as seen in the simulations, various substructures would form during the physical self-assembly process. These substructures are reflected in Figure 22. Components were able to disassemble, again mirroring the simulations, Figure 23. This was due to the physical realization of the breaks rule type. In the physical systems, magnets were placed within a nonmagnetic material, and therefore the magnets were not able to make

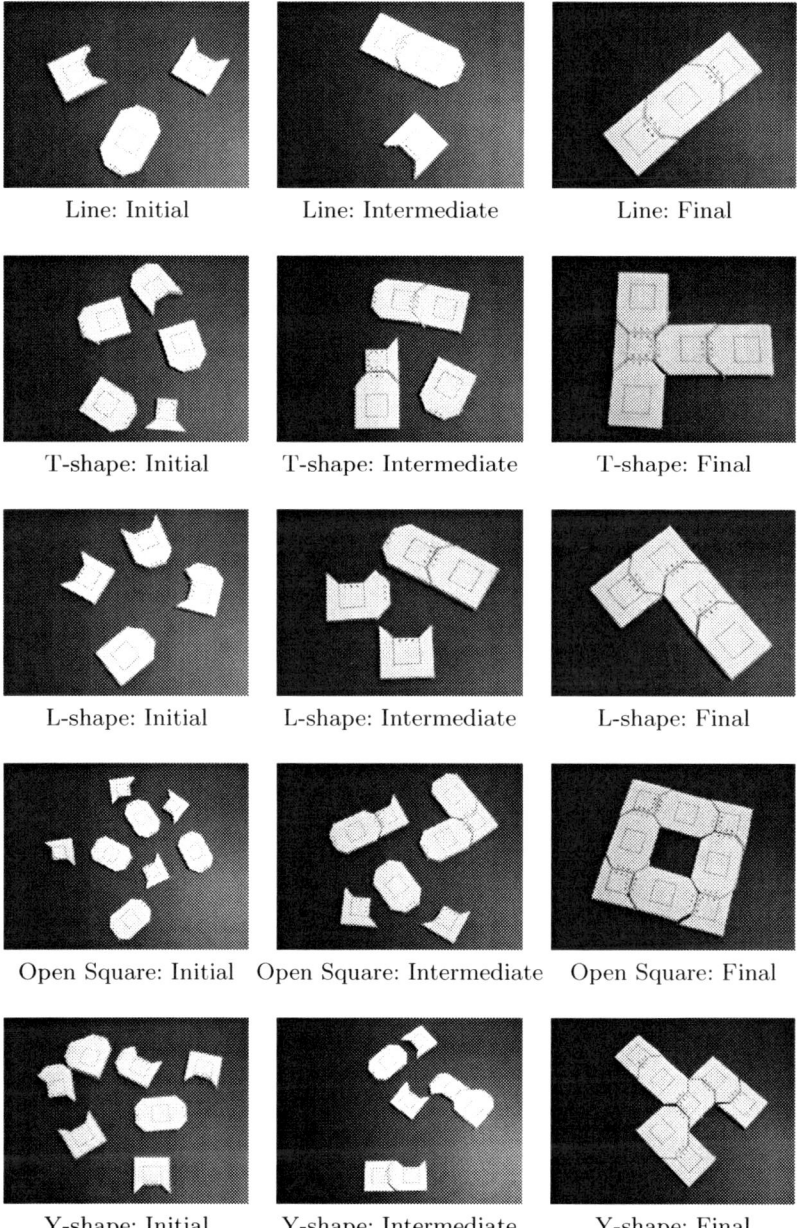

Figure 22. Results of the five physical systems

direct connections. As a result, components were able to break apart from collisions with each other and with the environment. With time, corresponding information associated with components were able to self-assemble again. Therefore, the physical systems are reversible, and are in keeping with the definition of self-assembly.

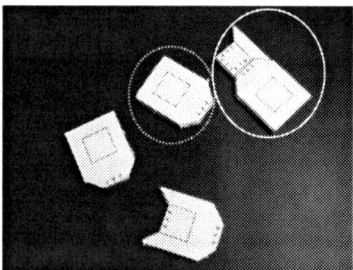 

Figure 23. Before configuration (left) and after configuration (right), where the structure within the solid oval is broken apart by the component within the dashed oval.

5.6   Experiments and results summary

The objective here was to determine if it was possible to create self-assembled desired entities by using a bottom-up rule-based approach to map virtual self-assembly rules to physical systems. For each system, a level one rule set was specified. Each level one rule set was tested using a level two simulation to determine if each system's rule set and components types could achieve its particular desired entity through self-assembly. Each virtual simulation experiment was successful. As a result, component types were translated to physical systems, based on the rule set of each system. This mapping procedure was done directly, without any information of the desired entities' global structures. Each physical system was successful in having its set of components self-assemble into their respective desired entity. With the successful results of the physical systems, it demonstrates that by using a bottom-up rule-based approach it is possible to create physical emergent systems capable of self-assembling into desired entities.

Investigating systems with a variety of rule types would allow for the possible creation of dynamic systems being able to create multiple structures. Such a rule set could include rules of the form A fits B, C+D fits E, and F breaks G+H. Research in this respect is ongoing by the authors.

In future applications, knowledge in regards to the required rule sets, component information configurations, and the physical design of components and their environment may not be entirely formulated. Utilizing this rule-based approach in the context of a process exploiting optimization techniques, such as evolutionary computation, to design and create self-assembling systems could address such instances [19] [20]. This would be in keeping with living organisms in nature, where evolution is acting on their rule set, their DNA. Through transcription to RNA and translation to proteins, these rules are mapped to physical shapes. Leveraging this three-level approach in this context is another area of research being pursued by the authors.

6 Conclusions

Designing and creating a set of components that can self-assemble into a desired entity continues to be a major challenge. Top-down techniques that utilize information from a desired entity's global structure to create a set of components will not scale effectively. This applies in particular to problems requiring the self-assembly process to create multiple structures. Such a case is likely in embodied computation where a computation may require multiple states (structures) to complete.

In nature, interacting components, governed by simple rules, self-assemble to create specific entities. The programs constituting these entities are based on the rules present in a given system and are executed on the physical and chemical encoded information comprising the components and their environment.

This rule-based methodology in nature is the basis to the three-level approach presented here. This approach encompasses specifying a set of rules, modeling these rules to determine the outcome of a specific system in software, and translating to a physical system based on the set of rules present. The benefit of this approach is that no knowledge of the end result is required to create the physical system, mirroring the bottom-up design process in nature.

Five experiments, based on an example implementation of this approach, were created to test the feasibility of the three-level approach. Each system varied in the number of rules, component types, and complexity of the structures of the desired entities. Each virtual system was successful in having a set of components self-assemble into their respective desired entities. The translations of these virtual systems to physical systems were also successful in having a set of components self-assemble into their respective desired entities. This demonstrates that by using a bottom-up rule-based approach it is possible to create emergent systems by mapping virtual self-assembly rules to physical systems.

Acknowledgements

Financial support for this research is provided by NSERC, the Natural Sciences and Engineering Research Council of Canada.

References

[1] Jones, C., Matari, M.J.: From local to global behavior in intelligent self-assembly. In: Proc. of the 2003 IEEE Int. Conf. on Robotics and Automation. Volume 1., IEEE Computer Society Press (2003) 721–726

[2] Klavins, ., Ghrist, R., Lipsky, D.: A grammatical approach to self-organizing robotic systems. In: IEEE Trans. Automat. Contr. Volume 51. (2006) 949–962

[3] Nagpal, R.: Self-organizing shape and pattern: From cells to robots. In: IEEE Intell. Syst. Volume 21. (2006) 50–53

[4] Rothmund, P.W.: Design of dna origami. In: Proceedings of the International Conference on Computer-Aided Design (ICCAD). (2005)

[5] Ball, P.: The Self-made Tapestry: Pattern Formation in nature. Oxford University Press (1999)

[6] Thompson, D.: On Growth and Form. Canto Series. Cambridge University Press (1961)

[7] Wolfram, S.: A New Kind of Science. Wolfram Media, Champaign, IL (2002)

[8] Whitesides, G., Gryzbowski, B.: Self-assembly at all scales. Science **295**(5564) (2002) 2418–2421

[9] Penrose, L., Penrose, R.: A self-reproducing analogue. Nature **4571** (1957) 1183–1184

[10] Adleman, L.: Molecular computation of solutions to combinatorial problems. Science **266**(5187) (1994) 10211024

[11] Wang, H.: Proving theorems by pattern recognition. Bell Systems Technical Journal **40**(II) (1961) 1–42

[12] Mao, C., LaBean, T., Reif, J., Seeman, N.: Logical computation using algorithmic self-assembly of dna triple-crossover molecules. Nature **407** (2000) 493–496

[13] Rothemund, P., Winfree, E.: The program-size complexity of self-assembled squares. In: ACM Symposium on Theory of Computing (STOC). (2000)

[14] Pelletier, O., Weimerskirch, A.: Algorithmic self-assembly of dna tiles and its application to cryptanalysis. In: Genetic and Evolutionary Computation Conference (GECCO-2002). (2002)

[15] Prusinkiewicz, P., Lindenmayer, A.: The Algorithmic Beauty of Plants. Springer, New York (1990)

[16] Jacob, C.: Illustrating Evolutionary Computation with Mathematica. Morgan Kaufmann Publishers, San Francisco, CA (2001)

[17] Bhalla, N., Jacob, C.: A framework for analyzing and creating self-assembling systems. In: Proc. of the 2007 IEEE Symposium Series on Computational Intelligence. (2007) 281–288

[18] Klein, J.: breve: a 3d environment for the simulation of decentralized systems and artificial life. In: ICAL 2003: Proceedings of the eighth international conference on Artificial life, Cambridge, MA, USA, MIT Press (2003) 329–334

[19] Bhalla, N., Bentley, P.J.: Working towards self-assembling robots at all scales. In: Proc. of the 3rd Int. Conf. on Autonomous Robots and Agents. (2006) 617–622

[20] Bhalla, N., Bentley, P.J.: Self-assembling systems. GB Patent: 061400.0 (2006)

148

Halting in quantum Turing computation

Willem Fouché[1], Johannes Heidema[2], Glyn Jones[3], and
Petrus H. Potgieter[4]

[1] Department of Decision Sciences, `fouchwl@unisa.ac.za`
[2] Department of Mathematical Sciences, `heidej@unisa.ac.za`
[3] Department of Physics, `joneseg@unisa.ac.za`
[4] Department of Decision Sciences, `php@member.ams.org`

University of South Africa, PO Box 392, Unisa 0003, Pretoria

Abstract The paper considers the halting scheme for quantum
Turing machines. The scheme originally proposed by Deutsch
appears to be correct, but not exactly as originally intended.
We discuss the result of Ozawa [1] as well as the objections
raised by Myers [2], Kieu and Danos [3] and others. Finally, the
relationship of the halting scheme to the quest for a universal
quantum Turing machine is considered.

Keywords: quantum Turing machine, halting scheme for quantum Turing machines, possibly universal quantum Turing machines, quantum computing

1 Introduction

The quantum logic circuit model, initiated by Feynman and Deutsch,
has been more prominent than quantum Turing machines in research
into quantum computing. However quantum Turing machines have been
studied for at least two major reasons:

(a) Quantum Turing machines form a class closely related to deterministic and probabilistic Turing machines, the basis for the theory of
classical computation which is well developed and quite well understood.

(b) Quantum logic circuits are devices purpose built for one specific
application whereas quantum Turing machines are devices that (like
ordinary Turing machines) can be used to operate on arbitrarily large
inputs and that can (possibly—this depends on the existence of a

universal device in the same class) be programmed. Programmability also allows a program-based notion of complexity.

Since a quantum Turing machine (QTM) could in principle operate for an indefinite number of steps, the interaction between the operator/observer and the apparatus is of crucial importance. Specifically, the operator needs to be able to tell when s/he may disturb the device and observe the output. This makes the halting of a quantum Turing machine a quite delicate issue (as also mentioned in [4] for example) and this paper examines the solution proposed by Deutsch, which is part of the standard description of the QTM.

2   Classical and probabilistic Turing machines

Since quantum Turing machines are based on the ordinary Turing machine, we start by reviewing the classical model. By the beginning of the twentieth century mathematicians had become quite interested in establishing a formal model of computability. In response, Alan Turing described an abstract device in 1936, now called a *Turing machine*, which follows a simple, finite set of rules in a predictable fashion to transform finite strings (input) into finite strings (output, where defined). The Turing machine (TM) can be imagined to be a small device running on a two-way infinite tape with discrete cells, each cell containing only the symbol **0** or **1** or a blank. The TM has a finite set of possible internal states and a head that can read the contents of the cell of the tape immediately under it. The head may also, at each step, write a symbol to the cell over which it finds itself. There are two special internal states: an *initial state* q_0 and a *halt state* q_H.

A TM has a finite list of instructions, or *transition rules*, describing its operation. There is at most one transition rule for each combination of cell content (under the head) and internal state. If the internal state is q_i and the head is over a cell with content S_j then the machine looks for a rule corresponding to (q_i, S_j). If no rule is found, the machine enters the halting state immediately. If a rule corresponding to (q_i, S_j) is found, it will tell the machine what to write to the cell under the head, whether to move left or right and which internal state to enter. There is no transition rule corresponding to the halting state. Sometimes we refer to the entire collection of individual rules for all the different (q_i, S_j) as *the transition rule* of the machine. A *computation* consists of starting the TM with the head over the first non-blank cell (which we may label position 0 on the tape) from the left of the tape (it is assumed that there is nothing but some finite *input* on the tape) and the machine in internal state q_0. Now the transition rules are simply applied until the machine enters the halting state q_H, at which point the content of the tape will be the *output* of the computation. If, for some input, the machine never halts then the output corresponding to that input is simply undefined. It is clear how every TM defines a (possibly, partial) function $f : \mathbb{N}_0 \to \mathbb{N}_0$ from the set of counting numbers to itself.

Turing machines are the canonical models of computing devices. No deterministic device, operating by finite (but possibly unbounded) means has been shown to be able to compute functions not computable by a Turing machine. In fact, one may view one's desktop computer as a Turing machine with a *finite* tape.

A probabilistic Turing machine (PTM) is identical to an ordinary Turing machine except for the fact that at each machine configuration (q_i, S_j) there is a finite set of transition rules (each with an associated probability) that apply and that a random choice determines which rule to apply. We fix some threshold probability greater than even odds (say, 75%) and say that a specific PTM computes $f(x)$ on input x if and only if it halts with $f(x)$ as output with probability greater than 75%.

3   Operation of a quantum Turing machine (QTM)

The *quantum Turing machine* (QTM) was first[5] described by David Deutsch [6]. The basic idea is quite simple, a QTM being roughly a probabilistic Turing machine (PTM) with complex transition amplitudes (the squared moduli of which add up to one at each application) instead of real probabilities. The QTM is related to the classical deterministic TM in much the same way as the PTM is.

In the following the *classical machine* is a machine with a two-way infinite tape, starting over position 0 on the tape as described above, that we use as a kind of template for the quantum Turing machine. The corresponding quantum Turing machine (QTM) might work as follows (based on the Deutsch description [6], Ozawa [1], Bernstein and Vazirani [7]).

I. The quantum state space of the machine is spanned by a basis consisting of states
$$|h\rangle\, |q_C\rangle\, |T_C\rangle\, |x_C\rangle$$
where $|h\rangle$ is the halt qubit, $h \in \{0, 1\}$ and (q_C, T_C, x_C) is a configuration of the corresponding classical machine, where x_C denotes

---

[5] Paul Benioff had related a similar idea somewhat earlier [5] but primarily in connection with presenting a possible physical basis for reversible computing.

the position of the head, q_C the internal state of the machine and T_C the non-blank content of the tape.

II. Special initial and terminal internal states have been identified (corresponding to the initial state and halting state of the classical machine).

III. The single transition rule is now a unitary operator U which, in each step, maps each basic $|h\rangle\,|q\rangle\,|T\rangle\,|x\rangle$ to a superposition of only finitely many $|h'\rangle\,|q'\rangle\,|T'\rangle\,|x'\rangle$, where

(a) the rule is identical for $|h\rangle\,|q\rangle\,|T_1\rangle\,|x\rangle$ and $|h\rangle\,|q\rangle\,|T_2\rangle\,|y\rangle$ when T_1 in position x and T_2 in position y have the same content, i.e. the rule depends only on the content of the tape under the head and the internal state q and not on the position of the head or on the content of the rest of the tape;

(b) T' and T differ at most in position x;

(c) $|x' - x| \leq 1$ (depending on whether the corresponding classical machine moves one position to the left, to the right, or not at all);

(d) $h' = 1$ if and only if q' is the halting state of the classical machine; and

(e) $T' = T$, $q' = q$ and $h' = h$ whenever $h = 1$.

Finitely many subrules

$$|h\rangle\,|q\rangle\,|T\rangle\,|x\rangle \quad \longmapsto \quad \sum_{i=1}^{n} c_i\,|h_i\rangle\,|q_i\rangle\,|T_i\rangle\,|x_i\rangle \qquad (1)$$

will determine U as there are, by the stipulations above, only finitely many possible—given that the alphabet of the tape (binary in our case) and the number of internal states are both finite. Note that the transitional rule ("program") will have a finite specification only if the transition amplitudes in the superposition of the $|h'\rangle\,|q'\rangle\,|T'\rangle\,|x'\rangle$ are all *computable* complex numbers, which we will of course assume to be the case throughout. The transition rule can also, obviously, be extended (linearly) to finite superpositions of $|h\rangle\,|q\rangle\,|T\rangle\,|x\rangle$.

IV. The machine is started with a finite superposition of inputs

$$|0\rangle\,|q_0\rangle\,|T\rangle\,|x\rangle\,.$$

Because of the form that the transition rule is allowed to take (and the fact that there are only finitely many internal machine states) the machine will be in the superposition of only *finitely many* basic states $|h\rangle\,|q\rangle\,|T\rangle\,|x\rangle$ at any step during the entire run[6] of computation.

---

[6] A more hazy concept than for classical Turing machines, as a QTM only really stops when one has observed the halt qubit and the content of the

The description of the machine given here differs from a classical reversible Turing machine in two obvious respects.

1. Transition rules are allowed to map a state of the machine to the superposition of several states. The crucial distinction with classical probabilistic machines is that the QTM goes to a quantum superposition of states whereas the classical PTM can be seen as either going to a classical probability distribution over states or to a specific state with some classical probability. Quantum computing, of course, uses superposition in an essential way[7]—as in the the algorithms of Shor or Grover.
2. The input is allowed to be a superposition of a finite number of "classical" inputs.

It is not immediately obvious why a finite collection of specifications of the form (1) should necessarily define a unitary U, however, just as it might not be apparent why a finite collection of rules

$$|h\rangle\,|q\rangle\,|T\rangle\,|x\rangle \quad \longmapsto \quad |h'\rangle\,|q'\rangle\,|T'\rangle\,|x'\rangle$$

for a classical machine would necessarily specify a reversible machine. Unitarity is, of course, a precondition for the quantum device to be feasible.

4   Physicality of the QTM

We start by assuming that the operator U described above is what is often called *well-formed*, i.e. that the subrules of the form (1) give rise to a unitary operator. Without loss of generality everything can be assumed to be coded in binary so that each position on the tape will correspond to a single qubit (quantum bit). A unit of quantum information, the qubit is a two level quantum mechanical system, whose state is described by a linear superposition of two basis quantum states, often labelled $|0\rangle$ and $|1\rangle$. The actual (quantum) state space of the machine will be a direct sum of n-qubit spaces (where n is an indication of how much tape has been used, each n-qubit space being the n-fold tensor of the single qubit space). The direct sum is, however, not a *complete* inner-product space (i.e. not a Hilbert space) and therefore—by the postulates of quantum

---

tape, so one may think of the transition rule being applied *ad infinitum*, step-by-step, unless the operator (physically, classically and externally) stops the machine.

[7] An very readable and accessible explanation of how and why this works can be found in [8].

mechanics—not a valid state space. However, the underlying Hilbert space can be taken to be the completion of the direct sum and a unitary operator U on the direct sum (see [7]) can be extended to a unitary operator $\widehat{U}$ on the Hilbert space. This completed space and operator will correspond to the physical system associated with the QTM, thereby taking care of the *physicality* of the QTM, provided that U was *well-formed*.

To see that well-formedness of U is not a triviality, consider a putative QTM with initial state q_0 and final state q_H. Let the following subrules define the machine:

$$|0\rangle\,|q_0\rangle\,|\ldots 0 \ldots\rangle\,|x\rangle \quad \longmapsto \quad \frac{1}{\sqrt{2}}\,|1\rangle\,|q_H\rangle\,(|\ldots 0 \ldots\rangle + |\ldots 1 \ldots\rangle)\,|x+1\rangle$$

$$|0\rangle\,|q_0\rangle\,|\ldots 1 \ldots\rangle\,|x\rangle \quad \longmapsto \quad |1\rangle\,|q_H\rangle\,|\ldots 1 \ldots\rangle\,|x+1\rangle$$

$$|1\rangle\,|q_H\rangle\,|T\rangle\,|x\rangle \quad \longmapsto \quad |1\rangle\,|q_H\rangle\,|T\rangle\,|x+1\rangle$$

These subrules specify that the machine starts out by replacing a 0 under the head by a superposition of 0 and 1, leaves a 1 initially under the head unchanged and then halts (i.e. enters a final state in which the tape head only may still move). Given a classical input, it could be said to operate reversibly as we can recover the input from the output. However, the putative QTM is not *well-formed* and therefore not a real QTM at all since

$$|0\rangle\,|q_0\rangle\,|\ldots 0 \ldots\rangle\,|x\rangle \quad \text{is perpendicular to} \quad |0\rangle\,|q_0\rangle\,|\ldots 1 \ldots\rangle\,|x\rangle$$

but

$$\frac{1}{\sqrt{2}}\,|1\rangle\,|q_H\rangle\,(|\ldots 0 \ldots\rangle + |\ldots 1 \ldots\rangle)\,|x+1\rangle$$

is certainly not perpendicular to $|1\rangle\,|q_H\rangle\,|\ldots 1 \ldots\rangle\,|x+1\rangle$ at all. A possibly correct version of the same machine could perhaps be given by

$$|0\rangle\,|q_0\rangle\,|\ldots 0 \ldots\rangle\,|x\rangle \quad \longmapsto \quad \frac{1}{\sqrt{2}}\,|1\rangle\,|q_H\rangle\,(|\ldots 0 \ldots\rangle + |\ldots 1 \ldots\rangle)\,|x+1\rangle$$

$$|0\rangle\,|q_0\rangle\,|\ldots 1 \ldots\rangle\,|x\rangle \quad \longmapsto \quad \frac{1}{\sqrt{2}}\,|1\rangle\,|q_H\rangle\,(|\ldots 0 \ldots\rangle - |\ldots 1 \ldots\rangle)\,|x+1\rangle$$

$$|1\rangle\,|q_H\rangle\,|T\rangle\,|x\rangle \quad \longmapsto \quad |1\rangle\,|q_H\rangle\,|T\rangle\,|x+1\rangle$$

for example, as the previous objection has been eliminated. The constraint of unitarity clearly excludes certain combinations of transition rules and directly implies that a great deal of care should be taken to ensure well-formedness of any machine that one may construct.

5   Time evolution of the QTM

If U is the operator that describes one application of the transition rule (i.e. one step in the operation) of the machine, then the evolution of an unobserved machine (where not even the halt bit is measured) for n steps is simply described by $V = U^n$. If the first measurement occurs after n_1 steps, and the measurement is described by an operator J_1 then the evolution of the machine for the first $n_1 + j$ steps is described by

$$U^j J_1 U^{n_1}$$

which is in general no longer unitary since the operator J_1 is a measurement (always in the computational basis). It is important to note that the machine evolves unitarily only when no measurement takes place at all. The quantum Turing machine should therefore not be seen as a pure quantum device but as a kind of hybrid device. Actually, as will be discussed below, the output from operating the machine for n steps without measurement of the halt bit is equivalent to operating it for n steps and observing the halt bit after every step. However, the QTM is envisioned with no explicit limit on the number of steps for which it may run, so the whole machine is evidently not equivalent to any single quantum experiment.

6   The halting scheme

The output of the machine on the tape is of course a superposition of basis states and should be read off after having measured the content of the halt bit and finding it in the state 1. The operator may at any time, or indeed between every two applications of U, measure the halt bit[8] in order to decide whether to read the tape content (and collapse the state of the machine to one of the basis states). The halt bit is intended to give the operator of the machine an indication of when an output may be read off from the tape (and by observation collapsing the system to an eigenstate) without interfering excessively with the computation.

It seems that Deutsch's original idea was that there would be no entanglement at all between the halt bit and the rest of the machine, but this cannot be guaranteed, as desribed by Myers [2]. The *output* of a QTM for some specific input x (which may be a superposition of classical inputs) is a probability distribution P_x over all possible contents of the tape at the time of observing the halt bit to have been activated. Actually, Miyadera and Ohya [9] have provided a simple proof that it

---

[8] The halt *qubit*, of course, until we measure it.

is not possible to effectively distinguish between those quantum Turing machines which have deterministic and those which have probabilistic halting behaviour.

6.1   Validity of the halting scheme

In an ordinary QTM the evolution of the machine continues even when the halt bit has been observed, and can continue after the halt bit has been observed without perturbing the probability distribution that has been defined to be the QTM's *output* (Ozawa [1]) since the observation projects one, in a certain sense, only into a specific ($h = 0$ or $h = 1$) branch of the computation. Let us consider this idea in slightly more detail.

Suppose U describes a QTM with a proper halting scheme, as above. By this we mean that U is unitary and that if

$$|1\rangle |q_H\rangle |T\rangle \sum_i c_i |x_i\rangle$$

is a halting configuration that occurs (maybe in a superposition) during the evolution of the machine for some valid input, then

$$U |1\rangle |q_H\rangle |T\rangle \sum_i c_i |x_i\rangle = |1\rangle |q_H\rangle |T\rangle \sum_j d_j |y_j\rangle$$

for some y. This simply means that U satisfies condition III(e) of Section 3 which is sufficient to guarantee this. Assume that the QTM is in a state

$$c_1 |1\rangle |\phi_1\rangle + c_2 |0\rangle |\phi_2\rangle$$

after $n > 0$ applications of U and (Scenario 1) that the halt bit is not measured at this point. One may assume $|\phi_1\rangle$ and $|\phi_2\rangle$ to be normalised, of course, and since $|1\rangle |\phi_1\rangle$ and $|0\rangle |\phi_2\rangle$ are orthogonal[9],

$$|c_1|^2 + |c_2|^2 = 1.$$

Apply U once more, to get

$$c_1 U |1\rangle |\phi_1\rangle + c_2 U |0\rangle |\phi_2\rangle$$

Since U is unitary and because of III(e)

$$U |1\rangle |\phi_1\rangle = |1\rangle |\psi_1\rangle$$

---

[9] Obviously, the probability of (Scenario 2) measuring an activated halt bit at this stage would be $|c_1|^2$.

and furthermore

$$U \left| 0 \right\rangle \left| \phi_2 \right\rangle = d_1 \left| 1 \right\rangle \left| \psi_2 \right\rangle + d_2 \left| 0 \right\rangle \left| \psi_3 \right\rangle$$

for some normalised ψ_i and d_i with

$$|d_1|^2 + |d_2|^2 = 1.$$

Therefore the state of the device is now

$$c_1 \left| 1 \right\rangle \left| \psi_1 \right\rangle + c_2 d_1 \left| 1 \right\rangle \left| \psi_2 \right\rangle + c_2 d_2 \left| 0 \right\rangle \left| \psi_3 \right\rangle .$$

It follows from unitarity of U that since $\left| 1 \right\rangle \left| \phi_1 \right\rangle$ is orthogonal to $\left| 0 \right\rangle \left| \phi_2 \right\rangle$ it must also be the case that $U \left| 1 \right\rangle \left| \phi_1 \right\rangle$ is orthogonal to $U \left| 0 \right\rangle \left| \phi_2 \right\rangle$. In other words

$$\left| 1 \right\rangle \left| \psi_1 \right\rangle \quad \text{is orthogonal to} \quad d_1 \left| 1 \right\rangle \left| \psi_2 \right\rangle + d_2 \left| 0 \right\rangle \left| \psi_3 \right\rangle .$$

But $\left| 1 \right\rangle \left| \psi_1 \right\rangle$ is orthogonal to $\left| 0 \right\rangle \left| \psi_3 \right\rangle$ and hence the three normalised states

$$\left| 1 \right\rangle \left| \psi_1 \right\rangle , \quad \left| 1 \right\rangle \left| \psi_2 \right\rangle , \quad \left| 0 \right\rangle \left| \psi_3 \right\rangle$$

are pairwise orthogonal. The probability of now measuring an activated halt bit is therefore

$$|c_1|^2 + |c_2 d_1|^2$$

which is greater than at the previous step, as one might expect. Unsurprisingly, if (Scenario 2) we had measured the halt bit one step earlier then we would have read off an activated state with probability $|c_1|^2$ and an inactive halt bit with probability $|c_2|^2$. Having measured an inactive halt bit would have collapsed the state of the machine to $\left| 0 \right\rangle \left| \phi_2 \right\rangle$ which, after another appplication of U would give

$$d_1 \left| 1 \right\rangle \left| \psi_2 \right\rangle + d_2 \left| 0 \right\rangle \left| \psi_3 \right\rangle .$$

Observing the halt bit now would show an activated state with probability $|d_1|^2$. Classical probability calculus, given that the events are mutually exclusive and using Bayes' formula, yields a probability of having observed an activated halt bit after the last step of Scenario 2 as

$$|c_1|^2 + |c_2|^2 \cdot |d_1|^2$$

which is then the same as the probability of measuring it after the last step of Scenario 1.

The preceding elementary exposition illustrates the idea developed by Ozawa in [1]. That paper considers the probability of observing the tape of the machine in a specific state T with the halt bit activated, either (i) by observing the halt bit after n_1 steps (and possibly reading of the content of the tape then, if the halt bit was activated) and then again after another n_2 steps or by (ii) just letting the machine run without observation and after $n_1 + n_2$ steps measuring the halt bit (and possibly the tape). Ozawa showed that the probability of observing the specific state T of the tape is identical in the two cases. It should perhaps be noted that in Ozawa's description the position of the head on the tape is considered part of the tape, whereas it has been explicitly separated in this contribution. This does not detract in any essential way from Ozawa's result.

6.2 Objections to the halting scheme

As mentioned earlier, Myers [2] pointed out that it must be possible to find a QTM in a superposition of basic states with the halt bit activated and basic states with the halt bit not (yet) active. Among other reasons, this is simply true because if x is an input for which the machine takes N_x steps to active the halt bit and y is an input for which it takes $N_y \neq N_x$ steps, then the machine, when the input is a superposition of x and y, will—between the N_x-th and N_y-th step—find itself in such a superposition. This emphasises the inherent probabilistic nature of quantum Turing machines, as also pointed out clearly by Miyadera and Ohya [9] and is not an objection *per se*, as Ozawa [1] has shown.

Kieu and Danos [3] purport to have shown the impossibility of any unitary operator U having the desired halting scheme. They argue as follows. Suppose there exists a state $|\psi\rangle$ of the machine such that

$$(\langle 0| \otimes I) \; |\psi\rangle = 1$$

where I is the identity operator, and there exists an $N > 0$ such that

$$(\langle 1| \otimes I) \; U^N \, |\psi\rangle > 0. \tag{2}$$

Let N further be the smallest number for which relation (2) holds. Such a $|\psi\rangle$ could constitute the QTM prepared with a valid input that eventually activates the halt bit—at least a little bit—and N would be the number of applications of U necessary for any chance of observing an activated halt bit. If no such $|\psi\rangle$ exists then the machine described by U has no input leading to a positive probability of an output and is, in other words, useless. Kieu and Danos proceed to observe correctly that (in our notation)

$$U^{N-1} \, |\psi\rangle \quad \text{is perpendicular to all} \quad |1\rangle \, |q_H\rangle \, |T\rangle \, |x\rangle$$

because of the minimality of N and therefore, since U is unitary,

$$U^N \, |\psi\rangle \quad \text{is perpendicular to all} \quad U \, |1\rangle \, |q_H\rangle \, |T\rangle \, |x\rangle \, .$$

This is then supposed to contradict (2). It would be true if, as stated in the Kieu and Danos paper, the subspace V spanned by the $|1\rangle \, |q_H\rangle \, |T\rangle \, |x\rangle$ were identical to the subspace $\tilde{V}$ spanned by the $U \, |1\rangle \, |q_H\rangle \, |T\rangle \, |x\rangle$. Kieu and Danos appear to presume that the restriction $U|_V$ of U to V is unitary, from which it would indeed follow that $V = \tilde{V}$, but this presumption is without proof. Of course, $U|_V$ does preserve the inner product but it is not necessarily unitary because its range need not span all of V. In fact, $U|_V$ being unitary (i.e. $V = \tilde{V}$) would *directly* imply that no $|\psi\rangle$ as above exists. In [10] the same assumption (that $U|_V$, denoted B there, is unitary) appears to have lead to the same conclusion.

Concerns about the halting scheme were raised also by Shi [11], albeit in the context originally introduced by Deutsch, where superpositions

$$c_1 \, |1\rangle \, |\phi_1\rangle + c_2 \, |0\rangle \, |\phi_2\rangle$$

with $|c_1 c_2| > 0$ are assumed not to occur. Shi had actually been examining the existence of universal quantum computers where the halting scheme is very relevant to the discourse.

6.3   The halting scheme and universality

Consider a general countable class of machines that compute partial functions, i.e. functions that are not necessarily defined for all inputs. Assume that each machine is fully described by a natural number. Let Φ_n denote the random variable (not simply function, since the machine is not assumed to be deterministic) computed by machine n and fix a reasonable

$$h : \mathbb{N}_0 \times \mathbb{N}_0 \to \mathbb{N}_0$$

which will be used for the encoding of programs and data as a single input.

Definition 6.1. *If there exists a number N such that*

$$\Phi_N \, (h(n, m)) = \Phi_n(m)$$

which means that the functions are either equal and both defined or both undefined if deterministic, and if not deterministic then the values have the same distribution, for all n and m, then the machine described by N is called a universal machine for the class.

Programmability follows from universality and this is why universality is such an important concept. A universal Turing machine, for example, can simulate all the Turing machines, and is thus programmable for the entire class of Turing machines. In spite of very powerful results by Bernstein and Vazirani [12], universality in the sense of the definition above has not (yet) been established [13]. The observation of the halt bit goes quite far in explaining why not. Bernstein and Vazirani showed that there exists a quantum Turing machine $\mathcal{U}$ such that

"for any well-formed[10] QTM M, any $\varepsilon > 0$, and any T, $\mathcal{U}$ can simulate M with accuracy ε for T steps with slowdown polynomial in T and $\frac{1}{\varepsilon}$."

The full Bernstein-Vazirani result could be summarised by the statement that

there exists a QTM $\mathcal{U}$ such that for each QTM M with finite description $\bar{M}$, n, ε and T there is a program $\mathcal{P}(\bar{M}, n, \varepsilon, T)$ and a function $f_{\bar{M}}(T, n, \frac{1}{\varepsilon})$ (both recursive in their inputs) such that running $\mathcal{U}$ on input $\left|\mathcal{P}(\bar{M}, n, \varepsilon, T)\right\rangle \otimes |x\rangle$ where $|x| = n$ for $f_{\bar{M}}(T, n, \frac{1}{\varepsilon})$ steps results—within accuracy ε—in the same distribution over observable states as running M on input $|x\rangle$ for T steps. [8].

So far there is no problem except if one wants to simulate the running of a given QTM using Berstein and Vazirani's $\mathcal{U}$ for an indefinite period of time, i.e. as long as it might take to obtain a result. In this case $\mathcal{U}$ would have to be run for $n = 100$, reset and run for $n = 200$ and so on, for an unknown but finite number of times. The problem is that after each run of $\mathcal{U}$ (for example, for $n = 500$) the resetting to the original input value is no longer necessarily possible—because the observation of the halt bit could actually have caused the familiar measurement-related collapse of the state of the machine[11] from

$$c_1 |1\rangle |\phi_1\rangle + c_2 |0\rangle |\phi_2\rangle$$

to either $|1\rangle |\phi_1\rangle$ or $|0\rangle |\phi_2\rangle$. Of course, at that point an operator could step up and reprogram the machine, but such an intervention would clearly violate the principle of autonomy of operation of the device in the same way that it would if a desktop computer were unable to perform

---

[10] Meaning that the time evolution operator is unitary, as discussed in Section 4.

[11] If not, recovering the initial state (i.e. the input) is as simple as applying the inverse operation n times.

the NOT operation and had to request the manual flipping of a switch from an operator each time the unary operator were required.

Concatenation is a crucial and closely related issue for universality in quantum computing which has also been addressed by Shi [11].

7   Conclusion

The halting scheme for quantum Turing machines, as proposed by Deutsch, is a valid idea — or appears, so far, to be—if one keeps in mind that the operation of the machine will be essentially probabilistic and not deterministic. Any classical reversible universal Turing machine, which uses the halt bit to identify its single terminal state and which after reaching the terminal state keeps moving the head in one direction, corresponds to a well-formed quantum Turing machine since its operation consists of a permutation of the basic states of the machine. This provides a simple example of one machine for which the halting scheme is obviously valid. However it is the view of the authors that more research is needed into the power of well-formed quantum Turing machines. Among other things, although a simulation procedure has been described for arbitrary QTMs, it has not been clearly shown whether any universal such machine (or a machine universal for some subclass) exists. The halting scheme for quantum Turing machines, while providing the means of using them in practice, seems to be a serious impediment in this regard.

References

[1] Ozawa, M.: Quantum Nondemolition Monitoring of Universal Quantum Computers. Phys. Rev. Lett. **80** (1997) 631–634

[2] Myers, J.M.: Can a Universal Quantum Computer Be Fully Quantum? Physical Review Letters **78**(9) (1997) 1823–1824

[3] Kieu, T.D., Danos, M.: A No-Go Theorem for Halting a Universal Quantum Computer. Acta Physica Hungarica A) Heavy Ion Physics **14**(1) (2001) 217–225

[4] Hirvensalo, M.: Computing with quanta — impacts of quantum theory on computation. Theor. Comput. Sci. **287**(1) (2002) 267–298

[5] Benioff, P.: The computer as a physical system: A microscopic quantum mechanical hamiltonian model of computers as represented by turing machines. J. Stat. Phys. **22** (1980) 563–591

[6] Deutsch, D.: Quantum theory, the Church-Turing principle and the universal quantum computer. Proc. R. Soc. Lond. A **400** (1985) 97–117

[7] Bernstein, E., Vazirani, U.: Quantum complexity theory. SIAM J. Comp. **26** (1997) 1411–1478

[8] Fortnow, L.: One complexity theorist's view of quantum computing. Electronic Notes in Theoretical Computer Science **31** (2000)

[9] Miyadera, T., Ohya, M.: On Halting Process of Quantum Turing Machine. quant-ph/0302051 (2003) Open Systems and Information Dynamics, Vol.12, No.3 261-264 (2005).

[10] Kieu, T.D., Danos, M.: The halting problem for universal quantum computers. quant-ph/9811001 (1998)

[11] Shi, Y.: Remarks on universal quantum computer. Physics Letters A **293**(5–6) (2002) 277–282

[12] Bernstein, E., Vazirani, U.: Quantum complexity theory. SIAM J. Comput. **26** (1997) 1411–1473

[13] Fouché, W., Heidema, J., Jones, G., Potgieter, P.H.: Deutsch's Universal Quantum Turing Machine (Revisited). quant-ph/0701108 (2007)

Design of DNA spike oscillator

Kohta Suzuki and Satoshi Murata

Tokyo Institute of Technology,
G3-53, 4259 Nagatsuta, Midori, Yokohama, 226-8502 Japan
murata@dis.titech.ac.jp

Abstract A model of DNA spike oscillator is proposed. The behavior of the proposed oscillator depends on DNA sequences; thus, the frequency and amplitude of the DNA oscillator can be determined a priori. If we use sets of orthogonal DNA sequences, we can realize many independent oscillators working in parallel in a single tube. The oscillator is a double feedback system implemented by three elementary reactions composed of restriction, enzyme, and DNA hybridization. We have built a numerical model of the oscillator to determine the parametric domain of oscillation and have verified each elementary reaction by PAGE experiment. This kind of oscillator will be useful for generating periodic clock signal for various DNA-based nanosystems.

1 Introduction

DNA nanotechnology has been gathering increased attention for its substantial successes in the past decade [1]. Its application ranges from the self-assembly of complicated nanostructure [2,3] to molecular robots driven by molecular logic gates. DNA molecular motors/actuators are especially important subjects of research; they are expected to become mechanical components that give motion to nanoscale robots. There are several types of DNA motors/actuators: The earliest type of DNA actuators is a nanomechanical device based on B-Z transition of DNA helix upon a change of ionic concentration [4]. It was very simple to implement, however, all the actuators in the tube responds to the input (the ionic concentration) in non-selective manner. Typical DNA actuator in another category is the molecular tweezers controlled by sequential addition of "fuel" DNA strands [5–7]. This is more advantageous than the non-selective device because it enables us to design sequence-specific activation of each actuator. The third type of the actuators is based on

DNA-enzyme [8]. The operation of DNA-enzyme based actuators can be autonomous, while more difficult to control.

In this paper, we propose a novel design of DNA spike oscillator that provides a driving signal for the DNA actuators in the second category. The oscillator changes the concentration of one specific DNA strand. The output DNA strand can be used as fuel strand for the actuator. This kind of oscillator may also be used in reaction-diffusion systems to execute spatio-temporal computation.

The proposed design of DNA oscillator has the following features:

1) The frequency and amplitude of oscillation are determined by sequence design of DNA.
2) Many oscillators with different properties can work simultaneously if the DNA sequences of these oscillators are orthogonal.

2   Model of spike oscillator

Karfunkel [9] found that Michaeris-Menten type enzyme reaction could generate spike oscillation by adding inhibitory reaction to the enzyme that is controlled by the product of the enzyme reaction itself. Sustainable oscillation occurs between concentrations of substrate and product (Fig. 1)

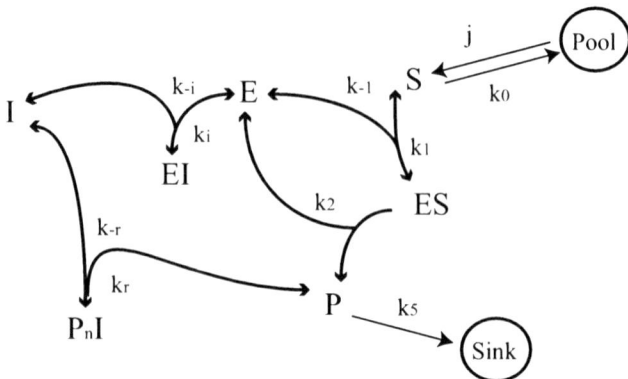

Figure 1. Spike oscillation of Michaeris-Menten type enzyme reaction

There are five components in the oscillator system; Substrate (S) that is provided at constant rate j from external environment, Enzyme (E)

that catalyzes reaction S→P, Product (P) that is discarded at constant rate k_5, and Inhibitor (I). The overall system consists of the following three reactions:

$$\left. \begin{array}{l} S + E \rightleftharpoons ES \rightarrow E + P \\ E + I \rightleftharpoons EI \\ nP + I \rightleftharpoons P_nI \end{array} \right\} \qquad (1)$$

The first reaction represents enzyme reaction. Then the enzyme is suppressed by inhibitor in the second reaction. In the third reaction, the inhibitor itself is suppressed by the product of the first reaction. Equilibrium constants for these reactions are expressed in terms of rate constant, respectively:

$$\left. \begin{array}{l} K_m = \dfrac{k_{-1} + k_2}{k_1} \\[2mm] K_I = \dfrac{E \cdot I}{EI} = \dfrac{k_{-i}}{k_i} \\[2mm] K_R = \dfrac{P^n \cdot I}{P_nI} = \dfrac{k_{-r}}{k_r} \end{array} \right\} \qquad (2)$$

This system oscillates under appropriate values of the rate constants and the equilibrium constants. Especially, valence n in the third reaction in (1) has an important role, namely the system oscillates only when $n \geq 2$.

The time course of the concentration of each component is calculated by numerically integrating the rate equations (3). Although the equations are highly nonlinear, we can simplify them by the steady state approximation $(dES/dt = 0)$ [9].

$$\left. \begin{array}{l} \dfrac{dS}{dt} = j - k_1 E \cdot S + k_{-1} ES \\[2mm] \dfrac{dP}{dt} = k_2 ES - k_5 P - n \cdot k_r P^n \cdot I + n k_{-r} P_nI \\[2mm] \dfrac{dES}{dt} = k_1 E \cdot S - (k_{-1} + k_2) ES \\[2mm] \dfrac{dEI}{dt} = k_i E \cdot I - k_{-i} EI \\[2mm] \dfrac{dP_nI}{dt} = k_r P^n \cdot I - k_{-r} P_nI \\[2mm] I_t = I + P_nI + EI \\[1mm] E_t = E + ES + EI \end{array} \right\} \qquad (3)$$

$$\frac{dS}{dt} = j - k_2 ES$$

$$\frac{dP}{dt} = k_2 ES - k_5 P$$

$$ES = -\frac{\sigma}{2}\left\{\frac{I_t - E_t}{1 + \sigma} + K_I(1 + \rho_n)\right\} +$$

$$\frac{\sigma}{2}\sqrt{\left\{\frac{I_t - E_t}{1 + \sigma} + K_I(1 + \rho_n)\right\}^2 + \frac{4K_I E_t(1 + \rho_n)}{1 + \sigma}} \qquad (4)$$

where, $\rho = S/K_m$, $\rho_n = p^n/K_R$, I_t and E_t denote initial concentrations of the Inhibitor and the Enzyme. Dimensions of these variables are as follows: [M] for S, E, ES, EI, P_nI, I_t, and Et; and [Ms^{-1}] for j; [M^{-1}s^{-1}] for k_1; [s^{-1}] for k_{-1}, k_2, and k_5; and [M] for K_m, K_I and K_R.

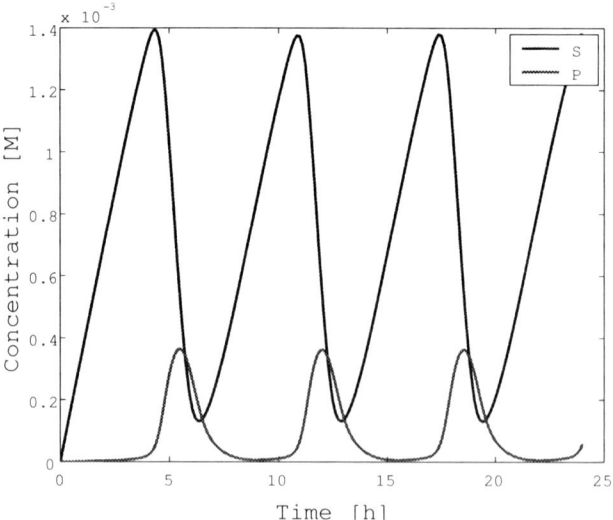

Figure 2. Dynamical behaviour of system. n=2, K$_m$=100[nM], k$_2$ = 1.0 × 10^{-2}[s^{-1}], K$_I$ = k_{-i}/k_i = 0.1[nM], K$_R$ = k_{-r}/k_r = 0.1[nM], E$_t$ = 60[μ M], I$_t$ = 70[μ M], j = 1.0× 10^{-6}[Ms^{-1}], k$_5$=1.0×10^{-2}[s^{-1}]

We did numerical simulation of the system for various parameter settings. One of the results is given in Fig. 2. In this case, the period of oscillation on S and P is about 2500[s]. The parameters for the simulation are the same to those of implementation model described later. Fig. 3 is the phase plot of the oscillator system. Dark gray curve is the null cline of S($dS/dt = 0$) and light gray one is the null cline of P($dP/dt = 0$). Vector field (dS/dt, dP/dt) is expressed as arrows. The oscillation (limit cycle) is shown by gray trajectory.

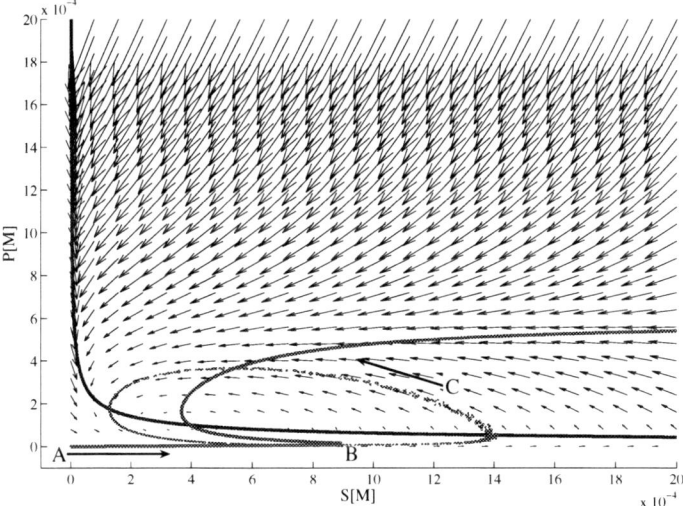

Figure 3. Phase plot of the oscillator system. n=2, K_m=100[nM], $k_2 = 1.0 \times 10^{-2}$[s^{-1}], $K_I = k_{-i}/k_i = 0.1$[nM], $K_R = k_{-r}/k_r = 0.1$[nM], $E_t = 60[\mu$ M], $I_t = 70[\mu$ M], j = 1.0$\times 10^{-6}$[Ms^{-1}], k_5=1.0$\times 10^{-2}$[s^{-1}]

The mechanism of the spike oscillation can be outlined as follows. Assume the initial condition of the system is at the origin $(S, P) = (0, 0)$. Here, most of the enzymes are in the form of EI, and thus inactive. Therefore, an increase in S is mostly limited by the constant inward flow of S. When P reaches its maximum (Fig. 3, B), the state of the system becomes unstable, resulting in rapid increase of P along the null cline (Fig. 3, C). At the same time, the concentration of product P becomes higher and suppresses the activity of enzyme E. Consequently, the state of the system is brought to near its origin, which initiates another cycle of oscillation.

3   Implementation by DNA

The implementation of the spike oscillator using DNA molecules is described in this section. The overall reaction system is illustrated in Fig. 4.
The DNA oscillator is composed of three reactions corresponding to the
reactions of the spike oscillator model. Labels in alphabets denote specific
DNA sequences. $\bar{x}$ means complementary sequence to x, and x* denotes
the partial sequence of x. Numbers in parentheses means the strength
of base pairing of that DNA segment. Hereafter, we assume that two
molecules bound by strength three or less does not hold the bond. This
assumption corresponds to a suitable combination of temperature of solution, the length of hybridized base pairs, and the concentrations of the
DNA molecules.

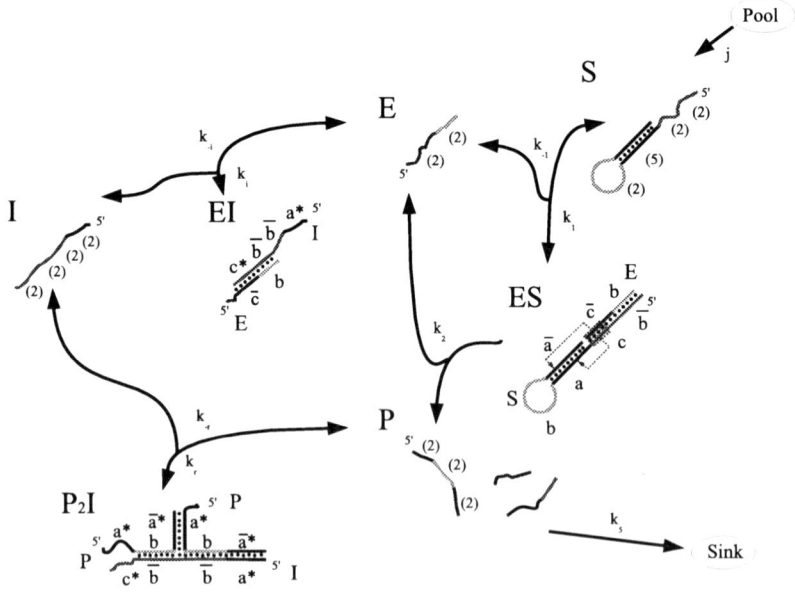

Figure 4. Implementation model of DNA spike oscillator

Three elementary reactions comprise the oscillator system:
$$\sharp\ S + E \rightleftharpoons ES \rightharpoonup E + P$$
$$\sharp\ E + I \rightleftharpoons EI$$
$$\sharp\ 2P + I \rightleftharpoons P_2 I$$

Molecule S is the substrate injected to the tube at constant rate. Molecule P is the product, which is decomposed at the rate proportional to its concentration. In the proceeding sections, each elementary reaction is explained in detail.

3.1 Elementary reaction ♯ 1 (Molecules E and S)

Elementary reaction♯ 1 (Molecule E and S comprise complex ES and produce P) is implemented by restriction enzyme. The restriction enzyme is an enzyme that cuts double-stranded DNA at specific recognition sequence. In this paper, we use a restriction enzyme Fok I by which the cleavage occurs not in the recognition sequence but 5~13 bp downstream of the recognition site.

The reaction ♯ 1 takes place under the existence of Fok I. The Enzyme E in the original model is represented by the combination of DNA molecule E and the restriction enzyme. In other words, the recognition site of Fok I is formed only when E and S hybridize.

The hairpin structure of S is stable because the bonding strength at its neck is more than three. When Fok I cleaves the double helix, it cannot hold the hairpin, and consequently S is opened (this opened molecule is called P). Then, some of the molecule P and two inhibitor molecules I form P_2I complex (see 3.3). The rest of molecule P is decomposed by another enzyme called Exonuclease, thus no longer affects the overall reaction. Here, rate constant k_1 denotes the association rate of E to S, and k_{-1} denotes dissociation rate. k_2 represents the enzymatic turnover number (the number of cleavages per unit time).

3.2 Elementary reaction ♯2 (Molecules E and I)

The oscillator system has double negative feedback loops. The first feedback loop is in the reaction where the inhibitor molecule I suppresses catalytic function of the molecule E by forming complex EI (Fig. 4). The bonding segment between E and I should slightly be (2~5bp) dislocated in order not to form another recognition site for Fok I. Reaction rate k_i is the association rate of E to I, and k_{-i} is the rate of reverse reaction.

3.3 Elementary reaction ♯3 (Molecules P and I)

The second negative feedback loop is in the reaction where molecule P forms complex with molecule I to inhibit I from bonding with E. As is mentioned in Section 2, it is necessary to consume two P molecules for

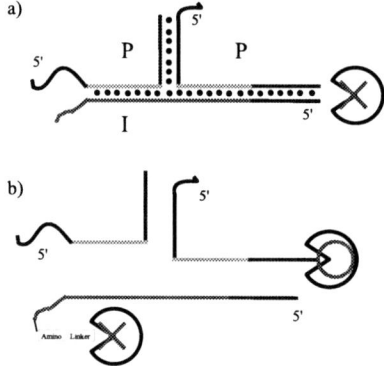

Figure 5. Excretion scheme for molecule P_2I

each I molecule. This is realized by using three-way junction structure formed by two Ps and one I (Fig. 5a). k_r is the forward rate of this hybridization reaction, and k_{-r} denotes the reverse rate.

3.4 Supply and excretion of waste molecules

In order to generate sustainable oscillation, we need to supply molecule S, and at the same time, excrete molecule P. Supplying S can simply be done by pipette injection. On the other hand, molecule P should be removed at a rate proportional to its concentration. This can be done with an enzyme called Exonuclease T (also called RNaseT), which specifically decompose single-stranded DNA (or RNA) from its free (non-modified) 3'-end. The Exonuclease T attacks all the free 3'-end, therefore, ssDNAs other than P should be protected by amino-linker at 3'-end (Fig. 5).

4 Experiments

We have verified each of elementary reactions by DNA. Note that these reactions are examined independently, and their reaction constants are not tuned to the parameters obtained in section 2. In this sense, the results are preliminary.

Elementary reaction ♯ 1 (Molecules E and S)

Selective activation of restriction enzyme Fok I by ssDNA E is examined. Fok I has recognition sequence 5' GGATG-3'/3'-CCTAC 5' (dark gray

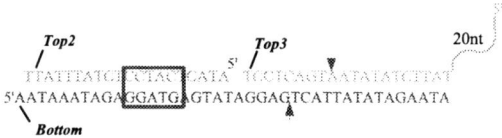

Figure 6. Activation control of restriction enzyme Fok I. Bottom(40nt) and Top2(20nt) correspond to hairpin substrate S. Existence of Top3(40nt) activate Fok I by forming recognition site.

box in Fig. 6) and cleaves 9/13 bases downstream (gray arrows in Fig. 6). Here, the molecule S is represented by two ssDNA, Top3 and Bottom. Existence of another ssDNA Top2 activates Fok I. The activation by adding strand Top2 (E) is verified by the result of gel analysis (Fig. 7).

4.1 Elementary reaction ♯ 3 (Molecules P and I, excretion of P)

The formation of complex P_2I (Fig. 8) is verified by gel analysis. P_2I^0 and PI^0 correspond to the top two bands in lane 4. The concentration of P is increased in Lanes 5 and 6, and we found that the upper band is always thicker than the lower. This finding implicates the formation of the desired complex P_2I^0.

This experiment also proves that the excretion of waste molecule P can be achieved through adding Exonuclease T (ExoT hereafter). I^0 is non-protected inhibitor DNA, while I^1 is protected by amino-linker (molecular weight 180) at 3' end. Lanes 7 and 8 show that unprotected complex P_2I^0 is decomposed by ExoT. The complex is protected by amino-linker in the gel analysis in Figs. 9 and 10 (Lane 5 and 7).

5   Conclusion

In this paper, we presented the novel design of DNA oscillator based on the spike oscillation of enzymatic reaction. The implementation of the oscillator by using DNA molecules is also proposed, and the basic behavior of three elementary reactions involved in the implementation model is verified by PAGE experiments.

These reactions should be combined into one. The issues in the next stage are:

1) To verify that there are no more unexpected intermediate complexes in the reaction network. SI and S_2I are some of such complexes that

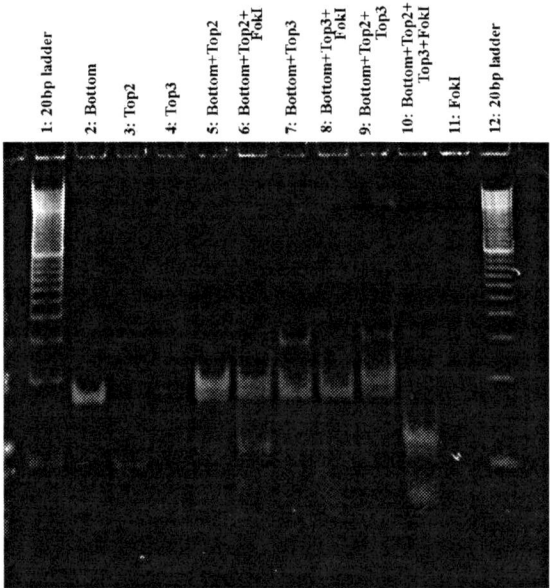

Figure 7. Non-Denaturing PAGE (8%) analysis of elementary reaction ♯ 1. Each sample is incubated at 37 °C for 60min in buffer solution (Tris-HCl:100mM, MgCl$_2$:100mM, Dithiothreitol:10mM, NaCl:500mM). Sample concentrations are 1 μM applied 4 units of Fok I.

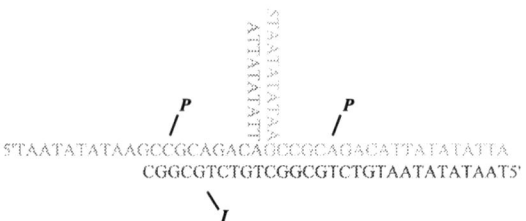

Figure 8. DNA sequence for P and I

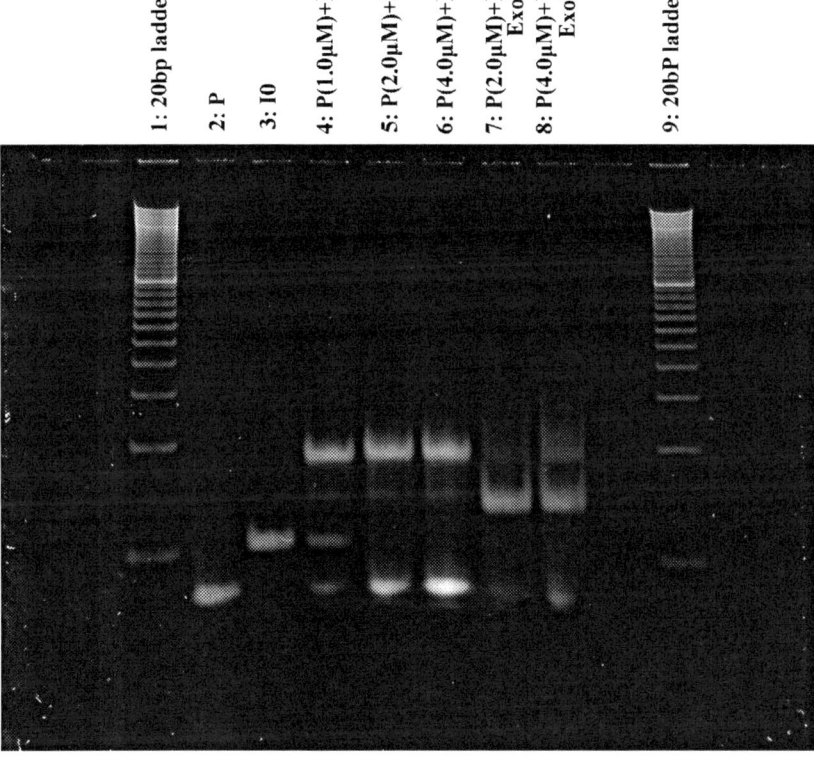

Figure 9. Non-Denaturing PAGE(8%) analysis of elementary reaction 3. Each sample is incubated at 37 °C for 120min in buffer solution (potassium acetate 50mM, Tris-acetate 20mM, magnesium acetate 10mM, DTT 1mM). Each I^0 concentration is 1.0μM and ExoT is 5 units.

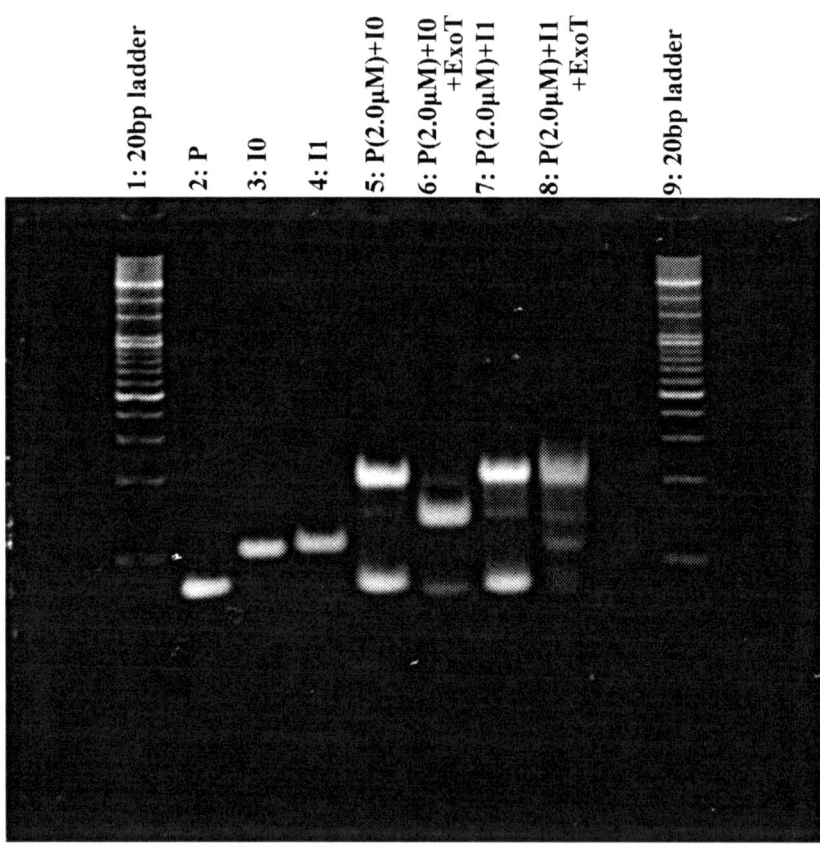

Figure 10. Non-Denaturing PAGE(8%) analysis of elementary reaction 3. Each sample is incubated at 37°C for 120min in buffer solution (potassium acetate 50mM, Tris-acetate 20mM, magnesium acetate 10mM, DTT 1mM). Each I^{0} and I^{1} concentration is 1.0μM and ExoT is 5 units.

current implementation model might have. They should be avoided by redesigning DNA sequences.

2) To measure rate constant and equilibrium constant of each elementary reaction by using FRET.

3) To reconstruct numerical model of DNA oscillator to examine its behavior under more realistic conditions.

References

[1] Chen, J., Jonoska, N., Rozenberg,G. (Eds.), *Nanotechnology: Science and Computation*, Springer-Verlag Berlin Heidelberg 2006.

[2] Winfree, E., Liu, F., Wenzler, L.A., Seeman, N.C., Design and self-assembly of two-dimensional DNA crystals, *Nature*, **394**(6693), 539-544.

[3] Rothemund,P.W.K., Folding DNA to create nanoscale shapes and patterns, *Nature* **440**, 297-302, 2006.

[4] Mao,C., Sun, W., Shen, Z., Seeman, N.C., A nanomechanical device based on the B-Z transition of DNA. *Nature* **397**, 144-146, 1999.

[5] Yurke, B., Tuberfield, A.J., Mills Jr, A.P., Simmel, F.C., Neumann, J.L., A DNA-fuelled molecular machine made of DNA, *Nature*, **406**, 605-608, 2000.

[6] Shin,J.S., Pierce N.A., A Synthetic DNA Walker for Molecular Transport, *J Am Chem Soc.*, **126**(35), 10834-10835(2004).

[7] Tian,Y., Mao,C., Molecular Gears: A Pair of DNA Circles Continuously Rolls against Each Other, *J. Am. Chem. Soc.*, **126**(37), 11410-11411(2004).

[8] Stojanovic,M.N. and Stefanovic,D., A deoxyribozyme-based molecular automaton, *Nat. Biotechnol.*, **21**(2003), 1069-1074.

[9] Kerfunkel, H.R., Seeling, F.F., Reversal of inhibition of Enzymes and the Model of Spike Oscillator, *J.Theor. Biol*, **36**, 237-253, 1972.

176

The computing power of structured molecules with gaps: Watson-Crick insertion systems

Kaoru Onodera

Department of Information Sciences, School of Science and Engineering, Tokyo Denki University, Ishizaka, Hatoyama-cho, Hiki-gun, Saitama 350-0394, JAPAN kaoru@j.dendai.ac.jp

Abstract   New operations based on DNA complementarity called *Watson-Crick insertion operations* are introduced in order to investigate the computational power of insertion operations in case of taking double strands into consideration through operation processes. A *Watson-Crick insertion system* makes it possible for a computation to proceed by using sticking for incomplete molecules with *gaps* (i.e. missing nucleotides) in one of the strand. The difference between the new type of computing models and the existing insertion systems is due to the data structures they handle, double stranded sequences with gaps in Watson-Crick insertion systems, while linear strings in insertion systems. By introducing a variety of types of insertion rules, our results suggest that restrictions on insertion rules as well as on computation processes influence on the computational powers of Watson-Crick insertion systems with those types of rules.

1   Introduction

Insertion and deletion systems have been considered in formal language theory as well as in theoretical research on molecular computing models. These operations, insertion and deletion, were proposed to model the behavior of biomolecules with context dependence and to investigate the computational capability of those molecules based on the biomolecular property of DNA complementarity. In the DNA framework, the existing insertion and deletion systems on strings enable to simulate behavior of molecules for restricted ones under suitable conditions [2] [3]. Moreover, Watson-Crick automata and sticker systems are known as a language recognition devices and language generating devices, respectively. Those computational devices operate on double strands without gaps which satisfy the complementarity, that is the matching letters [2].

On the other hand, a concise and precise notation for DNA molecules is proposed in [5], where all of the followings are considered: single-stranded DNA molecules, double-stranded DNA molecules containing gaps and nicks (missing phosphodiester bonds). In this paper, we consider incomplete DNA molecules containing gaps (missing nucleotides in one of the strands). Biologically, the DNA expressions characterize sets of bulge-forming DNA sequences as well as bulge-free ones.

The purpose of this paper is twofold: one is to define the DNA expressions and sticking operations for incomplete molecules with gaps based on complementarity. The other is to explore the computational power of an insertion operation working on incomplete molecules with gaps (in other words, the language generative power of an insertion operation over incomplete molecules with gaps through the computation) in terms of formal language theory. From insight of the known results on the insertion and deletion systems on strings [2], one would easily imagine that insertion operations together with deletion operations, when working on double strands, might be already powerful enough. This makes us consider using only insertion operations. From our results about the computational powers for a variety of types of Watson-Crick insertion systems, the restriction on insertion rules as well as on computational process (i.e. existence of gaps through computation) enhances the computing power of Watson-Crick insertion systems.

2   Preliminaries

We assume the reader to be familiar with the rudiments on DNA computing as well as basic notions in formal language theory (see, e.g., [2] [3]).

For a string x in V^*, $|x|$ is the length of x and $|x|_a$ is the number of occurrences of the symbol a in x.

2.1   Grammars and automata

A grammar $G = (N, V, P, S)$ is *context-sensitive* if N is a set of *non-terminal symbols*, V is a set of *terminal symbols*, S in N is the *initial symbol*, and P is a finite set of *productions* of the form $u_1 N_1 u_2 \to u_1 \beta u_2$, where $N_1 \in N$, $u_1, u_2, \beta \in (N \cup V)^*$, and $\beta \neq \epsilon$. For any x_1 and x_2 in $(N \cup V)^*$, if $x_1 = y_1 \alpha_1 y_2$, $x_2 = y_1 \alpha_2 y_2$, and $r : \alpha_1 \to \alpha_2 \in P$, then we write $x_1 \overset{r}{\Longrightarrow}_G x_2$. If there is no danger of confusion, we write $x_1 \Longrightarrow x_2$. The reflexive and transitive closure of $\Longrightarrow$ is denoted by $\Longrightarrow^*$.

A *language* $L(G)$ generated by a grammar G is defined by $L(G) = \{x \in V^* \mid S \Longrightarrow_G^* x\}$. A language L is a *context-sensitive language* if there is a context-sensitive grammar G such that $L = L(G)$.

A grammar $G = (N, V, P, S)$ is *context-free* if P is a finite set of *context-free productions* of the form $N_1 \to \alpha$, where $N_1 \in N$ and $\alpha \in (N \cup V)^*$. A language L is a *context-free language* if there is a context-free grammar G such that $L = L(G)$. For any context-free language $L \subseteq V^*$, there exists a context-free grammar $G = (N, V, P, S)$ such that each of whose productions in P is of the form $N_1 \to aN_2N_3$, $N_1 \to aN_2$, $N_1 \to a$, where $N_1, N_2, N_3 \in N$, $a \in V$. If the nonempty string ϵ is in L, then $S \to \epsilon$ is in P.

A *deterministic finite automaton* is a construct $M = (Q, V, q_0, F, \delta)$, where Q and V are disjoint alphabets, V is the alphabet of the automaton, $q_0 \in Q$ is the initial state, $F \subseteq Q$ is the set of final states, and $\delta : Q \times V \to Q$ is the transition mapping.

A relation $\vdash$ is defined on the set $Q \times V^*$: for $q, q' \in Q$, $a \in V$, $x \in V^*$, we write $(q, ax) \vdash (q', x)$ if $\delta(q, a) = q'$. By the definition, $(q, \epsilon) \vdash (q, \epsilon)$. If $\vdash^*$ is the reflexive and transitive closure of the relation $\vdash$, the language of the strings recognized by automaton M is defined by $L(M) = \{x \in V^* \mid (q_0, x) \vdash^* (q_f, \epsilon), q_f \in F\}$. A language L is regular if there is a deterministic finite automaton M such that $L = L(M)$.

Let FIN, REG, CF, and CS be the classes of finite languages, regular languages, context-free languages, and context-sensitive languages, respectively.

2.2 Watson-Crick molecules

For an alphabet V, $\rho \subseteq V \times V$ is a symmetric relation over V. We denote an element $(x, y) \in V^* \times V^*$ by $\binom{x}{y}$. The identity element $\binom{\epsilon}{\epsilon}$ is often identified with ϵ. For an alphabet V and a symmetric relation ρ, we define $V_\rho = V_+ \cup V_- \cup V_\pm$, where $V_+ = \{\binom{a}{\epsilon} \mid a \in V\}$, $V_- = \{\binom{\epsilon}{a} \mid a \in V\}$, $V_\pm = \{\begin{bmatrix} a \\ b \end{bmatrix} \mid a, b \in V, (a, b) \in \rho\}$. Elements in V_+ and V_- correspond to gaps in the lower strand and the upper strand, respectively. Let $V_\pm^*$ be the set of all *complete double stranded molecules over* V including $\begin{bmatrix} \epsilon \\ \epsilon \end{bmatrix}$.

The elements $\begin{bmatrix} a_1 \\ b_1 \end{bmatrix} \begin{bmatrix} a_2 \\ b_2 \end{bmatrix} \cdots \begin{bmatrix} a_n \\ b_n \end{bmatrix}$ in $V_\pm^*$ are also written in the form $\begin{bmatrix} x_1 \\ x_2 \end{bmatrix}$ for $x_1 = a_1 a_2 \cdots a_n$, $x_2 = b_1 b_2 \cdots b_n$. We call such elements $\begin{bmatrix} x_1 \\ x_2 \end{bmatrix} \in$

$V_{\pm}^*$ well-formed double stranded sequences, or simply *double stranded sequences*, or *molecules*. By the definition of $V_{\pm}^*$, $\begin{bmatrix} \epsilon \\ \epsilon \end{bmatrix}$ is also a molecule which identified with ϵ in a special case. The element $\begin{pmatrix} a_1 \\ \epsilon \end{pmatrix}\begin{pmatrix} a_2 \\ \epsilon \end{pmatrix}\cdots\begin{pmatrix} a_n \\ \epsilon \end{pmatrix}$ in V_+^* (resp. $\begin{pmatrix} \epsilon \\ b_1 \end{pmatrix}\begin{pmatrix} \epsilon \\ b_2 \end{pmatrix}\cdots\begin{pmatrix} \epsilon \\ b_n \end{pmatrix}$ in V_-^*) is also written in the form $\begin{pmatrix} x_1 \\ \epsilon \end{pmatrix}$ for $x_1 = a_1 a_2 \cdots a_n$ (resp. $\begin{pmatrix} \epsilon \\ x_2 \end{pmatrix}$ for $x_2 = b_1 b_2 \cdots b_n$).

Definition 2.1. An *incomplete molecule with gaps* over V_ρ is a string $\alpha = \alpha_1 \alpha_2 \cdots \alpha_n$ with $n \geq 1$, where
$$\text{for any } 1 \leq i \leq n, \quad \alpha_i \in V_\rho,$$
$$\text{for any } 1 \leq j \leq n-1, \text{ if } \alpha_j \in V_+, \text{ then } \alpha_{j+1} \notin V_-,$$
$$\text{if } \alpha_j \in V_-, \text{ then } \alpha_{j+1} \notin V_+.$$
That is, an incomplete molecule with gaps α over V_ρ satisfies $\alpha \in ((V_+^* \cup V_-^*)V_{\pm}^+)^+ (V_+^* \cup V_-^*) \cup V_+^+ \cup V_-^+$. Let $WG_\rho(V)$ denote the set of incomplete molecules with gaps over V_ρ.

Example 2.1. For $V = \{a, b, c, d, e, f\}$ and $\rho = \{(a,a), \cdots, (f,f)\}$, two elements $\alpha_1 = \begin{pmatrix} a \\ \epsilon \end{pmatrix}\begin{bmatrix} b \\ b \end{bmatrix}\begin{pmatrix} \epsilon \\ c \end{pmatrix}\begin{bmatrix} d \\ d \end{bmatrix}\begin{pmatrix} \epsilon \\ e \end{pmatrix}$ and $\alpha_2 = \begin{bmatrix} ab \\ ab \end{bmatrix}\begin{pmatrix} \epsilon \\ cde \end{pmatrix}\begin{bmatrix} f \\ f \end{bmatrix}$ are incomplete molecules with gaps. In a graphical representation, $\alpha_1 = \boxed{\begin{array}{c} a\ b\ \sqcup\ d\ \sqcup \\ b\ c\ d\ e \end{array}}$,

$\alpha_2 = \boxed{\begin{array}{c} a\ b\ \sqcup\ \ f \\ a\ b\ c\ d\ e\ f \end{array}}$.

Neither $\beta_1 = \begin{pmatrix} a \\ \epsilon \end{pmatrix}\begin{pmatrix} \epsilon \\ b \end{pmatrix}\begin{pmatrix} c \\ \epsilon \end{pmatrix}\begin{pmatrix} \epsilon \\ b \end{pmatrix}\begin{pmatrix} a \\ \epsilon \end{pmatrix}$ nor $\beta_2 = \begin{pmatrix} a \\ \epsilon \end{pmatrix}\begin{pmatrix} a \\ \epsilon \end{pmatrix}\begin{pmatrix} \epsilon \\ b \end{pmatrix}\begin{pmatrix} \epsilon \\ b \end{pmatrix}\begin{bmatrix} c \\ c \end{bmatrix}$ are in $WG_\rho(V)$.

Consider the prolongation of incomplete molecules with gaps α by β caused by mismatch annealing. In case that both α and β have a gapless region in $V_{\pm}^+$, Figure 1 below illustrates some of the cases for sticking of α and β. In case that α (resp. β) has a gapless region in $V_{\pm}^+$ and β (resp. α) has no gapless region in V_+^+ or V_-^+, Figure 2 (resp. Figure 3) illustrates some of the cases for sticking of α and β. If both α and β have no gapless region in V_+^+ or V_-^+, Figure 4 illustrates some of the cases for sticking of α and β.

Among the elements of $WG_\rho(V)$, we define a partial operation *sticking*, simulated the ligation or annealing operation as in Figures 1, 2, 3, 4. An incomplete molecule with gaps can be inserted by sticking into the single-stranded positions, having the corresponding positions being complete double-stranded molecules. The result of sticking should be an element in $WG_\rho(V)$.

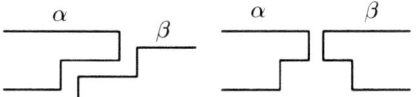

Figure 1. Both α and β have a gapless region

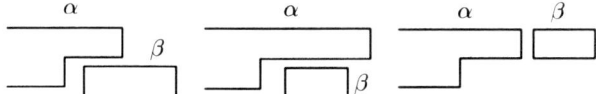

Figure 2. α has a gapless region, β has no gapless region

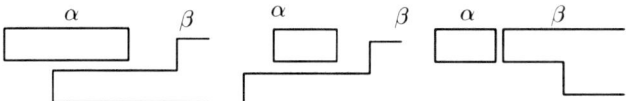

Figure 3. α has no gapless region, β has a gapless region

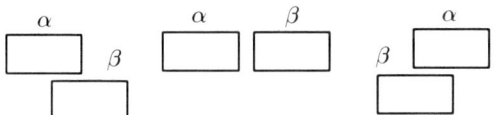

Figure 4. Neither α nor β has gapless region

Definition 2.2. The *sticking* of incomplete molecules with gaps $\alpha = \alpha_1\alpha_2$, $\beta = \beta_2\beta_1$ in $WG_\rho(V)$, denoted by $\mu(\alpha, \beta)$, is defined (in this order, non-commutatively) in the following cases:

1. for α_2 in V_+^*,

- $\alpha_2 = \begin{pmatrix} x_1 \\ \epsilon \end{pmatrix}$, $\beta_2 = \begin{pmatrix} y_1 \\ \epsilon \end{pmatrix}$ or $\begin{bmatrix} y_1 \\ y_2 \end{bmatrix}$ with $x_1, y_1 \in V^+$, $\begin{bmatrix} y_1 \\ y_2 \end{bmatrix} \in V_\pm^+$, $\alpha_1, \beta_1 \in V_\rho^*$,

$\mu(\alpha, \beta) = \alpha_1\alpha_2\beta_2\beta_1$, graphically, $\alpha_1 \boxed{x_1}{\boxed{\beta_2}\boxed{\beta_1}}$

- $\alpha_2 = \begin{pmatrix} x_1x_2x_3 \\ \epsilon \end{pmatrix}$, $\beta_2 = \begin{pmatrix} \epsilon \\ y_2 \end{pmatrix}$ with $x_1, x_3 \in V^*$, $x_2, y_2 \in V^+$, $\alpha_1 \in V_\rho^*$, $\beta_1 = \epsilon$,

$\mu(\alpha, \beta) = \{\alpha_1 \begin{pmatrix} x_1 \\ \epsilon \end{pmatrix} \begin{bmatrix} x_2 \\ y_2 \end{bmatrix} \begin{pmatrix} x_3 \\ \epsilon \end{pmatrix} \mid \begin{bmatrix} x_2 \\ y_2 \end{bmatrix} \in V_\pm^+\}$, graphically, $\alpha_1 \boxed{x_1\ x_2\ x_3}{\boxed{y_2}}$

- $\alpha_2 = \begin{pmatrix} x_2 \\ \epsilon \end{pmatrix}$, $\beta_2 = \begin{pmatrix} \epsilon \\ y_1y_2y_3 \end{pmatrix}$ with $y_1, y_3 \in V^*$, $x_2, y_2 \in V^+$, $\alpha_1 = \epsilon$, $\beta_1 \in V_\rho^*$,

$\mu(\alpha, \beta) = \{\begin{pmatrix} \epsilon \\ y_1 \end{pmatrix} \begin{bmatrix} x_2 \\ y_2 \end{bmatrix} \begin{pmatrix} \epsilon \\ y_3 \end{pmatrix} \beta_1 \mid \begin{bmatrix} x_2 \\ y_2 \end{bmatrix} \in V_\pm^+\}$, graphically, $\boxed{x_2}{\boxed{y_1\ y_2\ y_3}}\beta_1$

- $\alpha_2 = \begin{pmatrix} x_1 x_2 \\ \epsilon \end{pmatrix}$, $\beta_2 = \begin{pmatrix} \epsilon \\ y_2 y_3 \end{pmatrix}$ with $x_1, y_3 \in V^*$, $x_2, y_2 \in V^+$, $\alpha_1, \beta_1 \in V_\rho^*$,

$\mu(\alpha, \beta) = \{\alpha_1 \begin{pmatrix} x_1 \\ \epsilon \end{pmatrix} \begin{bmatrix} x_2 \\ y_2 \end{bmatrix} \begin{pmatrix} \epsilon \\ y_3 \end{pmatrix} \beta_1 \mid \begin{bmatrix} x_2 \\ y_2 \end{bmatrix} \in V_\pm^+ \}$, graphically, $\alpha_1 \boxed{\begin{matrix} x_1 & x_2 \\ y_2 & y_3 \end{matrix}} \beta_1$

- $\alpha_2 = \begin{pmatrix} x_2 x_3 \\ \epsilon \end{pmatrix}$, $\beta_2 = \begin{pmatrix} \epsilon \\ y_1 y_2 \end{pmatrix}$ with $x_3, y_1 \in V^*$, $x_2, y_2 \in V^+$, $\alpha_1 = \beta_1 = \epsilon$,

$\mu(\alpha, \beta) = \{ \begin{pmatrix} \epsilon \\ y_1 \end{pmatrix} \begin{bmatrix} x_2 \\ y_2 \end{bmatrix} \begin{pmatrix} x_3 \\ \epsilon \end{pmatrix} \mid \begin{bmatrix} x_2 \\ y_2 \end{bmatrix} \in V_\pm^+ \}$, graphically, $\boxed{\begin{matrix} x_2 & x_3 \\ y_1 & y_2 \end{matrix}}$

In the above four cases from the second case to the last one, we may consider that the sticking occurs for *any* decomposition which yields a gapless region $\begin{bmatrix} x_2 \\ y_2 \end{bmatrix}$.

2. for α_2 in V_-^*, in a similar way to α_2 in V_+^*, we can define the sticking.

3. for $\alpha_2 = \begin{bmatrix} x_1 \\ x_2 \end{bmatrix}$ in $V_\pm^+$ with $x_1, x_2 \in V^*$, $\alpha_1, \beta \in V_\rho^*$,

$\mu(\alpha, \beta) = \alpha_1 \begin{bmatrix} x_1 \\ x_2 \end{bmatrix} \beta$, graphically, $\alpha_1 \boxed{\begin{matrix} x_1 \\ x_2 \end{matrix}} \beta$

Example 2.2. For elements $\alpha = \begin{pmatrix} x_1 \\ \epsilon \end{pmatrix}$ in V_+^*, $\beta = \begin{bmatrix} y_1 \\ y_2 \end{bmatrix}$ in $V_\pm^+$, $\pi = \begin{pmatrix} \epsilon \\ z_1 \end{pmatrix}$ in V_-^*, the sticking $\mu(\mu(\alpha, \beta), \pi) = \mu(\alpha, \mu(\beta, \pi)) = \begin{pmatrix} x_1 \\ \epsilon \end{pmatrix} \begin{bmatrix} y_1 \\ y_2 \end{bmatrix} \begin{pmatrix} \epsilon \\ z_1 \end{pmatrix}$.

For elements $\alpha = \begin{bmatrix} x_1 \\ x_2 \end{bmatrix}$ in $V_\pm^+$, $\beta = \begin{pmatrix} y_1 y_2 \\ \epsilon \end{pmatrix}$ in V_+^+, $\pi = \begin{pmatrix} \epsilon \\ z_1 z_2 \end{pmatrix}$ in V_-^+,

if $\begin{bmatrix} y_2 \\ z_1 \end{bmatrix}$, $\begin{bmatrix} y_1 \\ z_2 \end{bmatrix} \in V_\pm^+$, then $\mu(\mu(\alpha, \beta), \pi) = \{ \begin{bmatrix} x_1 \\ x_2 \end{bmatrix} \begin{pmatrix} y_1 \\ \epsilon \end{pmatrix} \begin{bmatrix} y_2 \\ z_1 \end{bmatrix} \begin{pmatrix} \epsilon \\ z_2 \end{pmatrix} \mid \begin{bmatrix} y_2 \\ z_1 \end{bmatrix} \in$

$V_\pm^+ \}$, $\mu(\alpha, \mu(\beta, \pi)) = \{ \begin{bmatrix} x_1 \\ x_2 \end{bmatrix} \begin{pmatrix} \epsilon \\ z_1 \end{pmatrix} \begin{bmatrix} y_1 \\ z_2 \end{bmatrix} \begin{pmatrix} y_2 \\ \epsilon \end{pmatrix}, \begin{bmatrix} x_1 \\ x_2 \end{bmatrix} \begin{pmatrix} y_1 \\ \epsilon \end{pmatrix} \begin{bmatrix} y_2 \\ z_1 \end{bmatrix} \begin{pmatrix} \epsilon \\ z_2 \end{pmatrix} \mid \begin{bmatrix} y_1 \\ z_2 \end{bmatrix}, \begin{bmatrix} y_2 \\ z_1 \end{bmatrix}$

$\in V_\pm^+ \}$.

Note that Example 2.2 showed the non-associativity of μ.

2.3   Watson-Crick Insertion Systems

A *Watson-Crick insertion system* (abb. *WK-ins system*) is a 4-tuple $\gamma = (V, \rho, A, R)$, where V is a finite set of symbols, $\rho \subseteq V \times V$ is the symmetric relation on V, $A \subseteq (V_+^* \cup V_-^*)V_\pm^+(V_+^* \cup V_-^*)$ is a finite set of *axioms*, and R is a finite set of triples, called *insertion rules*, of the form $(\alpha, \pi, \beta) \in (V_+^* \times V_+^* \times V_+^*) \cup (V_-^* \times V_-^* \times V_-^*)$.

Intuitively, the meaning of (α, π, β) is that an incomplete molecule with gaps π can be inserted in between α and β, having α and β being complete double stranded molecules, that is, if $\alpha = \begin{pmatrix} \epsilon \\ x_2 \end{pmatrix}$, $\beta = \begin{pmatrix} \epsilon \\ y_2 \end{pmatrix}$, $\pi = \begin{pmatrix} \epsilon \\ z_2 \end{pmatrix}$, then an element $\begin{bmatrix} x_1 \\ x_2 \end{bmatrix} \begin{pmatrix} \epsilon \\ z_2 \end{pmatrix} \begin{bmatrix} y_1 \\ y_2 \end{bmatrix}$ is obtained from $\begin{pmatrix} x_1 \\ \epsilon \end{pmatrix} \begin{pmatrix} y_1 \\ \epsilon \end{pmatrix}$ by using the insertion rule (α, π, β), where x_1, y_1 in V^* satisfy $\begin{bmatrix} x_1 \\ x_2 \end{bmatrix}$, $\begin{bmatrix} y_1 \\ y_2 \end{bmatrix} \in V_{\pm}^*$,

graphically, $\boxed{x_1 \, y_1} \Longrightarrow_\gamma \boxed{\begin{array}{c} x_1 \;\underline{\quad}\; y_1 \\ x_2 \; z_2 \; y_2 \end{array}}$.

The following figures show how insertion proceeds by mismatching annealing. Figure 5 illustrates two single stranded DNA sequences of the form $5' - x_1 y_1 - 3'$ and $3' - x_2 z_2 y_2 - 5'$, where $\alpha_1 = \begin{pmatrix} x_1 \\ \epsilon \end{pmatrix}$, $\beta_1 = \begin{pmatrix} y_1 \\ \epsilon \end{pmatrix}$, $\alpha = \begin{pmatrix} \epsilon \\ x_2 \end{pmatrix}$, $\beta = \begin{pmatrix} \epsilon \\ y_2 \end{pmatrix}$, $\pi = \begin{pmatrix} \epsilon \\ z_2 \end{pmatrix}$, $\alpha_{\pm} = \begin{bmatrix} x_1 \\ x_2 \end{bmatrix}$, $\beta_{\pm} = \begin{bmatrix} y_1 \\ y_2 \end{bmatrix}$. The two DNA sequences are sticking together in Figure 6. We obtain the situation in Figure 7 by cutting the double stranded subsequence using a restriction enzyme.

Figure 5. Two incomplete molecules with gaps

Figure 6. Sticking together

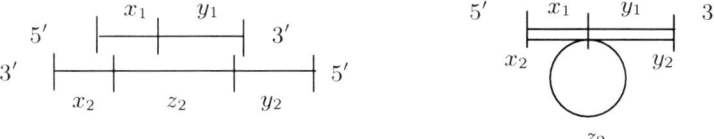

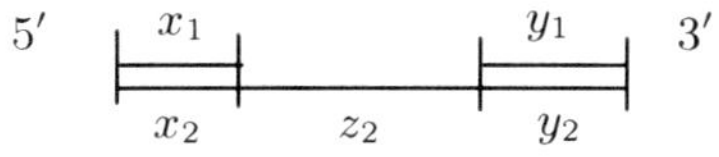

Figure 7. Insertion completed

Informally, for an insertion rule $(\begin{pmatrix} \epsilon \\ x_2 \end{pmatrix}, \begin{pmatrix} \epsilon \\ z_2 \end{pmatrix}, \begin{pmatrix} \epsilon \\ y_2 \end{pmatrix})$, we write $w_1 \Longrightarrow w_2$ if $w_2 \in WG_\rho(V)$ is obtained from $w_1 = w_{11} \begin{pmatrix} x_1 \\ \epsilon \end{pmatrix} \begin{pmatrix} y_1 \\ \epsilon \end{pmatrix} w_{12}$, by sticking elements w_{11}, $\begin{bmatrix} x_1 \\ x_2 \end{bmatrix}$, $\begin{pmatrix} \epsilon \\ z_2 \end{pmatrix}$, $\begin{bmatrix} y_1 \\ y_2 \end{bmatrix}$, w_{12} for μ. Similarly, we can define for the case $(\begin{pmatrix} x_1 \\ \epsilon \end{pmatrix}, \begin{pmatrix} y_1 \\ \epsilon \end{pmatrix}, \begin{pmatrix} z_1 \\ \epsilon \end{pmatrix})$.

Formally, $\Longrightarrow_\gamma$ is defined as follows: for incomplete molecules with gaps w_1 and w_2 in $WG_\rho(V)$, we write $w_1 \Longrightarrow_\gamma w_2$ if w_2 can be obtained from w_1 by using an insertion rule (α, π, β) in R such that $\alpha_\pm \in \mu(\alpha_1, \alpha)$, $\beta_\pm \in \mu(\beta_1, \beta)$ with $\alpha_\pm, \beta_\pm \in V_\pm^*$, and $w_1 \Longrightarrow_\gamma w_2$, where $w_1 = w_{11}\alpha_1\beta_1 w_{12}$, $w_{11}, w_{12} \in V_\rho^*$, $w_2 \in \mu(w_{11}, w_3) \cup \mu(w_4, w_{12})$ with $w_3 \in \mu(w_5, w_{12})$, $w_4 \in \mu(w_{11}, w_5)$ for $w_5 = \mu(\alpha_\pm, \mu(\pi, \beta_\pm))$.

We denote by $\Longrightarrow_\gamma^*$ the reflexive and transitive closure of $\Longrightarrow_\gamma$. If there is no confusion, we use $\Longrightarrow$ instead of $\Longrightarrow_\gamma$. A sequence $w_1 \Longrightarrow w_2 \Longrightarrow \cdots \Longrightarrow w_k$ with $w_1 \in A$ is called a *computation* in γ. A computation $w_1 \Longrightarrow^* w_k$ is *complete* if w_k is in $V_\pm^*$ (neither overhangs nor gaps are present in the last of the sequence).

Example 2.3. For an element $\begin{pmatrix} aaa \\ \epsilon \end{pmatrix}$ and an insertion rule $(\epsilon, \begin{pmatrix} \epsilon \\ c \end{pmatrix}, \epsilon)$ with $\begin{bmatrix} a \\ c \end{bmatrix} \in V_\pm$, we have $\begin{pmatrix} aaa \\ \epsilon \end{pmatrix} \Longrightarrow \begin{pmatrix} a \\ \epsilon \end{pmatrix}\begin{bmatrix} a \\ c \end{bmatrix}\begin{pmatrix} a \\ \epsilon \end{pmatrix}$, it is because $\begin{pmatrix} a \\ \epsilon \end{pmatrix}\begin{bmatrix} a \\ c \end{bmatrix}\begin{pmatrix} a \\ \epsilon \end{pmatrix}$ $\in \mu(\begin{pmatrix} a \\ \epsilon \end{pmatrix}, \begin{bmatrix} a \\ c \end{bmatrix}\begin{pmatrix} a \\ \epsilon \end{pmatrix})$ with $\begin{bmatrix} a \\ c \end{bmatrix}\begin{pmatrix} a \\ \epsilon \end{pmatrix} \in \mu(\begin{pmatrix} \epsilon \\ c \end{pmatrix}, \begin{pmatrix} aa \\ \epsilon \end{pmatrix})$ *for* $\begin{pmatrix} \epsilon \\ c \end{pmatrix} = \mu(\begin{bmatrix} \epsilon \\ \epsilon \end{bmatrix}, \mu((\ _c^\epsilon\ , \begin{bmatrix} \epsilon \\ \epsilon \end{bmatrix})))$.

Similarly, $\begin{pmatrix} aaa \\ \epsilon \end{pmatrix} \Longrightarrow \begin{bmatrix} a \\ c \end{bmatrix}\begin{pmatrix} aa \\ \epsilon \end{pmatrix}$, $\begin{pmatrix} aaa \\ \epsilon \end{pmatrix} \Longrightarrow \begin{pmatrix} aa \\ \epsilon \end{pmatrix}\begin{bmatrix} a \\ c \end{bmatrix}$, and $\begin{pmatrix} aaa \\ \epsilon \end{pmatrix} \Longrightarrow^* \begin{bmatrix} aaa \\ ccc \end{bmatrix}$.

Starting from an axiom and using insertion rules in R, we obtain a set of complete double stranded sequences in $V_\pm^*$. A set of molecules generated by γ called *molecular language* is defined by $LM(\gamma) = \{w \in V_\pm^* \mid w_1 \Longrightarrow^* w,\ w_1 \in A\}$. Furthermore, a *(string) language* $L(\gamma)$ *generated* by γ is a coding image of $LM(\gamma)$, i.e., the set of all *upper* components of the molecular language $LM(\gamma)$.

In this paper, we assume that there is a coding $h_\rho : V_\pm^* \to V^*$ such that
$$\begin{cases} h_\rho(a_\pm) = a_1 \text{ for } a_\pm \in V_\pm \text{ with } a_\pm = \begin{bmatrix} a_1 \\ a_2 \end{bmatrix}, a_1, a_2 \in V, \\ h_\rho(\epsilon) = \epsilon. \end{cases}$$

For a molecular language $LM(\gamma) \subseteq V_\pm^*$, $h_\rho(LM(\gamma))$ denotes the image of $LM(\gamma)$ by h_ρ, that is, $h_\rho(LM(\gamma)) = \{h_\rho(w) \mid w \in LM(\gamma)\} = L(\gamma)$.

We say that a WK-ins system $\gamma = (V, \rho, A, R)$ is *of weight* (n, m) if $n = max\{|\pi| \mid (\alpha, \pi, \beta) \in R\}$, $m = max\{|\alpha| \mid (\alpha, \pi, \beta) \in R \text{ or } (\beta, \pi, \alpha) \in R\}$.

The classes of molecular languages and that of string languages generated by WK-ins systems of weight (n', m') such that $n' \leq n$, $m' \leq m$

are denoted by $WINM_n^m$ and WIN_n^m, respectively. When one of the parameters m, n is not bounded, we replace it by $*$. Thus, the classes of all molecular languages and that of string languages generated by WK-ins systems are $WINM_*^*$ and WIN_*^*, respectively.

A WK-ins system γ is said to be *strong*, if for each (α, π, β) in R, we have $(\alpha, \pi, \beta) \in (V_+^+ \times V_+^* \times V_+^+) \cup (V_-^+ \times V_-^* \times V_-^+)$, that is, α and β are nonempty.

The classes of molecular languages and that of string languages generated by strong WK-ins systems of weight (n', m') such that $n' \leq n$, $m' \leq m$ are denoted by $WINM_n^m(s)$ and $WIN_n^m(s)$, respectively. In a similar way to the WK-ins systems, we use the notations $WINM_*^*(s)$ and $WIN_*^*(s)$.

A complete computation $w_1 \Longrightarrow w_2 \Longrightarrow \cdots \Longrightarrow w_k$ (hence $w_1 \in A$, $w_k \in V_\pm^*$) is said to be *without gaps* if for any $1 \leq i < k$, we have $w_i \in (V_+^* \cup V_-^*)V_\pm^+(V_+^* \cup V_-^*)$ (w_i is without gaps). We denote by $L_g(\gamma)$ the string language generated by γ at the end of computations of without gaps.

We denote by g-WIN_*^* the class of string languages $L_g(\gamma)$, where γ is an arbitrary WK-ins system.

2.4 Insertion systems on strings

An insertion system on strings is a construct $\gamma = (V, A, R)$, where V is a finite set of symbols, A is a finite set of strings over V, and R is a finite set of triples of the form (u, x, v) with $u, v \in V^*$, $x \in V^+$. We say that an insertion system on strings $\gamma = (V, A, R)$ is of weight (n, m) if $n = max\{|x| \mid (u, x, v) \in R\}$, $m = max\{|u| \mid (u, x, v) \in R$ or $(v, x, u) \in R\}$.

The class of string languages generated by insertion systems on strings of weight (n', m') such that $n' \leq n$, $m' \leq m$ are denoted by INS_n^m. In a similar way to the WK-ins systems, we use the notations INS_*^* and INS_*^*.

The following theorem concerning the class of string languages generated by insertion systems on strings is shown in [2].

Theorem 2.1. [2]

1. $FIN \subset INS_*^0 \subset INS_*^1 \subset \cdots \subset CS$.
2. CF is incomparable with all families INS_*^m, $m \geq 2$, and INS_*^*.
3. INS_2^2 contains non-semilinear languages.

3   Main results

In this section, we show the relationships between the classes of string languages generated by WK-ins systems as well as the relationships of these classes to the classes in the Chomsky hierarchy. From the definitions of the WK-ins systems, we obviously have the following inclusions.

Lemma 3.1.   *1. $WINM_n^m \subseteq WINM_{n'}^{m'}$, $WIN_n^m \subseteq WIN_{n'}^{m'}$ for any $0 \le n \le n'$, $0 \le m \le m'$.*
   2. $WINM_n^m(s) \subseteq WINM_{n'}^{m'}(s)$, $WIN_n^m(s) \subseteq WIN_{n'}^{m'}(s)$, for any $0 \le n \le n'$, $1 \le m \le m'$.
   3. $WINM_n^m(s) \subseteq WINM_n^m$, $WIN_n^m(s) \subseteq WIN_n^m$, for any $n \ge 0$, $m \ge 1$.
   4. $WINM_*^*(s) \subseteq WINM_*^*$, $WIN_*^*(s) \subseteq WIN_*^* \subseteq CS$.

At first, in the case of WK-ins systems in which any computation is without gaps, the following result holds.

Lemma 3.2. *g-$WIN_*^* \subseteq REG$.*

Proof. (sketch) In a computation without gaps, sequences prolong by using the overhangs in the right and left delay of incomplete molecules without gaps. We can always use the insertion rule which sticks to the existing overhanging ends, then the length of the overhanging ends are not longer than the bounded length. Remember simple one-sided sticker systems in [2]. Therefore, we can construct a deterministic automaton in which the state can control the process as the sticky ends do this.   □

3.1   WK-ins systems with gaps

In this subsection, we consider the generative powers of WK-ins systems with gaps, starting with examining the relationship between the class of languages generated by WK-ins systems with gaps and the class of regular languages.

Lemma 3.3. *$REG-WIN_*^0 \ne \emptyset$.*

Proof. Consider the regular language $L = \{ba^i b \mid i \ge 1\}$, and assume that $L = L(\gamma)$ for some WK-ins system $\gamma = (V, \rho, A, R)$ of weight $(n, 0)$ for $n \ge 0$. Since A is a finite set and L is an infinite language, there is a pair in R of the following forms: $(\begin{pmatrix} \epsilon \\ \epsilon \end{pmatrix}, \begin{pmatrix} a^i \\ \epsilon \end{pmatrix}, \begin{pmatrix} \epsilon \\ \epsilon \end{pmatrix})$ and $(\begin{pmatrix} \epsilon \\ \epsilon \end{pmatrix}, \begin{pmatrix} \epsilon \\ c^j \end{pmatrix}, \begin{pmatrix} \epsilon \\ \epsilon \end{pmatrix})$, where $i \ge 1$, $j \ge 1$, $(a, c) \in \rho$, that are used arbitrarily many times in the generation of strings in L of arbitrarily large length.

For some $n \geq 1$, a complete computation $\alpha \Longrightarrow^* \begin{bmatrix} x_1 \\ x_2 \end{bmatrix}$ with $\alpha \in A$, $x_1 = ba^n b$, $x_2 \in V^*$ can be continued as follows: $\alpha \Longrightarrow^* \begin{bmatrix} x_1 \\ x_2 \end{bmatrix} \Longrightarrow^* \begin{bmatrix} x_1 \\ x_2 \end{bmatrix} \begin{bmatrix} a^{ij} \\ c^{ji} \end{bmatrix}$, see Example 2.3. This is a complete computation, producing the string $ba^n ba^{ij}$ in the components of $L(\gamma)$, which is not in L, a contradiction. □

From Lemma 3.3, we obtain the following corollary.

Corollary 3.1. $CF{-}WIN_*^0 \neq \emptyset$.

Lemma 3.4. $WIN_2^0{-}REG \neq \emptyset$.

Proof. Let us consider a WK-ins system $\gamma = (\{\#, a, b\}, \rho, A, R)$, where

$$\rho = \{(\#, \#), (a, a), (b, b)\}, \qquad A = \{ \begin{bmatrix} \# \\ \# \end{bmatrix} \},$$

$$R = \{ (\begin{pmatrix} \epsilon \\ \epsilon \end{pmatrix}, \begin{pmatrix} ab \\ \epsilon \end{pmatrix}, \begin{pmatrix} \epsilon \\ \epsilon \end{pmatrix}), \ (\begin{pmatrix} \epsilon \\ \epsilon \end{pmatrix}, \begin{pmatrix} \epsilon \\ a \end{pmatrix}, \begin{pmatrix} \epsilon \\ \epsilon \end{pmatrix}), \ (\begin{pmatrix} \epsilon \\ \epsilon \end{pmatrix}, \begin{pmatrix} \epsilon \\ b \end{pmatrix}, \begin{pmatrix} \epsilon \\ \epsilon \end{pmatrix}) \}.$$

For any complete double stranded molecule $\begin{bmatrix} x \\ y \end{bmatrix}$ with $x, y \in \{\#, a, b\}^*$, $\begin{bmatrix} \# \\ \# \end{bmatrix} \Longrightarrow^* \begin{bmatrix} x \\ y \end{bmatrix}$, from the insertion rule $(\begin{pmatrix} \epsilon \\ \epsilon \end{pmatrix}, \begin{pmatrix} ab \\ \epsilon \end{pmatrix}, \begin{pmatrix} \epsilon \\ \epsilon \end{pmatrix})$, we obtain $|x|_a = |x|_b$, and for any prefix x' of x, $|x'|_a \geq |x'|_b$ holds.

Consider a regular language $L_R = \{\#a^i b^j \mid i, j \geq 0\}$ and Dyck language D_{ab} over $\{a, b\}$, where the symbol a (resp. b) is corresponding to a left (resp. right) parenthesis, which is not a regular language. We are interested in the intersection $L(\gamma) \cap L_R = \{\#a^i b^i \mid i \geq 0\}$, which is not a regular language. □

From Lemma 3.1 and Lemma 3.4, we obtain the following corollary.

Corollary 3.2. *For any $i \geq 0$, $j \geq 2$, $WIN_j^i{-}REG \neq \emptyset$.*

Together with Lemma 3.2, Corollary 3.2 suggests that a computation without gaps restricts a prolongation of incomplete molecules with gaps.

The following lemma shows that the parameter of insertion strings in Lemma 3.4 is optimal for WK-ins systems to generate non-regular languages.

Lemma 3.5. $WIN_1^0 \subset REG$.

Proof. (sketch) In a computation of γ in WIN_1^0, where any insertion rule is without context, one nucleotide (letter) is either sticking to sequences with the overhangs by complementarity or making a gap. In a similar way to Lemma 3.2, we can always use the adequate insertion rule, then the overhanging ends are not longer than the bounded length. Therefore, there is a deterministic automaton that accepts $L(\gamma)$. The strictness of the inclusion is obvious. □

Lemma 3.6. *Each regular language L can be written in the form $L = pr(L')$, where pr is a projection and $L' \in WIN_2^1$.*

Proof. Let us consider a regular language L, a deterministic finite automaton $M = (Q, T, \delta, q_0, F)$ such that $L = L(M)$, and a projection $pr : (T \cup Q \cup \{\#\})^* \to T^*$ on T: $pr(a) = a$, $a \in T$, $pr(q) = \epsilon$, $q \in Q$, $pr(\#) = \epsilon$.

We construct a WK-ins system $\gamma = (T \cup Q \cup \{\#\}, \rho, A, R)$ as follows:

$$\rho = \{(a,a) \mid a \in T \cup Q \cup \{\#\}\}, \qquad A = \{ \begin{bmatrix} \# \\ \# \end{bmatrix} \begin{pmatrix} q_0 \\ \epsilon \end{pmatrix} \} \cup \{ \begin{bmatrix} \# \\ \# \end{bmatrix} \mid \epsilon \in L \},$$

$$R = \{(\begin{pmatrix} \epsilon \\ q_1 \end{pmatrix}, \begin{pmatrix} \epsilon \\ aq_2 \end{pmatrix}, \begin{pmatrix} \epsilon \\ \epsilon \end{pmatrix}) \mid (q_1, a) \vdash (q_2, \epsilon),\ q_1, q_2 \in Q,\ a \in T \}$$

$$\cup \{(\begin{pmatrix} \epsilon \\ q \end{pmatrix}, \begin{pmatrix} \epsilon \\ a \end{pmatrix}, \begin{pmatrix} \epsilon \\ \epsilon \end{pmatrix}) \mid (q, a) \vdash (q_f, \epsilon),\ q \in Q,\ q_f \in F,\ a \in T \}$$

$$\cup \{(\begin{pmatrix} a \\ \epsilon \end{pmatrix}, \begin{pmatrix} \epsilon \\ \epsilon \end{pmatrix}, \begin{pmatrix} \epsilon \\ \epsilon \end{pmatrix}) \mid a \in T \} \cup \{(\begin{pmatrix} q \\ \epsilon \end{pmatrix}, \begin{pmatrix} q \\ \epsilon \end{pmatrix}, \begin{pmatrix} \epsilon \\ \epsilon \end{pmatrix}) \mid q \in Q \}.$$

Intuitively, the computation proceeds to simulate the computation of the finite automaton M. We will show that $(q_0, a_1 \cdots a_n) \vdash^n (q_n, \epsilon)$ if and only if

for $q_n \notin F$,
$$\begin{bmatrix} \# \\ \# \end{bmatrix} \begin{pmatrix} q_0 \\ \epsilon \end{pmatrix} \Longrightarrow \begin{bmatrix} \# \\ \# \end{bmatrix} \begin{bmatrix} q_0 \\ q_0 \end{bmatrix} \begin{pmatrix} \epsilon \\ a_1 q_1 \end{pmatrix} \Longrightarrow \begin{bmatrix} \# \\ \# \end{bmatrix} \begin{bmatrix} q_0 \\ q_0 \end{bmatrix} \begin{bmatrix} a_1 \\ a_1 \end{bmatrix} \begin{pmatrix} \epsilon \\ q_1 \end{pmatrix}$$

$$\Longrightarrow^* \begin{bmatrix} \# \\ \# \end{bmatrix} \begin{bmatrix} q_0 a_1 q_1 q_1 a_2 \cdots a_n \\ q_0 a_1 q_1 q_1 a_2 \cdots a_n \end{bmatrix} \begin{pmatrix} \epsilon \\ q_n \end{pmatrix},$$

for $q_n \in F$,
$$\begin{bmatrix} \# \\ \# \end{bmatrix} \begin{pmatrix} q_0 \\ \epsilon \end{pmatrix} \Longrightarrow^* \begin{bmatrix} \# \\ \# \end{bmatrix} \begin{bmatrix} q_0 a_1 q_1 q_1 a_2 \cdots a_n \\ q_0 a_1 q_1 q_1 a_2 \cdots a_n \end{bmatrix}$$

We can prove this by the induction on n and leave it out in this paper. Then, by using the projection pr, we can assert the claim. □

The following result has a rather interesting consequence in contrast to the characterization for the class of regular languages by using WK-ins systems in WIN_2^1 and a projection in Lemma 3.6.

Theorem 3.1. *Each context-free language L can be written in the form $L = pr(L')$, where pr is a projection and $L' \in WIN_3^1$.*

Proof. Let us consider a context-free language $L(G)$ with $G = (N, T, P, S)$ such that each of whose productions in P is of the form $N_1 \to aN_2N_3$,

$N_1 \to aN_2$, $N_1 \to a$, where $N_1, N_2, N_3 \in N$, $a \in T$. In case the empty string ϵ is in $L(G)$, let a production $S \to \epsilon$ be in P.

Let $pr : (T \cup N \cup \{\#\})^* \to T^*$ be a projection on T: $pr(a) = a$, $a \in T$, $pr(M) = \epsilon$, $M \in N$, $pr(\#) = \epsilon$.

We construct a WK-ins system $\gamma = (T \cup N \cup \{\#\}, \rho, A, R)$ as follows:

$$\rho = \{(a,a) \mid a \in T \cup N \cup \{\#\}\}, \qquad A = \{\binom{S}{\epsilon}\begin{bmatrix}\#\\\#\end{bmatrix}\} \cup \{\begin{bmatrix}\#\\\#\end{bmatrix} \mid \epsilon \in L(G)\},$$

$$R = \{(\binom{\epsilon}{\epsilon}, \binom{\epsilon}{aN_2N_3}, \binom{\epsilon}{N_1}), (\binom{\epsilon}{\epsilon}, \binom{aN_2N_3}{\epsilon}, \binom{N_1}{\epsilon})) \mid N_1 \to aN_2N_3 \text{ in } P,$$
$$a \in T, \ N_1, N_2, N_3 \in N\}$$

$$\cup \{(\binom{\epsilon}{\epsilon}, \binom{\epsilon}{aN_2}, \binom{\epsilon}{N_1}), (\binom{\epsilon}{\epsilon}, \binom{aN_2}{\epsilon}, \binom{N_1}{\epsilon})) \mid N_1 \to aN_2 \text{ in } P,$$
$$a \in T, \ N_1, N_2 \in N\}$$

$$\cup \{(\binom{\epsilon}{\epsilon}, \binom{\epsilon}{a}, \binom{\epsilon}{N_1}), (\binom{\epsilon}{\epsilon}, \binom{a}{\epsilon}, \binom{N_1}{\epsilon})) \mid N_1 \to a \text{ in } P, a \in T, N_1 \in N\}$$

$$\cup \{(\binom{\epsilon}{\epsilon}, \binom{\epsilon}{\epsilon}, \binom{a}{\epsilon}), (\binom{\epsilon}{\epsilon}, \binom{\epsilon}{\epsilon}, \binom{\epsilon}{a})) \mid a \in T \}.$$

For any string in $L(G)$, there is the left most derivation $S \Longrightarrow^k a_1 \cdots a_n N_1 \cdots N_i$, where $a_1, \cdots, a_n \in T$, $N_1, \cdots, N_m \in N$ if and only if there is a computation

$$\binom{S}{\epsilon}\begin{bmatrix}\#\\\#\end{bmatrix} \Longrightarrow^* \begin{bmatrix}a_1\\a_1\end{bmatrix}\begin{bmatrix}M_1\\M_1\end{bmatrix}\begin{bmatrix}a_2\\a_2\end{bmatrix} \cdots \begin{bmatrix}a_n\\a_n\end{bmatrix}\begin{bmatrix}M_n\\M_n\end{bmatrix} N_1' \begin{bmatrix}M_{n+1}\\M_{n+1}\end{bmatrix} N_2' \cdots N_i' \begin{bmatrix}M_{n+i}\\M_{n+i}\end{bmatrix}\begin{bmatrix}\#\\\#\end{bmatrix},$$

where for $1 \le m \le n+j$, $M_m \in N^*$, for $1 \le j \le i$, $N_j' \in \binom{N_j}{\epsilon} \cup \binom{\epsilon}{N_j}$.

We can prove this by the induction on k and leave it out in this paper. Then, by using the projection pr, we can assert the claim. $\qquad\square$

3.2 Strong WK-ins systems

Concerning the class of languages generated by strong WK-ins systems, the relationships with the classes of languages in the Chomsky hierarchy are considered as follows.

Lemma 3.7. $WIN_2^1(s) - REG \ne \emptyset$.

Proof. Let us consider a strong WK-ins system $\gamma = (\{\#, a, b\}, \rho, A, R)$, where

$$\rho = \{(\#, \#), (a, a), (b, b)\}, \qquad A = \{\begin{bmatrix}\#\\\#\end{bmatrix}\binom{ab}{\epsilon}\},$$

$$R = \{(\binom{a}{\epsilon}, \binom{ab}{\epsilon}, \binom{b}{\epsilon}), (\binom{\epsilon}{a}, \binom{\epsilon}{ab}, \binom{\epsilon}{b}), (\binom{a}{\epsilon}, \binom{\epsilon}{\epsilon}, \binom{b}{\epsilon}), (\binom{\epsilon}{a}, \binom{\epsilon}{\epsilon}, \binom{\epsilon}{b})\}.$$

For any complete double stranded molecule $\begin{bmatrix} x \\ y \end{bmatrix}$ with $x, y \in \{\#, a, b\}^*$ and $\begin{bmatrix} \# \\ \# \end{bmatrix} \begin{pmatrix} ab \\ \epsilon \end{pmatrix} \Longrightarrow^* \begin{bmatrix} x \\ y \end{bmatrix}$, from the definition of R, obviously, $L(\gamma) = \{\#a^i b^i \mid i \geq 1\}$, which is not a regular language. The parsing of such a sequence $\begin{bmatrix} x \\ y \end{bmatrix}$ proceeds as follows:

$$\begin{bmatrix} \# \\ \# \end{bmatrix} \begin{pmatrix} ab \\ \epsilon \end{pmatrix} \Longrightarrow \begin{bmatrix} \# \\ \# \end{bmatrix} \begin{bmatrix} a \\ a \end{bmatrix} \begin{pmatrix} \epsilon \\ ab \end{pmatrix} \begin{bmatrix} b \\ b \end{bmatrix} \Longrightarrow^* \begin{bmatrix} \# \\ \# \end{bmatrix} \begin{bmatrix} a^{2n+1} \\ a^{2n+1} \end{bmatrix} \begin{pmatrix} \epsilon \\ ab \end{pmatrix} \begin{bmatrix} b^{2n+1} \\ b^{2n+1} \end{bmatrix}$$

$$\Longrightarrow \begin{bmatrix} \# \\ \# \end{bmatrix} \begin{bmatrix} a^{2n+1} \\ a^{2n+1} \end{bmatrix} \begin{bmatrix} ab \\ ab \end{bmatrix} \begin{bmatrix} b^{2n+1} \\ b^{2n+1} \end{bmatrix}$$

$$\begin{bmatrix} \# \\ \# \end{bmatrix} \begin{pmatrix} ab \\ \epsilon \end{pmatrix} \Longrightarrow^* \begin{bmatrix} \# \\ \# \end{bmatrix} \begin{bmatrix} a^{2n} \\ a^{2n} \end{bmatrix} \begin{pmatrix} ab \\ \epsilon \end{pmatrix} \begin{bmatrix} b^{2n} \\ b^{2n} \end{bmatrix} \Longrightarrow \begin{bmatrix} \# \\ \# \end{bmatrix} \begin{bmatrix} a^{2n} \\ a^{2n} \end{bmatrix} \begin{bmatrix} ab \\ ab \end{bmatrix} \begin{bmatrix} b^{2n} \\ b^{2n} \end{bmatrix} \quad \square$$

From Lemma 3.1 and Lemma 3.7, we obtain the following corollary.

Corollary 3.3. *For any $i \geq 1$, $j \geq 2$, $WIN^i_j(s) - REG \neq \emptyset$.*

The following lemma shows that the parameter of insertion strings in Lemma 3.7 is optimal for strong WK-ins systems to generate infinite languages.

Lemma 3.8. $WIN^1_1(s) = FIN$.

Proof. The inclusion $FIN \subseteq WIN^1_1(s)$ is obvious, thus we consider the other one. Suppose that there is an infinite language L such that $L = L(\gamma)$ for some WK-ins system $\gamma = (V, \rho, A, R)$ of weight $(1, 1)$. Since L is an infinite language, there is an insertion rule (α, π, β) in R with $\alpha, \beta, \pi \in V_+ \cup V_-$ that is used arbitrarily many times in the generation of strings in L of arbitrarily large length. Assume that there is a complete computation $w_1 \Longrightarrow^* w_2 \alpha_1 \beta_1 w_3 \Longrightarrow w_2 \alpha_\pm \pi \beta_\pm w_3 \Longrightarrow^* w$, where $\alpha_\pm \in \mu(\alpha_1, \alpha)$, $\beta_\pm \in \mu(\beta_1, \beta)$ with $\alpha_\pm, \beta_\pm \in V_\pm$, $w_1 \in A$, $w_2, w_3 \in V^*_\rho$, $w \in V^*_\pm$, in which (α, π, β) is used. There is no insertion rule in R which can apply for π in $V_+ \cup V_-$. Therefore, no complete double stranded molecule is obtained from $w_2 \alpha_\pm \pi \beta_\pm w_3$, which is a contradiction. $\square$

From Lemmas 3.4, 3.5, 3.7, 3.8, the parameters of insertion strings must be at least two in order to generate non-regular languages in WK-ins systems as well as strong WK-ins systems.

In contrast to the known results for insertion systems on strings in Theorem 2.1, the following result for strong WK-ins systems shows that nonempty context-checking properly reduces the computational powers of WK-ins systems.

Theorem 3.2. $WIN_*^*(s) \subseteq CF$.

Proof. Let $\gamma = (V, \rho, A, R)$ be a given strong WK-ins system. Let $l = max\{ |\alpha\beta|, |\alpha_1|, |\alpha_3| \mid (\alpha, \pi, \beta) \in R, \alpha_1\alpha_2\alpha_3 \in A$ with $\alpha_2 \in V_\pm^+$ and $\alpha_1, \alpha_3 \in V_+^* \cup V_-^*\}$. Construct a context-free grammar $G = (N, V, P, S)$ defined from γ as follows:

$N = \{N_{\alpha\beta} \mid 2 \leq |\alpha\beta| \leq l, \ \alpha\beta \in V_+^+ \cup V_-^+\}$,

$P = \{N_{\alpha\beta} \to x_1 N_{\pi_1} \cdots N_{\pi_n} y_1 \mid (\alpha_2, \pi_1 \cdots \pi_n, \beta_2) \in R$, for any $1 \leq i \leq n, 2 \leq |\pi_i| \leq l$,

$\begin{bmatrix} x_1 \\ x_2 \end{bmatrix} \in \mu(\alpha, \alpha_2), \begin{bmatrix} y_1 \\ y_2 \end{bmatrix} \in \mu(\beta, \beta_2)$ with $x_1, x_2, y_1, y_2 \in V^*\}$

$\cup \{N_{\alpha\beta} \to x_1 y_1 \mid (\alpha_2, \epsilon, \beta_2) \in R$,

$\begin{bmatrix} x_1 \\ x_2 \end{bmatrix} \in \mu(\alpha, \alpha_2), \begin{bmatrix} y_1 \\ y_2 \end{bmatrix} \in \mu(\beta, \beta_2)$ with $x_1, x_2, y_1, y_2 \in V^*\}$

$\cup \{S \to N_{\alpha_{11}} \cdots N_{\alpha_{1i}} x_1 N_{\alpha_{31}} \cdots N_{\alpha_{3j}} \mid \alpha_1\alpha_2\alpha_3 \in A$ with

$\alpha_1, \alpha_3 \in V_+^* \cup V_-^*, \alpha_2 = \begin{bmatrix} x_1 \\ x_2 \end{bmatrix} \in V_\pm^+, \alpha_{11} \cdots \alpha_{1i} = \alpha_1, \alpha_{31} \cdots \alpha_{3j} = \alpha_3$,

for any $1 \leq i' \leq i, 1 \leq j' \leq j, 2 \leq |\alpha_{1i'}| \leq l, 2 \leq |\alpha_{3j'}| \leq l\}$.

We will show that for any $k \geq 0$, $S \Longrightarrow_G^{k+1} z_1 \cdots z_n$, where for each $1 \leq i \leq n$, z_i is in V^* with $2 \leq |z_i| \leq l$ or z_i is in N, if and only if $\alpha \Longrightarrow_\gamma^k w_1 \cdots w_n$, where $\alpha \in A$, for each $1 \leq i \leq n$,

$$\begin{cases} w_i = \pi & \text{for } z_i = N_\pi \text{ in } N, \\ w_i = \begin{bmatrix} x_{i1} \\ x_{i2} \end{bmatrix} & \text{for } x_{i1}, x_{i2} \in V^*. \end{cases}$$

We can prove this by the induction on k. $\qquad\square$

4   Concluding remarks

In this paper, WK-ins systems have been introduced by using incomplete molecules with gaps and sticking operations for those molecules. By exploring the computational powers of WK-ins systems, we have given a new characterization for the insertion computation using not only incomplete molecules without gaps but also the ones with gaps which suggest bulge-forming DNA sequences of biomolecules. The results we have shown include that string languages generated by WK-ins systems using only incomplete molecules without gaps are in the class of regular languages. Furthermore, a computational power of strong WK-ins systems has been investigated in which an insertion operation proceeds by checking nonempty context. For example, the class of string languages generated by strong WK-ins systems of weight $(1, 1)$ is equivalent to the class of finite languages, besides, the class of context-free languages includes the class of string languages generated by arbitrary strong WK-ins systems. It is remarkable to compare these results with existing results for insertion systems of linear strings, concerning the characterizations of the class of regular languages, such as any regular language is the coding of a language in the family INS_*^1 and the family INS_*^* properly includes the class of regular languages [2]. Those results show that the gaps of DNA sequences and context-checking clearly enhance the computing powers of insertion systems.

Acknowledgements

The author would like to express her thanks to T.Yokomori for useful suggestions. The author is also grateful to H.Katsuno for the helpful comments. This work is supported in part by a Grant from the Research Institute for Science and Technology of Tokyo Denki University with no.Q07J-05.

References

[1] Hoogeboom, H.J. and Vugt, N.V. Fair sticker languages, *Acta Informatica*, **37**, 213–225 (2000).

[2] Păun, Gh., Rozenberg, G. and Salomaa, A. : *DNA Computing. New Computing Paradigms.*, Springer (1998).

[3] Rozenberg, G. and Salomaa, A. (Eds.) *Handbook of Formal Languages*, Springer (1997).

[4] Salomaa, A. *Turing, Watson-Crick and Lindenmayer: Aspects of DNA Complementarity*, In *Unconventional Models of Computation*, Springer, 94–107 (1998).

[5] Vliet, R. van, Hoogeboon, H.J. and Rozenberg, G. *Combinatorial Aspects of Minimal DNA Expressions*, Pre-proc. In *Tenth International Meeting on DNA Computing*, Univ. of Milano-Bicocca, Italy, 84–96 (2004).

Physical hypercomputation

Mike Stannett

Department of Computer Science, University of Sheffield,
Regent Court, 211 Portobello Street, Sheffield S1 4DP, United Kingdom
m.stannett@dcs.shef.ac.uk

Abstract We argue that observable values and the physical systems that generate them need not have representations belonging to the same computational class. It is possible both for computable systems to generate uncomputable values, and also for uncomputable systems to generate computable values. In particular, while quantum wave functions are typically uncomputable, the relationships between quantum theoretical operators are generally computable. At the same time we clarify an apparent temporal anomaly in recent attempts to provide a category theoretical semantics for quantum theory.

1 Introduction

In this paper, we consider the central question of physical hypercomputation [26], viz. *are uncomputable constructs physically realisable?*

Physics concerns itself on the one hand with *measurable values*, and on the other with *behaviours and their inter-relationships*, and since values and the behaviours that generate them are very different in nature, conclusions concerning the one cannot necessarily be carried across to the other. It is not sensible, therefore, to ask whether physics is 'computable' without first declaring which aspect of physical enquiry we are talking about. In this paper we consider the following questions:

Question 1.1. Does physics support the construction of uncomputable values?

Question 1.2. Does physics support the construction of uncomputable behaviours?

Notice that our interest goes beyond the mere physical *existence* of uncomputable values and behaviours, since these are easily seen to exist in standard—e.g. Newtonian, quantum and relativistic—models of physics (since at most countably many reals are computable, the probability that a randomly generated distance is uncomputable is 1). We wish to know whether such values and behaviours can be constructed *deliberately* [27].

1.1   Structure of the paper

The answers to Questions 1.1 and 1.2 will generally depend upon our underlying model of physics. In Sect. 2, we consider in turn the nature of computation in quantum, Newtonian and relativitic models of physics. By looking at various examples, we extract principles to help us define what we mean by physical computability and physical hypercomputation.

In the remainder of the paper we focus on quantum physics and the standard model. We argue that quantum theory is best represented not in terms of *functions*, but *processes*, and investigate a basic process calculus semantics for quantum theory.

The similarity between quantum theoretical principles and the rules of calculi like CCS and the π-calculus is striking, and has been noted in the literature over many years (e.g. [23]). Gay and Nagarajan [8] recently developed the connection by describing a language (CQP) for modelling communicating quantum processes; in an attempt to provide a coherent description at an appropriate level of abstraction, they introduce new primitives for measurement and state transformation. While this is a sensible step towards the development of a programming language for practical quantum computations, it is unhelpful for our own purposes. We therefore retain the pure calculus in our discussions below.

Our work is also related to recent attempts to provide a category theoretical semantics for quantum theory [2,3], but has the advantage of being clearly computational rather than mathematical; this suggests that certain identifiable quantum relationships (though not necessarily all observables) may be computable. In passing, we discuss an apparent temporal anomaly in the category-theoretic approach.

2   Computation in quantum, Newtonian and relativistic models of physics

2.1   The view from quantum physics

Let us suppose first (as appears to be experimentally the case) that quantum theory provides an acceptable model of the physical world. An important equation in quantum theory is the time-dependent version of Schrödinger's equation (TDSE),

$$-\frac{\hbar^2}{2m}\nabla^2\psi + V(\mathbf{r})\psi = i\hbar\frac{\partial\psi}{\partial t}$$

(where V gives the potential energy at each location $\mathbf{r}$). As the right hand term reminds us, quantum theory makes considerable use of complex numbers. Does this imply that complex numbers are in some sense *real*? Not at all; it simply means that the underlying mathematics of physical behaviours can conveniently be described using complex numbers.* This causes no problems in practice, because whenever we want to extract 'real world' information from such equations, the procedures used invariably result in real-number values. In other words:

Observation 2.1. It can be appropriate to use mathematical constructs to model the physical world, even when those constructs have no obvious physicality of their own.

To find physical values associated with steady-state solutions, the time-independent version of Schrödinger's equation (TISE) is more appropriate: $H\psi = E\psi$, where E is the total steady-state energy of the solution being described, and H, the *Hamiltonian* of the system, is the operator $H = -(\hbar^2/2m)\nabla^2 + V$. Feynman [7, §20–2] explains that the operator equation avoids having to focus on any particular choice of basis. He continues, by analogy with conventions in vector analysis,

It's similar to the difference between writing

$$\mathbf{c} = \mathbf{a} \times \mathbf{b}$$

instead of

$$c_x = a_y b_z - a_z b_y,$$
$$c_y = a_z b_x - a_x b_y,$$
$$c_z = a_x b_y - a_y b_x.$$

The first way is much handier. When you want results, however, you will eventually have to give the components with respect to some set of axes. Similarly, if you want to be able to say what you mean by [some abstract quantum theoretical operator] $\hat{A}$, you will have to be ready to give the [corresponding, worked-out] matrix A_{ij} in terms of some set of base states.

---

* Nonetheless, Penrose and Rindler have suggested that complex manifolds may have a real significance in spinor-based cosmologies [19, 20].

This idea underpins operator-based approaches to physics: specific physical values are only required when you want to test specific predictions about specific experiments; they are not generally needed to *describe* or *reason about* the patterns underpinning those phenomena. Indeed, this is an inevitable consequence of modern physics: first we reason about the probable existence of observables, then we perform measurements to test our theory.

Observation 2.2. If you want to reason about—that is, to *understand*—physical behaviours, it is typically more appropriate to use abstract representations than concrete ones.

While Obs. 2.1 justifies our use as tools of potentially unphysical mathematical representations, Obs. 2.2 introduces the important *computational* caveat that measured values on the one hand, and physical behaviours on the other, are likely to be described using rather different representations. Just as behaviours described using complex numbers or quaternions can give rise to measurements that are necessarily real, so it is entirely reasonable that uncomputable behaviours might give rise to computable observables, or computable behaviours to uncomputable observables.

Observation 2.3. There is no a priori reason for supposing that observable values and the behaviours that generate them should belong to the same computational class.

In fact, we shall argue that while wave functions ψ are presumably *uncomputable*, the relationships between operators—i.e. the properties of operators that we use to reason about and understand physical systems— are nonetheless generally *computable*. In large part this last claim is obvious, since physicists perform such reasoning on paper, blackboard and laptop every day; but more than this, we suggest that the relationships between operators can be represented within the π-calculus [15,28], whence these relationships are certainly computable (Table 1).

The structure of the TISE also illustrates another important point. Consider, for example, the TISE for a spherically symmetrical neutral hydrogen atom:

$$(\hbar^2/2m)\nabla^2 + e^2/r = -E \tag{1}$$

where e is the electronic charge. Equation (1) is clearly a finite specification expressed using finite means, and a single hydrogen atom is certainly a finitely resourced system. Nonetheless, (1) has embedded within it the potential for distinctly uncomputable behaviour, since it describes an

Table 1. Physical systems, their behaviours and experimental observations –
quantum theoretical vs. computational representations.

	Quantum Theoretical Representation	Computational Status
Systems	wave functions (complex-valued)	uncomputable
Behaviours	manipulation of operators (essentially algebraic)	computable?
Observations	observables (real-valued)	???

infinite set of energy levels for the atom's single electron. Since the electron can move non-deterministically between these levels, the TISE is
an effective specification of an infinitely non-deterministic system, the
behaviour of which is necessarily non-recursive [24, 25].

Observation 2.4. A finitely specified, finitely resourced, system need not
have computable behaviour.

On the other hand, it is well known that energy levels E_n charac-
terised by (1) are easily described: we have $E_n \propto \frac{1}{n^2}$. Clearly, if we take
the E_1 to be our unit of energy, each of the values E_n is computable.
Thus, the status of the TISE is rather subtle. It describes a system
which can evolve uncomputably between states which are themselves
computably describable.

Observation 2.5. A finitely specified, finitely resourced, system can dis-
play both computable and uncomputable behaviours at the same time,
depending on the particular aspect of behaviour in which we are interested.

2.2   Relativistic computation

Although the unconventional computing community has focussed (and
understandably so) on the potential for quantum computation, it should
not be forgotten that relativistic theories also give rise to models of computation that differ significantly from those of standard Newtonian/Eucli-
dean models. Morgenstein and Kreinovich [16] have shown, for example,
that the complexity of problems depends upon the underlying geometry
of space-time, with certain problems being easier in the hyperbolic plane.
Since quantum computation also offers exponential speed-up for certain
algorithms [21, 22], it would seem inevitable that the same should be
true in any consistent synthesis of quantum gravity. Indeed, if this is not

the case, it would imply that quantum and relativistic physics contain elements that 'exactly cancel out' each other's super-Turing potential – an intriguing result, indeed!

Observation 2.6. Both quantum physics and relativistic physics support systems that can outperform the standard Turing model of computation. It is likely that any consistent synthesis of quantum gravity will do likewise.

Moreover, a class of exotic (so-called *Malament-Hogarth*) spacetimes [10,11] supports the existence of singularities with properties that enable the computation of non-recursive functions. In particular, suppose we wish to compute the Turing halting function

$$h(n) = \begin{cases} 1 \text{ if } P_n \text{ halts} \\ 0 \text{ otherwise} \end{cases}$$

where $P_0, P_1, P_2, \ldots$ is some fixed Gödel-numbering of programs. A computer falling into a Malament-Hogarth singularity experiences infinite proper time, but its entire trajectory can be observed, in finite proper time, by an external observer. The observer can therefore determine in finite proper time whether or not any given program P_n running on that computer eventually terminates (the computer can be programmed to launch a rocket to rendezvous with the observer on termination of the program; if no rocket arrives to meet the observer at a pre-determined space-time location, the program never terminated).

Such space-times would seem to violate cosmic censorship [12], but there is currently no evidence that censorship is itself physically valid. Moreover, any attempt to implement such an algorithm for h would involve insuperable practical difficulties [6], but this not undermine the scenario's theoretical relevance. Its significance for our purposes lies rather in the signature of $h: \mathbb{N} \to \{0, 1\}$, for although h is well known to be uncomputable, it nonetheless transforms each (computable) input n into one of the (computable) outputs 0 or 1.

Observation 2.7. It is possible for an uncomputable physical behaviour to generate computable measurements.

2.3   Newtonian computation

While our main focus in this paper is on non-classical physics, we should not forget that Newtonian physics *also* supports the theoretical existence of super-Turing and non-Turing behaviours. Again, there are serious practical difficulties in implementing either of the schemes described below, but this should not blind us to their theoretical significance. As I have discussed elsewhere [27], three important findings are those due to Pour-El and Richards [18], Xia [30], and Myhill [17].

Myhill's work demonstrated that a certain function f is both smooth and computable, and yet has non-computable derivative. It is unclear to what extent this result can be applied in practice, but one can envisage a theoretical scenario in which f is computed using Turing-style equipment, and the output channeled through a (computably specified and designed) analog circuit to perform the subsequent differentation.

Observation 2.8. It is feasible that uncomputable values can be generated by applying computably generated analog equipment to computable inputs.

Pour-El and Richards looked at systems specified by the standard wave equation, and found that it was possible to provide computable initial conditions, set the system running, observe its amplitude at a computably selected location after a computable amount of time, and obtain an uncomputable result. Their method has been called into question on practical grounds, since the initial conditions could never be configured *precisely* by finite means, but this is not important for our argument, since we are concerned here only with the theoretical structure of physical systems, and its implications for computational power.

Observation 2.9. It is feasible for a computably specified system with computable initial conditions to be observed in a computable manner and yet return an uncomputable result.

Finally, Xia showed in his studies of the many-body problem that a finite gravitational system can propel a finite mass to infinity in finite time. The body's trajectory takes it past a family of double-star systems in an accelerating sequence of fly-bys, and using these rendezvous as 'decision events' in a suitably recast version of the Turing model allows us to complete an infinite supertask in finite time [27].*

Observation 2.10. The inability of Turing machines to perform supertasks is purely mathematical; it is not a consequence of the Newtonian model in which computation is traditionally considered to take place. Rather, the Newtonian n-body problem has solutions which permit the completion of infinite decision procedures in finite time.

---

* The relationship between Church's Thesis and Xia's work on the n-body problem seems to have been discussed first by Warren Smith in unpublished work from 1993 (revised 2005). See **http://math.temple.edu/~wds/ homepage/works.html**.

3   Computational semantics and quantum physics

Modern physics is firmly based in the tradition of experimental verification; for a theory to be scientifically relevant, it must be capable
of making predictions that can be tested empirically. To some extent,
this empirical emphasis forces physicists to focus on values and their
measurement; it is not enough that a physical theory be internally consistent - it must also explain why particles have the masses they do, why
space-time occupies a certain fixed number of dimensions, and so on.
Nonetheless, the focus on values is essentially secondary; the true focus
of mathematical physics is to understand the *processes* by which physical systems evolve, and by which they transform one observable value
into another. Performing measurements is important for verification purposes, but more important is the formulation of the underlying theories:
without theories, we cannot know what measurements are worth performing, nor understand what they tell us.

   This suggests that physics might usefully be represented using compu-
tation-theoretic models. Abramsky and Coecke have recently developed a
category theoretical semantics for quantum theory, which allows them to
represent protocols like quantum teleportation and entanglement swapping via simple graphical representations [2,3], an approach which mirrors that used by Penrose and Rindler in the 1980s to discuss spinor
theory [19]. However, category theory is a mathematical discipline, and
does not typically pre-suppose any particular computability requirements (rather, it is used to give a semantics to computation). It is,
however, entirely possible for an uncomputable behaviour to be given
a categorical semantics, whence the existence of categorical semantics
for quantum theory does not directly inform our debate as to the feasibility of physical hypercomputation.

   Given the process-based nature of mathematical physics, we would
therefore prefer to give a process-theoretic representation expressed in
CCS [13] or the π-calculus [15, 28]. These are standard computational
tools for modelling distributed processes (in the case of π-calculus, we
can also model process mobility), which makes them ideal candidates
for modelling the non-local reality [4] inherent in quantum theory. We
will assume for the remainder of this section that the reader is familiar
with basic π-calculus (as discussed in e.g. [15]), and present only a basic
summary to fix our notation.

3.1   The π-calculus

In the π-calculus, we assume a set $\mathcal{N}$ of *names* ranged over by a and
b, together with a corresponding set $\overline{\mathcal{N}} = \{\overline{a} \mid a \in \mathcal{N}\}$ of *co-names*; we

define $\overline{\overline{a}} = a$. Together with the *silent action* τ, these constitute the set $\mathcal{L} = \mathcal{N} \cup \overline{\mathcal{N}} \cup \{\tau\}$ of *labels*, ranged over by α and β. The notation $\widetilde{a}$ denotes a (possibly empty) vector of distinct names. We often write $\widetilde{a}$ as a list $a_1 \ldots a_n$ of names, where it is assumed implicitly that the a_i are distinct.

The set of π-calculus *process expressions*, ranged over by E, F and G, is defined in Table 2, and their semantics in Table 3.

Table 2. Syntax of the π-Calculus.

Action prefixes
$$\pi ::= b(\widetilde{a}) \qquad \text{input } \widetilde{a} \text{ on } b \text{ (binds all } a \in \widetilde{a})$$
$$| \quad \overline{b}\langle\widetilde{a}\rangle \qquad \text{output } \widetilde{a} \text{ on } b$$
$$| \quad \tau \qquad \text{unobservable action}$$

Process expressions
$$E, F, G ::= \mathbf{0} \qquad \text{nil}$$
$$| \quad \pi.E \qquad \text{prefix (we say that } \pi \text{ } guards \text{ } E)$$
$$| \quad E | F \qquad \text{composition}$$
$$| \quad E + F \qquad \text{sum (or } choice)$$
$$| \quad (\boldsymbol{\nu}a)(E) \quad \text{restrict names in } \widetilde{a} \text{ (binds all } a \in \widetilde{a})$$
$$| \quad !E \qquad \text{replication}$$

An important feature of the π-calculus is that it is Turing complete; any recursive function can be expressed via representation in the calculus. In order to illustrate this principle, we show briefly how to model *polymorphic binary trees*, defined in Haskell by the polymorphic algebraic datatype `Tree a`,

```
data Tree a   = Empty
            | Node { left  :: Tree a,
                     val   :: a,
                     right :: Tree a }
```

The method we describe is essentially that sketched by Milner in [15], but differs from his earlier exposition [14].

Note first that each 'data' process must have an specific *interface channel*, i, upon which 'functions' can synchronise in order to perform 'computations.' It follows that the same must be true of constructors, since a data-process is simply a tree of constructors in which every channel except i is restricted (see figure 1).

Table 3. Semantics of the π-Calculus.

Structural Equivalence Rules: These rules explain what it means for two process expressions to be algebraically identical. In particular, we adopt the usual computational convention (rule **Alpha**) that α-*equivalent* processes are identical i.e. processes that are equivalent up to renaming of bound variables. The *free names* of a process E are denoted fn E.

$$\text{SumComm}\ \frac{}{E + F \equiv F + E} \qquad \text{Repeat}\ \frac{}{!E \equiv E\,|\,!E} \qquad \text{ParComm}\ \frac{}{E\,|\,F \equiv F\,|\,E}$$

$$\text{SumId}\ \frac{}{E + 0 \equiv E} \qquad \text{Alpha}\ \frac{E \equiv_\alpha F}{E \equiv F} \qquad \text{ParId}\ \frac{}{E\,|\,0 \equiv E}$$

$$\text{SumAssoc}\ \frac{}{E + (F + G) \equiv (E + F) + G}$$

$$\text{ParAssoc}\ \frac{}{E\,|\,(F\,|\,G) \equiv (E\,|\,F)\,|\,G} \qquad\qquad \text{RestrNil}\ \frac{}{(\nu a)\,(0) \equiv 0}$$

$$\text{RestrPar}\ \frac{}{(\nu a)\,(E\,|\,F) \equiv E\,|\,(\nu a)\,(F)}\ \ x \notin \text{fn}\ E$$

Derivation Rules: These rules explain when it is possible for one process E to evolve into another process E' by performing an action α, denoted $E \xrightarrow{\alpha} E'$. For example, **Sum** tell us that, if E can evolve into E' by performing α, then so can $E + F$, for any process E'. In particular, **Sync** tells us that corresponding input/output pairs can synchronise; the reason for the synchronisation is regarded as *unobservable* outside the combined system, and is denoted τ.

$$\text{Sum}\ \frac{E \xrightarrow{\alpha} E'}{E + F \xrightarrow{\alpha} E'} \qquad \text{Act}\ \frac{}{\alpha.E \xrightarrow{\alpha} E} \qquad \text{Par}\ \frac{E \xrightarrow{\alpha} E'}{E\,|\,F \xrightarrow{\alpha} E'}$$

$$\text{Sync}\ \frac{E \xrightarrow{\overline{a}\langle \tilde{x} \rangle} E' \qquad F \xrightarrow{a(\tilde{y})} E'}{E\,|\,F \xrightarrow{\tau} E'\,|\,F'[\tilde{x}/\tilde{y}]}\ (\text{len}(\tilde{x}) = \text{len}(\tilde{y}))$$

$$\text{Restr}\ \frac{E \xrightarrow{\alpha} E'}{(\nu b)(E) \xrightarrow{\alpha} (\nu b)(E')}\ a \neq b \qquad \text{Struct}\ \frac{E \equiv E' \xrightarrow{\alpha} F' \equiv F}{E \xrightarrow{\alpha} F}$$

- To represent **Empty**, we need to specify the interface channel i on which we wish it to be located.

- To represent **Node**, we need to specify both the interface channel i, and also the *argument channels* l, v and r.

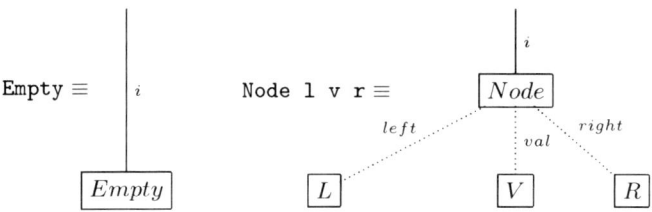

Figure 1. Flow graphs for the representations of typical binary trees `Empty` and `Node l v r`. The latter is recursively defined, with $V\langle val\rangle$ representing the value `v :: t` and $L\langle left\rangle$ and $R\langle right\rangle$ representing the trees `l` and `r`. Restricted channels are indicated using dotted arrows.

Let `Y` be some type whose values can already be represented, and suppose $F\langle io\rangle$ represents some function `f :: Tree a → Y`. Using pattern-matching techniques, the definition of `f` can be given generally as

```
f  Empty        =  y
f (Node l v r)  =  g l v r
```

where `y :: Y`, and `g :: Tree a → a → Tree a → Y` is some auxiliary function.

Suppose $[\![t]\!]\langle i\rangle$ represents `t :: Tree a`, and consider the problem of evaluating $[\![f\ t]\!]\langle o\rangle \equiv (\boldsymbol{\nu}i)\,([\![f]\!]\langle io\rangle\,|\,[\![t]\!]\langle i\rangle)$. From $[\![f]\!]$'s point of view, it needs to synchronise with $[\![t]\!]$, but from our point of view (since we know $[\![t]\!]$'s internal structure) $[\![f]\!]$ is actually synchronising with one of $[\![t]\!]$'s constructors. Of course, $[\![f]\!]$ has no a priori way of knowing whether it is synchronising with $[\![\texttt{Empty}]\!]\langle i\rangle$ or some $[\![\texttt{Node}]\!]\langle ilvr\rangle$ instance, so it engages in the following dialogue with the constructor.

- $[\![f]\!]\langle i\rangle$ creates two new channels e and n, and shares them with the constructor by sending them through i.
- The constructor receives e and n, and then replies to $[\![f]\!]$.

  - $[\![\texttt{Empty}]\!]\langle i\rangle$ identifies itself by replying along e.
  - $[\![\texttt{Node}]\!]\langle ilvr\rangle$ replies along n, and shares the private channels l, v and r. In this way it both identifies itself, and gives F access to its own arguments so that $[\![f]\!]$ can pass them to the process representing `f '`.

It follows that the representations of `Empty`, `Node` and `f` are the processes

$$\llbracket \texttt{Empty} \rrbracket \langle i \rangle \equiv i(en).\overline{e}$$
$$\llbracket \texttt{Node} \rrbracket \langle ilvr \rangle \equiv i(en).\overline{n}\langle lvr \rangle$$
$$\llbracket \texttt{f} \rrbracket \langle io \rangle \equiv (\boldsymbol{\nu} en)\left(\overline{i}\langle en \rangle.(e.Y\langle o \rangle + n(lvr).F'\langle olvr \rangle)\right)$$

where $Y\langle o \rangle$ represents y at o, and $F'\langle olvr \rangle$ represents $\texttt{f'}$ on o.

3.2   General structure of constructor and function processes

Ignoring type constraints for the time being, an algebraic data type definition has the form

$$\texttt{data T} = \texttt{C}_1 \; \texttt{T}_{1,1} \; \ldots \; \texttt{T}_{1,r_1}$$
$$| \quad \ldots$$
$$| \; \texttt{C}_n \; \texttt{T}_{n,1} \; \ldots \; \texttt{T}_{n,r_n}$$

where the type T may be polymorphic, in which case the various parameter types $\texttt{T}_{j,k}$ may depend on the parameters to T. The corresponding definition of a general (unary) function defined on objects of type T is then of the form

$$\texttt{f} \; (\texttt{C}_1 \; \texttt{v}_{1,1} \; \ldots \; \texttt{v}_{1,r_1}) \; = \; \texttt{d}_1$$
$$\ldots \qquad\qquad = \ldots$$
$$\texttt{f} \; (\texttt{C}_n \; \texttt{v}_{n,1} \; \ldots \; \texttt{v}_{n,r_n}) = \; \texttt{d}_n$$

Arguing as above, it follows that T will be represented by defining n different processes; the process $\llbracket \texttt{f} \rrbracket$ emits a name vector $\widetilde{c}$ of length n, and waits for the lead constructor of x to identify itself by replying on the appropriate channel c_m. To this end we define

$$\llbracket \texttt{f} \rrbracket \langle io \rangle = (\boldsymbol{\nu} c)\left(\overline{i}\langle \widetilde{c} \rangle . \sum_m c_m(\widetilde{p}).D_m\langle o\widetilde{p} \rangle\right)$$

where the data processes D_m are defined recursively, and

$$\llbracket \; \texttt{C}_m \; \rrbracket \langle i\widetilde{p} \rangle = i(\widetilde{c}).\overline{c_m}\langle \widetilde{p}_m \rangle.\mathbf{0} \; .$$

3.3   Polymorphism

The π-calculus is naturally polymorphic to some extent; for example, the representation procedure described above generates the process $i(ab).\overline{a}.\mathbf{0}$ as a representation of both the natural number Zero and the Boolean value `False`. Nonetheless, it is clear that the representation only works if the underlying data type is presented as an unconstrained algebraic data type. Unfortunately, it is not clear how we might give a finite representation of real numbers as an algebraic data type, and without a representation of reals, a representation of complex numbers is inaccessible (clearly, each of the fields $\mathbb{R}$ and $\mathbb{C}$ can be defined in terms of the other).

More generally, a limited form of polymorphism can be implemented if we are prepared to accept a more indirect representation. In languages like Haskell, it is possible to impose *type constraints*, so that the type `T a b c ...` is only inhabited in the event that the argument types `a`, `b`, `c`, ... can be shown to satify the assigned constraints. In particular, suppose we define a type class

```
class Incr a where
   incr :: a -> a
```

encapsulating the idea that an *incrementable value* is one whose type `a` comes equipped with a suitable function `incr` (which we can specify elsewhere as having the functional behaviour of adding 1 to its argument). We can then declare `Nat` to be such a class by declaring `incr ::` `Nat` $\rightarrow$ `Nat` to be the function `incr n = Succ n`.

Rather than attempt a direct representation of $\mathbb{R}$ and $\mathbb{C}$, we leave their construction to some oracle, and ask whether we can nonetheless represent the incrementing of such values in a generic way, just as we were able to determine the length of a list regardless of the value-types it contains.

Observation 3.1. It is possible to manipulate an object recursively even if it is defined in terms of components that are uncomputable.

To see that this is possible more generally, we can mirror the way that polymorphic functional languages like Haskell make use of *libraries*. We regard `Incr` as a (typically finite, though recursively enumerable is sufficient) list of pairs associating each relevant incrementable type `t` with its associated `incr` implementation. A type `T` is then a valid instance of `Incr`, if and only if the associated pair `(T, incr)` has been added to this list.

Armed with such a list, it is clear that a π-calculus representation of the incrementation class can be generated, provided that type identification is available. For suppose `x` is a value of some type `T`. In order to increment `x` we set $F = \{(\mathtt{T}, \mathtt{incr}) \mid \mathtt{x} :: \mathtt{T}\}$. The set F identifies all those instances of `incr` that can meaningfully be applied to x. We now declare all of the associated values `incr(x)` to be valid increments of x, and ask whether these values can be unified. If so, the result is the answer we seek. (There are various options available to us if the results cannot be unified; the precise choice is not relevant to our discussion.) This approach is entirely general. So, for example, while we cannot compute all complex values, we can nonetheless consider such functions as $\lambda z.z^*$ to be computable.

Observation 3.2. By using libraries, we can introduce sufficient polymorphism into π-calculus representations of functions to enable type constraints to be expressed in a generic way.

The use of libraries and polymorphism is, of course, only one approach. In other situations, we can achieve polymorphism very simply indeed. For example, consider the identity function `id :: Nat → Nat`. In our representation this becomes the process

$$Id\langle io\rangle = (\boldsymbol{\nu}zs)\left(\bar{i}\langle zs\rangle.(z\,.Zero\langle o\rangle + s(p).Succ\langle op\rangle)\right)$$

which moves any natural number n located on channel i to the new channel o. More generally, `id :: T → T`, becomes

$$Id\langle io\rangle = (\boldsymbol{\nu}c)\left(\bar{i}\langle\tilde{c}\rangle.\sum_{c_m\in\tilde{c}}c_m(\tilde{p}).C_m\langle o\tilde{p}\rangle\right)\quad.\tag{2}$$

Observation 3.3. A polymorphic process can operate on values of *any* type `T`. Some polymorphic processes can be implemented using libraries, while others can be implemented directly. If a behaviour is represented by such a process, it can be regarded as being *computable*, even if the values it operates upon are not.

3.4   Modelling quantum physical behaviours

Although the syntax and semantics of π-calculus are traditionally interpreted in computational terms, they are also relevant from a physical standpoint provided physical objects and systems are regarded not as an entities, but as a collection of quantum information channels. The fact that channels are identified solely by name means that geographical location is not an issue; actions-at-a-distance are entirely natural in the π-calculus.

Example 3.1. Consider the electromagnetic interaction shown in Fig. 2, in which a virtual photon is exchanged between two electrons.

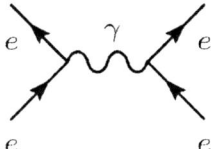

Figure 2. Electromagnetic interaction.

Since the virtual photon in this exchange acts as an information carrier, we model it as an information channel γ shared by all electrons. For the purposes of this example, we shall assume that certain names represent momenta, and that we can meaningfully 'add' such names; in effect, we will be using the exchange of names as a form of value-passing. The process $E_{\mathbf{a}}\langle\gamma\rangle$ defined in (3) represents an electron with momentum $\mathbf{a}$

$$E_{\mathbf{a}}\langle\gamma\rangle = \gamma(\mathbf{b}).E_{\mathbf{a}+\mathbf{b}}\langle\gamma\rangle + \sum_{\mathbf{b}\in M} \overline{\gamma}\langle\mathbf{b}\rangle.E_{\mathbf{a}-\mathbf{b}}\langle\gamma\rangle \tag{3}$$

which is capable of exchanging any momentum from some pre-defined set M. In this example, the choice of M is largely irrelevant, but for quantised systems, the ability to constrain such quantities is clearly important.

Suppose then that we have two electrons with momenta $\mathbf{x}$ and $\mathbf{y}$, respectively, so that the system as a whole can be represented as the composite process

$$E_{\mathbf{x}}\langle\gamma\rangle \,|\, E_{\mathbf{y}}\langle\gamma\rangle$$

The semantic axiom $\texttt{Par}$ ensures that $E_{\mathbf{x}}\langle\gamma\rangle \,|\, E_{\mathbf{y}}\langle\gamma\rangle$ and $E_{\mathbf{x}}\langle\gamma\rangle \,|\, E_{\mathbf{y}}\langle\gamma\rangle$ are indistinguishable, while $\texttt{Sync}$ ensures that momentum exchange occurs as required, since, given any available momentum $\mathbf{m} \in M$, we have the possible synchronisation instance

$$\cfrac{\cfrac{E_{\mathbf{x}}\langle\gamma\rangle \xrightarrow{\overline{\gamma}\langle\mathbf{m}\rangle} E_{\mathbf{x}-\mathbf{m}}\langle\gamma\rangle \qquad E_{\mathbf{y}}\langle\gamma\rangle \xrightarrow{\gamma(\mathbf{b})} E_{\mathbf{y}+\mathbf{b}}\langle\gamma\rangle}{E_{\mathbf{x}}\langle\gamma\rangle \,|\, E_{\mathbf{y}}\langle\gamma\rangle \longrightarrow E_{\mathbf{x}-\mathbf{m}}\langle\gamma\rangle \,|\, E_{\mathbf{y}+\mathbf{b}}\langle\gamma\rangle\{\mathbf{m}/\mathbf{b}\}}\;\texttt{Sync}}{E_{\mathbf{x}}\langle\gamma\rangle \,|\, E_{\mathbf{y}}\langle\gamma\rangle \longrightarrow E_{\mathbf{x}-\mathbf{m}}\langle\gamma\rangle \,|\, E_{\mathbf{y}+\mathbf{m}}\langle\gamma\rangle}\;\texttt{Struct}$$

Discussion How accurate is the representation in this example? Since

$$E_{\mathbf{x}}\langle\gamma\rangle \,|\, E_{\mathbf{y}}\langle\gamma\rangle \equiv E_{\mathbf{x}}\langle\gamma\rangle \,|\, E_{\mathbf{y}}\langle\gamma\rangle \ ,$$

our model correctly suggests that elecrons are indistinguishable . Unfortunately, this is not the end of the story – for while electrons are indeed indistinguishable, they are also fermions. Consequently, if they are exchanged, the system wavefunction should change sign. Depending on the system we are attempting to model this may be irrelevant, in which case the representation presented above is adequate. But what if the two-electron system is part of a larger system, for which the sign of the wave-function matters? How can we model this in our representation?

Clearly, we cannot represent the change of sign by amending our definitions of $E_\mathbf{x}$ and $E_\mathbf{y}$ – we have already seen that swapping these processes under the parallel composition operator cannot change their composite behaviour. Instead, we need to adopt a *higher-order* approach, regarding the interaction as being *mediated* through a third process (in the same way that quantum field theory introduces a third actor, the field, to mediate action at a distance). For example, a very simple representation might include a process $Field_\theta\langle ie_1e_2\rangle$, defined to interact with (a modified version of) E, in such a way that

$$(\boldsymbol{\nu} e_1e_2)\,(Field_\theta\langle ie_1e_2\rangle \mid E_\mathbf{x}\langle e_1\gamma\rangle \mid E_\mathbf{y}\langle e_2\gamma\rangle)$$

and

$$(\boldsymbol{\nu} e_1e_2)\,(Field_\phi\langle ie_1e_2\rangle \mid E_\mathbf{y}\langle e_1\gamma\rangle \mid E_\mathbf{x}\langle e_2\gamma\rangle)$$

are behaviourally equivalent, for any desired (fixed) phase change relationship $\delta_{\mathbf{x},\mathbf{y}}\colon \theta \mapsto \phi$. Since π-calculus is Turing-complete, we believe that suitable mediating processes can always be defined; the issue, however, is how best to do this, and at what level of complexity. This requires further research.

3.5   Which representation is most suitable?

It could be argued that π-calculus is too powerful to use for assigning a semantics to quantum theory, and that a simpler value-passing variant of CCS is sufficient (in effect, (3) is itself defined in such a system).

Indeed, the computational power of the calculus presents real difficulties. Clearly, (2) gives a general polymorphic representation $Id\langle io\rangle$ of the identity function, and it is easy to generate a copying function in the same way; for example, given $\mathbf{n}$ :: $\mathtt{Nat}$ on channel i we can generate a duplicate of $\mathbf{n}$ on channel o (without destroying the original) using the process

$$Copy\langle io \rangle = (\boldsymbol{\nu}zs)\,\bar{i}\langle zs \rangle.$$
$$(\; z\,.(Zero\langle i \rangle \mid Zero\langle o \rangle)$$
$$+ \; s(p).(Succ\langle ip \rangle \mid (\boldsymbol{\nu}q)\,(Copy\langle pq \rangle \mid Succ\langle oq \rangle)))$$

or more generally, Copy $::$ T $\rightarrow$ T can be represented recursively using the process

$$Copy\langle io \rangle = (\boldsymbol{\nu}\widetilde{c})\,\bar{i}\langle \widetilde{c} \rangle.$$

$$\sum_{c_m \in \widetilde{c}} c_m(\widetilde{p}).\left(C_m\langle i\widetilde{p} \rangle \mid (\boldsymbol{\nu}q)\left(C_m\langle o\widetilde{q} \rangle \mid \prod_{q_j \in \widetilde{q}} Copy\langle p_j q_j \rangle \right) \right)$$
$$(4)$$

where the local name vector $\widetilde{q}$ should be of the same length as the received vector $\widetilde{p}$. The existence of such functions presents severe difficulties to any π-calculus representation of quantum theory, since it is essential that such a function should *not* be available fully polymorphically – it would violate the no-cloning theorem [5, 29]!

Is this problem an unavoidable consequence of process-theoretic methods, and if so, does this mean that process-theoretic methods cannot safely be used to represent quantum theoretic systems? Not necessarily. In the first place, the fairly simple definition of $Copy\langle io \rangle$ given in (4) conceals a number of subtleties. For example, having carried out the input prefix $c_m(\widetilde{p})$, the $Copy\langle io \rangle$ process needs to generate a new local name vector $\widetilde{q}$ of the same length as $\widetilde{p}$ – but how can it do so?

All that (4) really shows is that processes can be cloned *provided channels can also be cloned*. But nothing in the syntax or the semantics of the π-calculus suggests that this is the case, and we must conclude that the definition of $Copy\langle io \rangle$ is not well-formed; and this is indeed a reasonable claim, since we are attempting to use the input vector $\widetilde{p}$ as a general purpose polymorphic input variable, whose length and sorting is immaterial, but have no justification for doing so. In physical terms this is precisely the kind of limitation that is required, because, as Example 3.1 illustrates, channels correspond to *virtual exchange particles*, which we would not expect to be clonable.

Observation 3.4. Communication channels in our process-theoretic representation correspond (amongst other things) to virtual exchange particles. The inability to clone the latter is reflected in the inability to clone channel names, and this in turn throws in doubt the ability to clone arbitrary processes using a single polymorphic *Copy⟨io⟩* process. This is as it should be: cloning should *not* be generally available.

3.6   Graphical representations of quantum semantics

The various process examples we have considered above share a common feature: the process we describe has an internal structure that mirrors an existing diagrammatic representation of the system in question. Thus, for example, our representation of `Tree a` mirrors the diagrams in Fig. 1, and likewise, the processes in Example 3.1 have the composite flow diagram

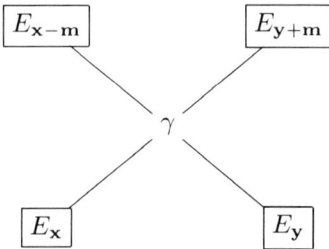

which is structurally identical to the Feynman diagram of Fig. 2.

This similarity is no coincidence, but reflects the shared algebraic structure of abstract computation and quantum theory, and likewise the structural similarity between these theories, the categorical constructions used by Abramsky and Coecke [3], and the graphical notation described by Penrose and Rindler to represent general cosmological tensor operations [19, App. 1] – it is easy to interpret the axioms of Table 3 in physical terms.

Nonetheless, we need to be alert to anomalies. Consider, for example, Abramsky and Coecke's graphical notation, as presented recently by Abramsky [1]. Abramsky considers the map $f : z \mapsto w$ obtained by chaining together various two-input unitary transformations as shown in Fig. 3(a). Given a transformation f, Abramsky and Coecke show how to define its *adjoint* $(f^\dagger)$ in purely categorical terms, and show that

$$f = c \circ d \circ b^\dagger \circ c^\dagger \circ a \circ b \tag{5}$$

In other words, we need to 'follow the route' shown in Fig. 3(b). Each time we enter a box from the left hand side, we apply the corresponding function, and each time we enter on the right, we apply the adjoint. While this is a neat result, it is also somewhat paradoxical, since it implies that the various functions are being applied in the 'wrong' order.

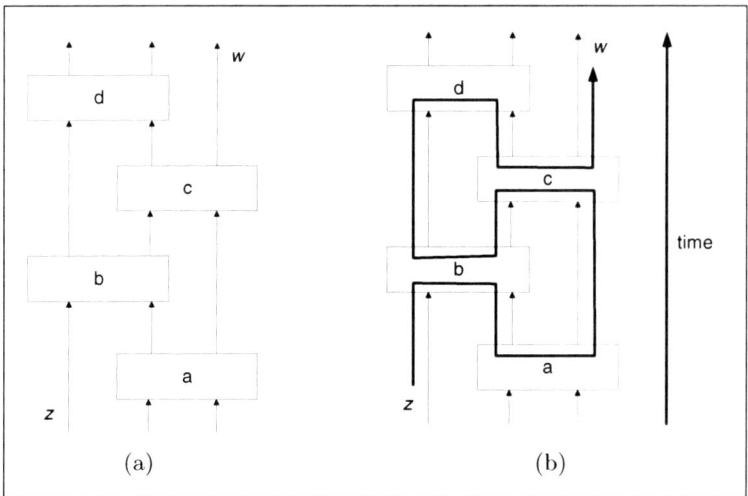

Figure 3. How can the map $f\colon z \mapsto w$ be characterised in terms of the component transformations shown? (based on [1]).

For example, a is actually applied before b in strictly temporal terms, but occurs after b in the route map. So while he diagrammatic approach proposed by Abramsky and Coecke seems reasonably natural, and ties up neatly both with the diagrammatic tensor approach, and our own process-theoretic approach to quantum semantics, it is reasonable to suspect that something is wrong!

We end this paper with a brief discussion of this problem, which suggests a future research programme. Our solution is rather straightforward – we simply rotate the route through $90°$, as shown in Fig. 4. If we trace the route now, it is clear that positive applications of a function f correspond to motion *in the positive spatial direction*, regardless of the temporal direction in which one is moving. Likewise, moving against the 'direction of space' requires the adjoint to be applied instead. The similarity with Feynman diagram conventions is clear; just as an anti-particle can be regarded as a particle moving backwards in time, so the adjoint corresponds to applying a transformation 'against the flow of space'.

The idea that 3-dimensional space has a preferred direction (as opposed to orientation) is rather unusual, and hard to explain in standard terms. However, we can offer a tentative explanation by comparison with ideas from superstring theory [9], where *dual models* are commonplace. In such models, we frequently encounter Feynman diagrams containing both s-channels (shown by horizontal lines, e.g. the photon in Fig. 2)

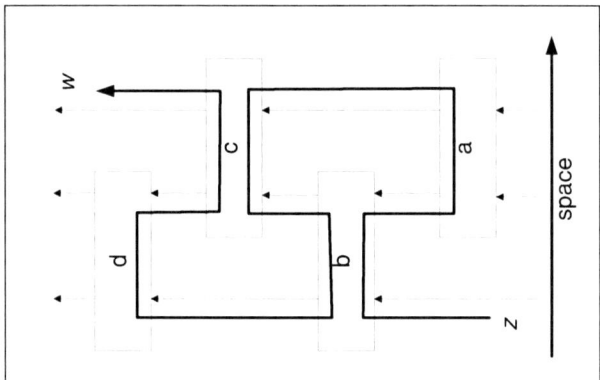

Figure 4. Re-interpretation of Fig. 3(b) in spatial terms.

and t-channels (shown by vertical lines), and any given diagram can be decomposed into a sum of t-channel components, or dually, of s-channel components. The fact that Abramsky and Coecke's model relates adjoints to negative *spatial* motion suggests that it may be inherently based around an underlying (though unobvious) s-channel model. It seems sensible, therefore, to investigate whether or not a dual model can be formulated, in which adjoints correspond to movement against the flow of *time* rather than *space*. If so, what does this imply for their category theoretic semantics of quantum theory?

4   Summary and conclusions

We have presented a wide-ranging discussion of the way physics underpins the scope of physical hypercomputation. In each of the three physical paradigms we considered (quantum, Newtonian and relativistic) we found scope for hypercomputation. However, there were often severe practical difficulties preventing the exploitation of hypercomputational features in the immediate future. Nonetheless, there seems little doubt that systems can and will be built (witness developments in quantum computation) that will surpass the capabilities of the standard Turing model of computation.

Given that the standard representation of quantum physics relies crucially upon the properties of complex numbers, which are uncomputable in general, we also asked to what extent quantum theory is, and to what extent it is not, computable. Answering this question requires us to state carefully which aspect of physics we are interested in: values or processes. The properties of physical systems would appear to be in some cases computable and in others uncomputable.

We argued that physics should be viewed as a study of *processes*, and that process-theortic semantics might therefore be worth investigating, rather than the more mathematical semantics offered by recent forays into category theory. We illustrated a standard representation of functions as processes, which is defined for data belonging to algebraic data types. We observed that our reprsentation cannot, and *should not* be complete, since this would break the no-cloning theorem. This suggests strongly that individual quanta cannot be represented by computable functional semantics, since these functions are themselves included in our representation.

Nonetheless, a polymorphic typing for the π-calculus may still offer scope for modelling certain quantum processes. In particular, any system that can be represented polymorphically, e.g. a list, can be regarded as computable regardless of the value-type upon which it operates. Moreover, any model in which interactions are mediated by a *finite* (or at least recursively enumerable) set of virtual exchange particles can be treated polymorphically by defining libraries of oracular methods. While measurable values may turn out to be nonrecursive, the methods used to manipulate them may nonetheless be computable.

5   Acknowledgements

We are grateful to Susan Stepney and an anonymous referee for comments that served to crystallise some of our early thinking in this matter, and to Frank Wilson and Simon Foster for several useful suggestions concerning types and polymorphism in the π-calculus.

References

[1] S. Abramsky. Information is Physical, but Physics is Logical, 9 May 2007. Research seminar, University of Sheffield, Department of Computer Science.

[2] S. Abramsky and B. Coecke. A Categorical Semantics of Quantum Protocols. In *Proceedings of the 19th Annual IEEE Symposium on Logic in Computer Science: LICS 2004*, pages 415–425. IEEE Computer Society, 2004.

[3] S. Abramsky and B. Coecke. Abstract Physical Traces. *Theory and Applications of Categories*, 14:111–124, 2005.

[4] J. Bell. On the Einstein-Podolsky-Rosen Paradox. *Physics*, 1:195–200, 1964.

[5] D. Dieks. Communication by EPR devices. *Physics Letters A*, 92(6):271–272, 1982.

[6] J. Earman and J. Norton. Infinite Pains: The Trouble with Supertasks. In A. Morton and S. Stich, editors, *Benacerraf and his Critics*, pages 231–261. Blackwell, Cambrdige, MA, 1996.

[7] R. Feynmann. *Lectures on Physics*, volume III. Addison Wesley, 1965.

[8] S. Gay and R. Nagarajan. Communicating quantum processes. In *POPL '05: Proceedings of the 32nd ACM SIGPLAN-SIGACT Symposium on Principles of Programming Languages*, pages 145–157, New York, NY, 2005. ACM Press.

[9] M. Green, J. Schwarz, and E. Witten. *Superstring theory*, volume 1, Introduction of *Cambridge Monographs on Mathematical Physics*. Cambridge University Press, Cambridge, 1987.

[10] M. Hogarth. Does General Relativity Allow an Observer to View an Eternity in a Finite Time. *Foundations of Physics Letters*, 5:73–81, 1992.

[11] M. Hogarth. Deciding Arithmetic using SAD Computers. *British Journal for the Philoophy of Science*, 55:681–691, 2004.

[12] S. Hawking and R. Penrose. *The Nature of Space and Time*. The Isaac Newton Institute Series of Lectures. Princeton University Press, Princeton, NJ, 1996.

[13] R. Milner. *Communication and Concurrency*. Prentice Hall International, London, 1989.

[14] R. Milner. Functions as processes. *Journal of Mathematical Structures in Computer Science*, 2(2):119–141, 1992.

[15] R. Milner. *Communicating and Mobile Systems: the π-calculus*. Cambridge University Press, Cambridge, UK, 1999.

[16] D. Morgenstein and V. Kreinovich. Which algorithms are feasible and which are not depends on the geometry of space-time. *Geocombinatorics*, 4(3):80–97, 1995.

[17] J. Myhill. A recursive function, defined on a compact interval and having a continuous derivative that is not recursive. Michigan Mathematical Journal, 18:97–98, 1971.

[18] M. Pour-El and J. Richards. The Wave Equation with Computable Initial Data such that its Unique Solution is not Computable. *Advances in Mathematics*, 39:215–239, 1981.

[19] R. Penrose and W. Rindler. *Spinors and Space Time*, volume 1, Two-spinor calculus and relativistic fields. Cambridge University Press, Cambridge, 1984.

[20] R. Penrose and W. Rindler. *Spinors and Space Time*, volume 2, Spinor and Twistor Methods in Space-Time Geometry. Cambridge University Press, Cambridge, 1984.

[21] P. Shor. Polynomial-Time Algorithms for Prime Factorization and Discrete Logarithms on a Quantum Computer. *SIAM Journal on Computing*, 26(5):1484–1509, 1997.

[22] D. Simon. On the power of quantum computation. In *Proc. 35th Annual Symp. on Foundations of Computer Science*, pages 124–134, Los Alamitos, CA, 1994. IEEE Computer Society Press.

[23] M. Stannett. An Introduction to post-Newtonian and non-Turing computation. Technical Report CS-91-02, University of Sheffield, Department of Computer Science, 1991. Re-issued 2000.

[24] M. Stannett. Hypercomputation is Experimentally Irrefutable. Technical Report CS-2001-04, University of Sheffield, Department of Computer Science, 2001.

[25] M. Stannett. Computation and Hypercomputation. *Minds and Machines*, 13(1):115–153, 2003.

[26] M. Stannett. Industrial Hypercomputation. In S. Stepney, editor, *The Grand Challenge in Non-Classical Computation International Workshop: 18–19th April 2005*. York University, 2005. Online: `http://www.cs.york.ac.uk/nature/workshop/papers.htm`.

[27] M. Stannett. The Case for Hypercomputation. *Applied Mathematics and Computation*, 178:8–24, 2006.

[28] D. Sangiorgi and D. Walker. *The π-calculus: A Theory of Mobile Processes*. Cambridge University Press, Cambridge, 2001.

[29] W. Wootters and W. Zurek. A Single Quantum Cannot be Cloned. *Nature*, 299:802–803, 1982.

[30] Z. Xia. The existence of noncollision singularities in the n-body problem. *Annals of Mathematics*, 135(3):411–468, 1992.

Collective perception of absolute brightness from relative contrast information — an emergent pattern formation approach

Jeff Jones and Mohammed Saeed

Department of Computer Science, University of Chester, Parkgate Road, Chester, U.K. CH1 4BJ
`jeff.jones@chester.ac.uk, m.saeed@chester.ac.uk`

Abstract The human visual system is able to effortlessly perceive absolute brightness in a scene even though the output of retinal processing is a relative spatial encoding of local edge contrast information. Neural filling-in processes are one suggested mechanism of reconstructing the original global surface brightness values from the local contrast edges but controversy continues as to whether such mechanisms are necessary, possible and biologically plausible. A low-level emergent pattern formation approach is described using a simple, reactive multi-agent system whose components perform a collective emergent perception of the scene. The population is able to generate absolute global brightness levels from local contrast based stimuli and can accurately perceive both real world imagery and classical illusory brightness phenomena. Recent brightness illusions that were introduced specifically to rule out low-level mechanisms of brightness perception can also be collectively perceived with the framework. The behaviour of the framework is analogised to the distortion of the background grey (eigengrau) by the diffusive flux of brightness and darkness information across brightness interfaces.

*K*eywords: Brightness perception, emergent pattern formation, illusory phenomena, multi-agent, collective perception

1 Introduction — brightness perception

The human visual system (HVS) is a remarkable adaptive system that is able to assess the lighting levels in different scenes where luminance can differ by nine orders of magnitude. In any given scene there are often areas under different illumination, complicated by shadows and areas of different reflectance and texture. The HVS is effortlessly able to parse the luminance and reflectance information so that we can make sense of the world, resulting in the perceived brightness of the environment. Brightness perception is but one of many tasks performed by the HVS, among them depth perception, motion perception, binocular vision, object tracking, visual attention, and the integration of vision with other senses. Although the HVS functions in both scotopic (dim light), photopic (daylight) and mesopic (where the two lighting environments overlap) conditions and is sensitive to colour frequencies (via different classes of cone photoreceptor and opponent process mechanisms), most research into brightness perception pertains to achromatic photopic brightness perception, i.e. daylight perception in the light to dark (white to black) range.

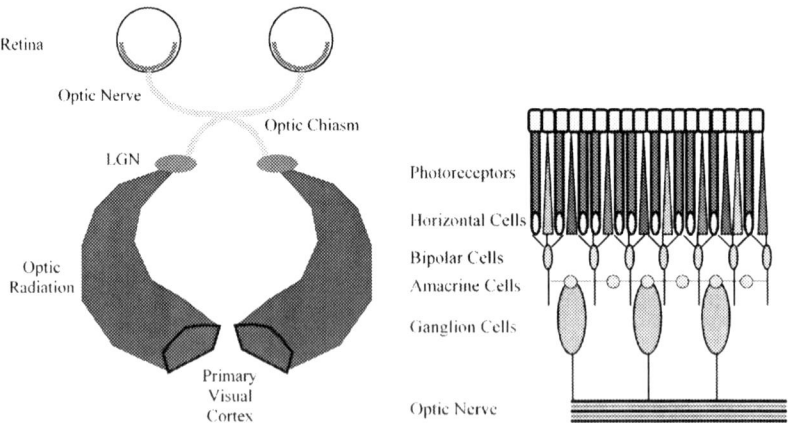

Figure 1. Architecture of the human visual system (left), retinal information processing (right)

The mechanisms of brightness perception start with the sensing and transduction of the light signal by retinal photoreceptors and progress through retinal processing, transport of signals via the optic nerve to the lateral geniculate nucleus (LGN), and on to the visual cortex (Fig. 1). The HVS system is thought to behave in a hierarchical manner — building upon low level processes such as local contrast (at the retinal / LGN level) and extracting higher level features at the visual cortex.

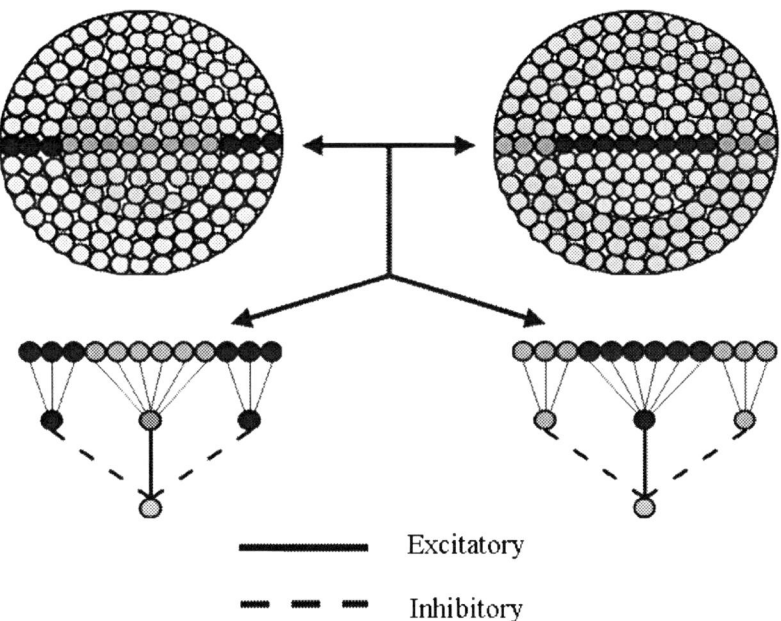

Figure 2. Schematic arrangement of the differing receptive field response

A central question in brightness perception, and the source of much historical and current debate, is the matter of how surface brightness is perceived. The sampling mechanisms in the early part of the HVS are the retinal photoreceptor neurons (approximately 100-130 million photoreceptors at the retina, most densely populated at the fovea, the most sensitive region of the retina), and the networks of neurons that the photoreceptor signals converge upon. The amount of light falling on the photoreceptors modulates their release of glutamate. The glutamate released from the photoreceptors modulates the behaviour of bipolar cells. Bipolar cells occur in two separate populations, either inhibited (ON type) by the photoreceptor response or excited (ON) by the response. Bipolar cell behaviour is further modulated by lateral connections from nearby horizontal and amacrine cells before converging on the ganglion cell neurons, whose action potentials are transmitted via the optic nerve to the later stages of the visual system (approximately 1.2 million axons in the optic nerve, signifying a large convergence from the photoreceptor inputs). The convergence of photoreceptor signals, the two types of bipolar call and their lateral connections generates receptive fields that affect an individual ganglion cell response. The most common types of receptive field are circular fields with an internal disk shape surrounded by an annulus surround. The two types of receptive field are shown in Fig. 2 and are described as on-centre/off-surround (ON fields) and off-centre/on-surround (OFF fields).

Ganglion cells with ON type receptive fields have their firing rate increased by light falling on their centre. Their firing rate is decreased by light falling on their surround. Off cells have their firing rate increased by light falling at their periphery and their rate decreased by light falling at their centre [1]. Neither type has their resting rate significantly changed by homogeneous input or by slow luminance gradient changes and are sensitive to luminance contrast within their receptive field [2]. The effect of lateral inhibitory wiring acts to amplify the contrast responses. Current models of retinal processing consider only local and relative changes in light intensity and the output from the retina may be considered as a spatial encoding of contrast responses (Fig. 3).

By having suitable large receptive fields it is possible to view centre-surround processing as a series of low-pass filters which may later be reconstructed (by additive weighting of separate scales) to provide a reference to absolute luminance values [3], [4]. An example of the multi-scale approach is the brightness perception model used by Blakeslee and McCourt that is able to predict a wide variety of illusory stimuli [5]. Blakeslee and McCourt use an oriented version of the difference of Gaussian (DOG) kernel to approximate the behaviour of the receptive

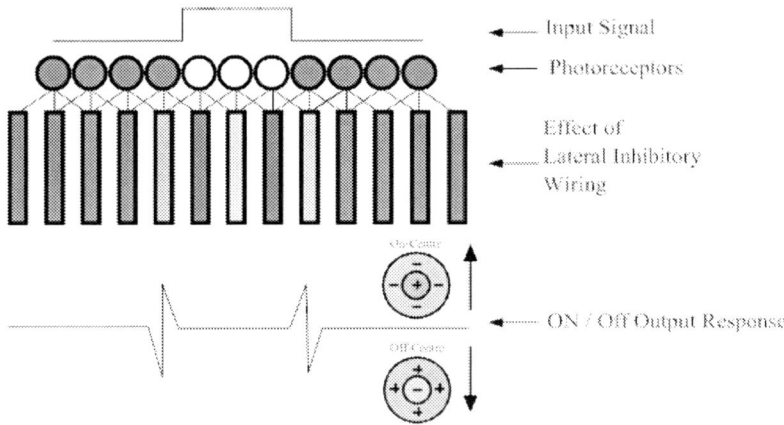

Figure 3. Retinal output as a localised spatial contrast response via lateral inhibition and ON/OFF channels. ON response shown as upward signal and OFF response shown as downwards signal

fields converging on the ganglion cells. There is some question as to whether is actually possible to encode at least some reference to the absolute luminance in the retinal encoding. Putative specialised cells that respond to luminance levels include the biplexiform cells in the retina [6] and the luminance sensitive ganglion cells [7] although the functions of these cells may merely be to affect the pupillary reflex and circadian rhythm rather than high level brightness perception. There is some evidence that local absolute luminance information may be able to modulate the contrast responses of centre-surround receptive fields (via an additional external surround), enabling the simultaneous encoding of luminance and contrast information [8, 9].

Recovering Absolute Brightness from Local, Relative Contrast Information

The response of retinal behaviour to light stimulus from the environment results in a local spatial encoding of relative contrast generated at the boundaries of edge and contour discontinuities. An illustration of what retinal output might globally 'look like' is shown in Fig. 4. The grey background represents the Eigengrau ('intrinsic grey' — representative of the resting firing rate of both ON and OFF fields [10]) which is distorted upwards and downwards by the ON and OFF contrast signals respectively.

Figure 4. Absolute surface brightness (left) encoded as by early visual process-ing as relative spatial contrast (right). ON channel contrast is represented by lighter signals above the eigengrau background grey and OFF channel contrast represented by darker signals than the background grey.

In real-life perception we obviously consciously perceive scenes as they are on the left side of Fig. 4. If the retina records only relative contrast information then the absolute surface brightness percepts must be inferred, recovered, or reconstructed from the relative contrast cues at later stages of the visual system. Isomorphic filling-in theories of sur-face brightness perception have been suggested to account for the special cases of filling-in of information at the location of retina blind spot cre-ated by the passage of the optic nerve from the retina [11], and to account for recovering missing visual details from patients with retinal scotomas [12]. As well as being mechanisms to cope with special cases, evidence for isomorphic filling-in behaviour for surface brightness perception has been seen in both psychophysical and neurological experiments [13], [14].

Gerrits and Vendrik [15] presented a theoretical model of filling-in where the separate ON and OFF retinal outputs were used to laterally spread the contrast information by diffusion and distort the eigengrau upwards (ON channel filling-in) and downwards (OFF channel filling-in). Grossberg, along with many later contributions, ([16],[17],[18]) presented a neural mechanism to instantiate the filling in hypothesis. Grossberg's system (part of an ongoing and much larger framework that considers many wider aspects of visual perception) is known as the BCS/FCS (the Boundary Contour System and Feature Contour System). The BCS/FCS is a hierarchical system comprising six layers and a simplified version that considers its use for achromatic brightness perception and illusory perception is fully detailed in [19]. The BCS/FCS system works by generating a map of contrast boundaries within the image using mechanisms analogous to the behaviour of retinal ganglion cells with receptive fields. The orientations of the stimuli are in turn recognised by mechanisms approximating the behaviour of 'complex 'cells in the visual cortex. The output from the BCS stage (separate retinoptic maps of light and dark boundaries) is passed to the FCS stage where the contrast signals diffuse from the boundaries to generate the absolute brightness percept. The boundaries themselves act to gate the flow of diffusion (isotropic diffusion based upon the heat equation). Grossberg suggested a syncytium map of cortical neurons to generate the speed of propagation necessary for real-time. The BCS/FCS mechanism explained many illusory brightness percepts and has become known as the standard filling-in model. Hong and Grossberg have presented a modified version of standard model that uses a multi-scale approach to signal propagation that reportedly sped up the diffusion speed by 1000 times [4]. One important feature of the standard filling-in model is that the boundary completion stage must be completed before the filling-in process can start, suggesting that a strict hierarchy of activity occurs in the visual cortex.

Variations on the standard model have been presented that attempt to overcome some of the model's weaknesses on certain stimuli. Arrington suggested a Directional Filling-In (DFI) addition for both lightness and darkness filling-in that overcame the BCS/FCS limitations on perceiving transitive stimuli (images with incremental plateaus of luminance) but the DFI addition has, as yet, only been presented for one dimensional data [20]. Confidence based filling-in [21] was suggested as a means of overcoming the 'bowing' of surface brightness seen in the middle of uniformly lit shapes (caused by greater activity at the borders than at the centre of the object). Keil suggested modifications to filling-in, including a non-linear diffusion operator and different methods of representing the boundary structures that overcame limitations of previous filling-in models (including the problems of fogging, blurring and activity trapping in the final percept) [22]. Keil's method included a means of multiplexing the retinal code (using a surround external to the receptive field) to include information about local luminance in the retinal output [9].

The mechanism in this report presents a method of absolute brightness perception from local contrast cues by means of emergent pattern formation. The mechanism works via a mass behaviour effect that utilises diffusion of a population of agents in their environment to affect population changes, resulting in differences in the activity in the landscape and, ultimately, differences in the brightness of the emergent percept. The framework is able to operate using a single sensory scale, using only relative contrast information and with an extremely simple behavioural algorithm for the agents' sensory, cognitive and motor behaviours.

2   Multi-agent systems and the PixieDust framework

PixieDust is a multi-agent system framework developed by the authors to explore a collective approach to image enhancement, image perception, and related problems where both the problem definition and its solution may be represented in terms of spatial patterns (such as the simulation of chemotaxis in diffusive environments for path planning). Previously Ramos has exploited the behaviour of ant-based systems that deposit a pheromone trail as the agents move and use the strength of the trail to affect the movement direction [23] forming a collective perception of the environment. Although at first glance the PixieDust framework is similar to Ant Colony System inspired approaches, the PixieDust framework actually attempts to use an even simpler approach to perception, inspired by the behaviour of very simple physical systems that collectively behave in a 'particle-like' rather than an 'entity-like' manner. Smolka used a particle-like model of image enhancement using a random walk based on Markov processes, recording the output as the frequency of visiting particular pixel sites [24]. Toffoli also suggested the idea of using massive iteration with fine grained systems ('programmable matter', cellular automata in his examples) to discover the patterns within complex data sets, giving the analogy of a phase locked loop for texture recognition [25]. The PixieDust framework attempts to invert the classical, 'program-centric', model of computation where the data is treated as fodder to be processed by a complex program. In The PixieDust framework the data is treated as a rich potential field of information that is used to influence the behaviour of thousands of very simple processors. The hidden patterns in the data are visualised by the emergent historical patterns formed by the collective behaviours of the agents [26]. Physical mechanisms of data discovery (where the content of one hidden system affects the behaviour of another homogeneous system) are commonly used in the physical sciences - DNA electrophoresis, X-ray imaging, fingerprint discovery, cell staining - all are examples where previously hidden data is visualised by the interactions of the data with a homogeneous medium. The system comprises of a topographic landscape consisting of a digitized image (binary, greyscale or colour, though only greyscale images are considered in this report). Each landscape site corresponds to a pixel location and the height of each site is dependent on the pixel brightness: for an 8-bit representation zero represents the lowest landscape level and

255 represents the highest. A population of simple mobile agents resides on the landscape. The base agent class provides basic functionality and is extensible so that agents with different behaviours may be specified. This report considers an equally distributed mixture of two types of agent that are either light-attracted or dark-attracted. Only one agent may exist on any particular landscape site. The agents are oriented in a particular direction (for example 8-way, compass point notation) and reactive in their actions - sensing their environment and modifying their behaviour (specifically orientation) in response to the local landscape configuration, usually a 3x3 window centred about the agent's location.

Each agent operates in a sensory-cognitive-motor manner: The local environment is sampled and the configuration affects the agent's choice of direction. If the agent receives sufficient stimulation from the local environment, the agent changes direction (for example, orienting the agent by rotating to face the stimulus) and, in the motor stage, attempts to move forwards in its current direction. The agent will attempt to move forwards at every system step and if the forward movement is successful a unit of 'trail' (specifiable by the parameter depT) will be deposited onto (light-attracted agents), or excavated from (dark-attracted agents) a separate trail map structure whose geometry corresponds to the image landscape structure. If the agent is affected by a stimulus above a certain threshold the agent will become 'activated' and deposit a unit of 'mark' (depM parameter) on another separate map that records mark stimuli points if the next step is successful, the mark stimulus map also corresponding to the geometry of the image landscape. The framework operates in a massively iterative manner where simple innate behaviours generate mass behaviours and global emergent patterns of activity, and the population stabilises onto a dynamic equilibrium of activity. The dynamic nature of the equilibrium is influenced by the forward biased nature of the agent sampling of the environment and the fact that, at every step, each agent attempts to move forwards. The structure and scheduling of the PixieDust framework is shown in Fig. 5.

The agent population is initialised on the landscape with random locations and random initial directions, facing in one of 8 possible directions, relative to the regular image pixel structure. The structure of an individual light-attracted agent with respect to the landscape is shown in Fig. 6 with the simple pseudocode governing the agent's sensory, cognitive and motor behaviour. The dark-attracted agent is structurally equivalent to the light-attracted type with the difference that the dark agent will attempt to orient itself towards darker pixels.

The size of the agent population may be specified with the parameter %p which specifies the population size as a percentage of image landscape

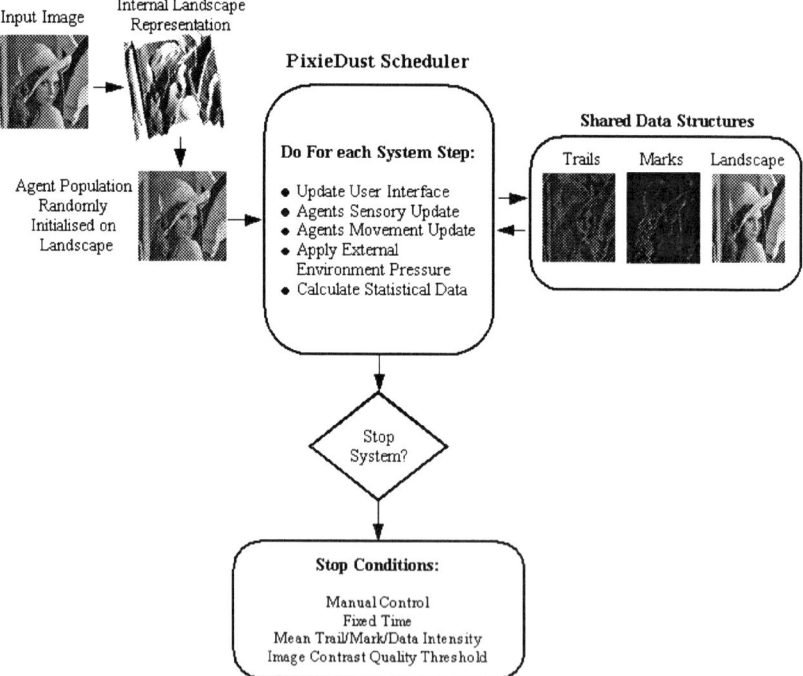

Figure 5. Structures and scheduling in the PixieDust multi-agent framework.

```
// Results from previous movement attempt
        If (activated AND moved_forwards_successfully)
                LEAVE A STIMULUS MARK IN CURRENT LOCATION
// sensory behaviour - sample the landscape environment
ACTIVATED = FALSE
getDirectionalLocalNeighbourhoodData()
// cognitive behaviour
Stage 1:
If (F>(FL+sMin)) AND (F>(FR+sMin))
        {
        ACTIVATED = TRUE
        Return
        }
Stage 2:
If (FL>(F+sMin)) AND (FR>(F+sMin))
        {
        Randomly turn
        45 degrees left
        Or 45 degrees right
        ACTIVATED = TRUE
        Return
        }
Stage 3:
If (FR > (FL+sMin))
        {
        Turn right 45°
        ACTIVATED = TRUE
        Return
        }
Stage 4:
If (FL > (FR+sMin))
        {
        Turn left 45°
        ACTIVATED = TRUE
        Return
        }
```

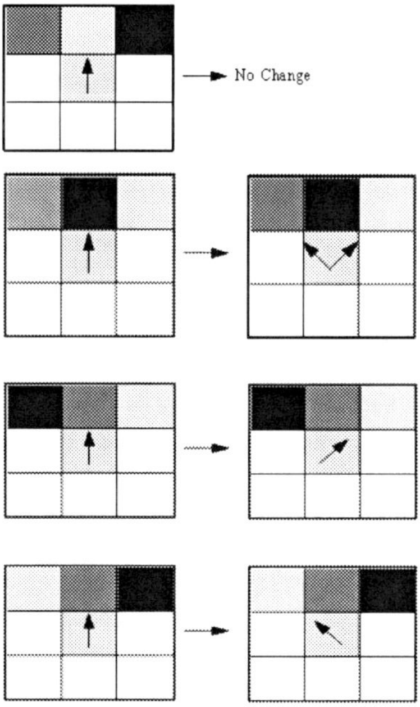

Figure 6. Structure of agent and simple light-attracted agent cognitive behaviour. Arrowed cell represents agent's current position and direction.

areas. Smaller values of %P will result in less agent-agent collisions and greater convergence into local landscape maxima and minima. Higher agent populations will result in the greater probability of agent-agent collisions and at levels of approximately 50% and greater, the emergent trail pattern will gradually be the inverse in brightness of the original landscape values due to high collision rates. The relative comparison between the front, front-left and front-right neighbouring cells can be modified with a sensitivity parameter, sMin, so that the agent will only be activated by a stimulus difference greater than sMin. The agents may also be subject to random influences on their direction (pCD parameter) and their position on the landscape (pTP parameter). The pCD specifies a probability that the agent will randomly change direction. The probability is sampled at every movement step. The pTP parameter specifies a probability that the agent will randomly jump ('teleport') to a randomly chosen, currently unoccupied location on the landscape. The effect of increasing the pCD parameter is to reduce straight line artifacts on the emergent trail patterns that are most noticeable with small population sizes. The effect of the pTP parameter is to reduce the population's convergence into local maxima (light-attracted) or local minima (dark-attracted) areas on the image landscape. The pTP parameter also affects the length of each agent's pseudo random walk on the landscape (an analysis of the framework's main parameters are given in [26]). The complete sensory, cognitive and motor step for all the members of the population is designated as a single system step. The system is massively iterative and as the system evolves, emergent trail and mark patterns are generated on the respective maps as a consequence of the combined simple behaviours of the individual agents in the population. At every system step the image data, trail map and mark map may also be subjected to a selection pressure such as simple erosion or diffusion that acts to reduce the historical length of time of pattern persistence.

The emergent trail pattern levels are interpreted as pixel brightness values and the trail pattern acts as a collective perception of the original image landscape stimulus. Examples of collective perception on conventional imagery are shown in Fig. 7.

It is important to note that the emergent trail pattern shown in Fig. 7c is not a processed version of the original image on the left. The trail pattern is simply an emergent historical record of agent movement across the image and was originally a homogeneous landscape initialised at the intensity of 127 (halfway in brightness for an 8-bit greyscale image). The homogeneous trail landscapes were deformed by the activity of accretive light-attracted and excavating dark-attracted agents respectively. The trail patterns represent an accurate collective perception of

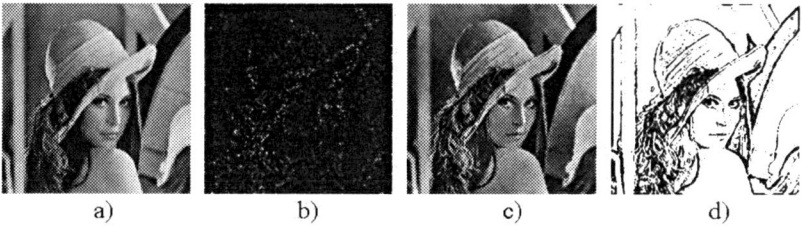

Figure 7. Collective perception of the Lena image by emergent pattern formation. a) Original Lena image, b) Agent distribution after 2000 system steps, c) Emergent trail pattern, d) Emergent mark stimulus pattern.

the original images by agents that conform to the behaviours seen in emergent systems: simplicity, locality and parallelism. Furthermore, the absolute brightness values in the emergent trail patterns arose from the consideration of only relative local contrast cues (see agent pseudocode).

The resulting trails, although perceiving the original image, appear somewhat artificial and poster-like because all the image features below the sensitivity threshold were 'missed' by the agents' sensory/cognitive behaviour and therefore are not perceived in the emergent trail patterns. By randomly modulating the sensitivity of each agent at each step (for example by randomly selecting the sMin parameter from the range of 0-60) a much more accurate and 'realistic' percept is generated (Fig. 8).

The random modulation of agent sensitivity acts to consider the relative strength of the contrast border in the attraction and corralling of agents — with the result that the agents are more likely to be affected by borders with a high contrast. The result effectively alters the 'permeability' of the border (brightness interface) to agent movement. By including low thresholds in the random modulation of the contrast sensitivity, the effects of small contrast changes (often providing the more subtle features of perception) are considered. By setting a relatively high maximum value for the highest possible random sensitivity the agents are less likely to become trapped into local maxima/minima, thus reducing the need to set the pTP and pCD parameters to high levels.

2.1 Mechanism of collective perception — Absolute brightness perception via mass flux across brightness interfaces

The mechanism of collective perception by the emergent trail patterns of the PixieDust framework is that of agent flux across the interface between light and dark areas. The behaviour of the light-attracted agent

Figure 8. Random modulation of agent sensitivity results in more accurate collective perception. Left column: original image. Middle column: collective perception with fixed sensitivity. Right column: collective perception with randomly modulated sensitivity.

will be considered since the dark-attracted agent is functionally equivalent, but opposite. Agent deposit a uniform amount of trail whenever the agent successfully moves to a new cell so the difference in intensity in the trail pattern must be caused by a population disparity in different areas of the image. Since the agent population is initially randomly distributed the population disparity arises due to stimulation from contrast changes above a certain threshold. These locations of stimuli can be seen on the mark map in Fig 7d which is, in effect, a recording of the locations of stimulus points deposited by the agents. It is at these points where the landscape influences the agent to change direction in favour of directions that are locally greater in intensity. Over time, the agents will be drawn into the areas of local maxima - increasing the population density in these areas and thus increasing the levels of trail being deposited in these areas. Conversely, the relative de-population in darker areas will reduce the trail accreted in darker areas. Furthermore, once agents become attracted to lighter areas they are statistically less likely to leave such areas because their innate behaviour is to be attracted towards relatively lighter areas. A small proportion of agents will, however, be able to leave the lighter area due to random direction changes caused by the pCD parameter or agent-agent collisions and also due to the fact that the particular alignment of landscape features in a particular agent orientation may not provide the agent with sufficient stimuli to keep the agent within the local area.

The attraction of light-attracted agents into areas of local maxima at stimulus points that correspond to large differences in contrast may be described as flux across a brightness interface. Enclosed brighter regions (such as the vertical bar at the left of the Lena image) will tend to 'corral' agents within the brighter regions. The opposite behaviour occurs for dark-attracted agents: dark agents are too influenced by large differences in contrast and change their orientation to face towards areas of local minima. Dark agents, with their innate 'preference' for darker areas of the landscape will tend to remain confined to such areas. The excavation of trail by darker agents serves to amplify the distinction between the accretion in the light areas and the absolute brightness levels of the trail pattern are further stretched from their initial homogeneous levels.

The mass flux across brightness interfaces explains how the agent population can perceive absolute brightness using only relative contrast cues. The behaviour of the population also explains how long range effects may emerge via only local stimuli. It is the long range (temporal as well as spatial) patterns of behaviour that enables the generation of trail patterns that show accurate absolute brightness or the original image.

2.2   Mass behaviour effects — from isotropy to anisotropy

The macroscopic behaviour of the flux of agents across brightness interfaces can be more simply understood if the brightness interfaces are completely removed, i.e. a landscape that is completely homogeneous (topographically flat). For simplicity, only a single agent type will be considered — the light-attracted agent. The flat landscape presents no stimuli to the population and the agents are behaving in a pseudo random walk. The length of the walk (in terms of pixels covered) if influenced by the pTP parameter. The pCD parameter affects the straightness of the walk (higher values becoming the so-called drunken walk) and the distance (in terms of the Euclidean distance between origin and destination). If neither the pTP or pCD parameter are set, the only deviations from straight walks will occur when agent/agent collisions occur (an agent tries to move to a site occupied by another agent). The default behaviour for an agent/agent collision is a random change in direction. As each agent is given an initially random site on the landscape and an initially random direction, the population movement tends to exhibit isotropic diffusion if no brightness interfaces are present, resulting in an even accretion of trail in the landscape. The emergent trail patterns on homogeneous (flat) landscapes have the lowest variance at population levels of approximately 45-50% of the landscape area.

When the landscape contains fluctuations in height (brightness), stimuli will be presented to the agent population. Since the light-attracted agent is attracted towards areas of local maxima, the agents will orient themselves towards the brightness interface when the interface is within range of the agent's sensory apparatus. The subsequent movement across the interface to the locally brighter area, when exhibited on a macroscopic scale, represents a shift in global population movement from isotropy to anisotropy. The population differences caused by agent migration across brightness interfaces results in differences in trail deposition at different areas of the landscape. This ultimately results in brightness differences in the collective perception of the landscape. When agents are not directly stimulated by their local landscape, for example when on a plateau, the local agent movement becomes (statistically, and over time) isotropic. As the anisotropic movement is initiated at the brightness interface, the different positions of brightness and darkness (i.e. which side of the interface) plays a critical role in the net direction of agent flux. Fig. 9 shows the result of flux in different directions across brightness interfaces.

The direction of the population flux is dependent on the orientation of the interface. When the lighter area is inside the darker area, the net population behaviour is a contracting movement of the population. This

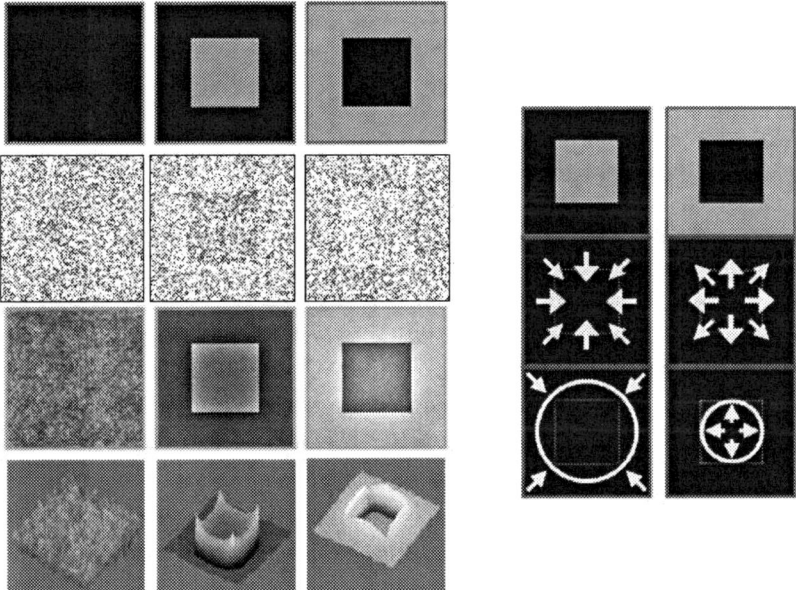

Figure 9. Isotropy, anisotropy and flux direction in light-attracted agents on homogeneous and varying landscapes. Left (top to bottom): Original images, agent distributions, emergent trail pattern, 3d plot of emergent trail pattern. Inset Right: Original images (top), direction of diffusion (middle), mass diffusion as a contractile or expansive phenomenon.

can be visualised as a circle contracting around the square border with the result that the corners of the square are more likely to be encountered by the population before the sides. This results in a stronger perception of the corners (as shown in the 3d plot) than the sides. When the square is bordered by a lighter surround, the net population behaviour is a movement outwards across the interface. This can be likened to a circular expansion of agent movement with the likelihood that the sides of the square will be encountered first. This too is borne out by the resulting trail pattern and 3d plot.

It should be noted that the circular contraction/expansion of agent flux is not innate to the agent behaviour, it is a macroscopic behaviour that emerges statistically over time - individual agents are only stimulated at the brightness interface and exhibit semi-random isotropic behaviour when not stimulated. The anisotropic movement only occurs at the brightness interface. The transport of agents across the interface, however, leaves a relatively depopulated space into which other (isotropically moving) agents are more likely to move into (the depopulated areas will statistically result in less agent-agent collisions). The net result is macroscopic movement, akin to the physical phenomenon of induced flow when pressure is reduced. The direction of the macroscopic movement (expansive or contractile) is governed by the landscape configuration.

2.3   Agent mass behaviour and the similarity to retinal processing

The behaviour of light-attracted agents and dark-attracted agents may be analogised to the behaviour of on-centre/off-surround and off-centre/on-surround receptive fields in the retina and LGN. The behaviours of both agent types in response to simple landscape stimuli show the characteristic overshoot/undershoot response at brightness interfaces in the emergent trail patterns and mark stimulus patterns (Fig. 10). These patterns were first noted by Hartline and Ratliff in their experiments on the encoding of luminance information in Limulus [27].

The trail patterns from both light and dark agents show the characteristic overshoot and undershoot response to boundary stimuli seen in lateral inhibition in early retinal processing. The mark stimulus point patterns show the location of the stimulus points at the brightness interfaces. When both light and dark agents are present on the image landscape the stimulus point cross-section profile is that of the chart at the bottom centre of Fig 10. This chart indicates that the positions of light and dark stimuli are recorded at different points — the light stimulus is on the 'inside' and the dark stimulus on the 'outside' of the interface.

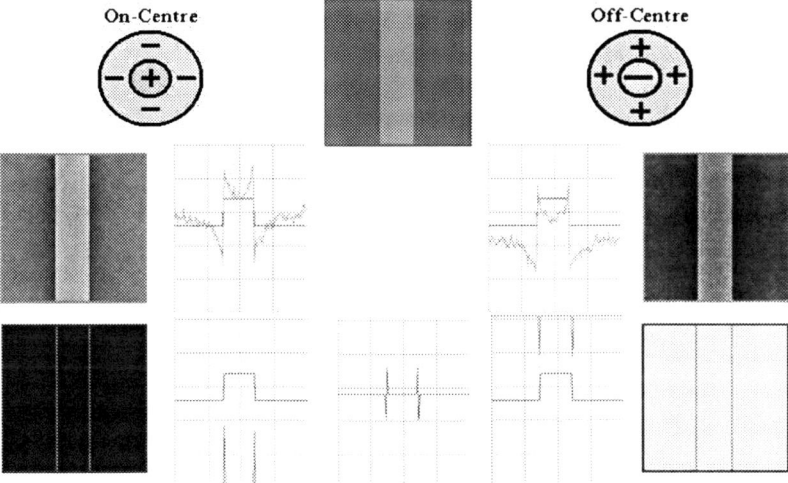

Figure 10. Analogous behaviour of light and dark-attracted agents to receptive field processing in the early mammalian visual system. Simple stimulus image (top centre) and the emergent trail (upper) and mark (lower) patterns formed by the responses of both light-attracted accretion based agents (left) and dark-attracted excavation based agents (right).

The mark map shows the boundaries in the image, whilst the trail map shows the 'filling' in-between the boundaries, i.e. a facsimile of the surface brightness of the original image. There is a correspondence between the mark map and the lateral inhibitory output of retinal processing, and also a correspondence between the trail map and post-retinal processing where surface brightness is thought to be recovered. There is also a similarity between the mark and trail maps and the outputs of Grossbergs's BCS/FCS model of brightness perception. The stimulus map appears to be similar to the boundary contour map (BC stage) and the trail pattern is similar to the final syncytium output of the Grossberg model. Unlike the BCS/FCS however, there is no distinct hierarchy in the PixieDust output — The BCS/FCS specifically requires that the boundary compartments be completed before the filling-in process can occur whereas, with the PixieDust framework, the filling-in process and boundary generation are simultaneous (in fact the visualization of the stimulus mark map was a secondary development in order to see the location of the stimulus points that guided the emergent trail pattern formation).

3   Collective perception of achromatic brightness illusions

Illusory brightness perception occurs when the HVS perceives the scene in such a way that the perceived brightness of parts of the scene differ from the actual photometric brightness. This apparent 'fooling' of the visual system may provide insights into how the HVS parses scenes under uneven, uncertain or changing lighting conditions. To attain the maximum impact, brightness illusions are often carefully constructed geometric figures with specific target patches of interest. Although the target patches are often the focus of measurement (and often the focus of the surprising nature of the percept) it should be noted that illusory perception can affect the entire image under scrutiny. This section presents the results of collective perception of brightness illusion images by the PixieDust framework.

3.1   Simple brightness illusions

Simple brightness illusions include simultaneous brightness contrast effects (Fig. 11a), where the identically bright target patches appear to differ in brightness when placed upon different backgrounds. The grey target on a dark background appears brighter than the grey patch on the lighter background. Grating induction (Fig. 11b) occurs when a sinusoidal gradient is overlaid with horizontal bar of uniform brightness. The gradient waveform appears to 'induce' a perceived brightness in the evenly lit bar that is opposite in phase to the main waveform. The mach band effect (Fig. 11c) appears at the start and end of a luminance ramp, the effect manifested as apparent vertical lines at the start and end of the luminance ramp. The Chevreul staircase illusion (Fig. 11d)) consists of a luminance staircase consisting of panels of increasing brightness, each panel uniformly bright. The illusory effect is a 'scalloping' of the edges where the panels join — an increase in brightness is seen at the left of a panel when it joins a darker panel and a corresponding decrease in darkness is also seen in the panel at the right side where the panel adjoins its brighter neighbour. The pyramid illusion features a nested series of squares, increasing in brightness from the outside inwards (Fig. 11e). There appear to be diagonal radiating lines of increased brightness at the corners of the steps of the pyramid even though the pyramid steps are uniform in brightness. The Hermann-Hering grid illusion (11f) shows a regular series of black square 'blocks' separated by white 'streets'. When casting the eyes over the image, there appear to be dark spots at the intersections of the streets even though the streets are of a uniform brightness. When fixating on a particular intersection, however, the illusory spot disappears.

Simple brightness illusions are commonly explained by the activity of low-level processes such as lateral inhibition via receptive field activity, for example in the Baumgartner model of the Hermann-Hering grid illusion, see [28]). The lateral inhibition explanation states that the target areas in brightness illusions differ in perceived brightness because the surrounds of the receptive fields receive different amounts of inhibition at different parts of the image. Fig. 12 shows the receptive field explanations for both the pyramid illusion and the Hermann-Hering grid illusion.

The ganglion cell whose receptive field centre is located at the corner of the pyramid step receives less inhibition from its surround (because most of its surround is in a darker area) than a ganglion cell whose receptive field centre is located on the side of a step. The reduced amount of lateral inhibition at the corner step is, according to the receptive field hypothesis, responsible for the illusory percept of the bright diagonal lines. The lateral inhibition explanation for the Hermann-Hering grid illusion states that the ganglion cell whose receptive field centre is in the street intersection (left), receives more inhibition from its surround than the ganglion cell whose receptive field centre lies in the middle of the street (right). The increased inhibition of the cell at the centre of the

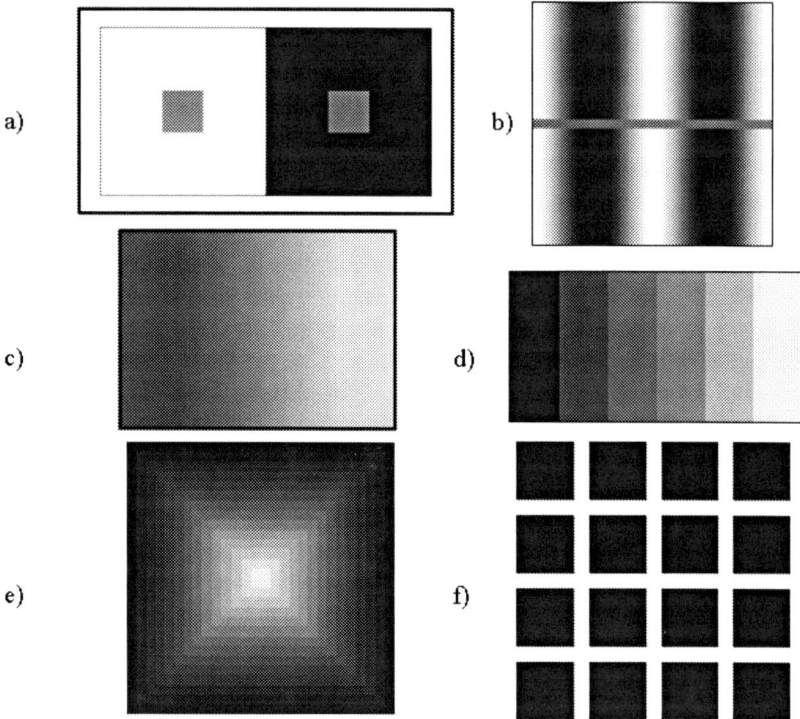

Figure 11. Simple brightness illusions.

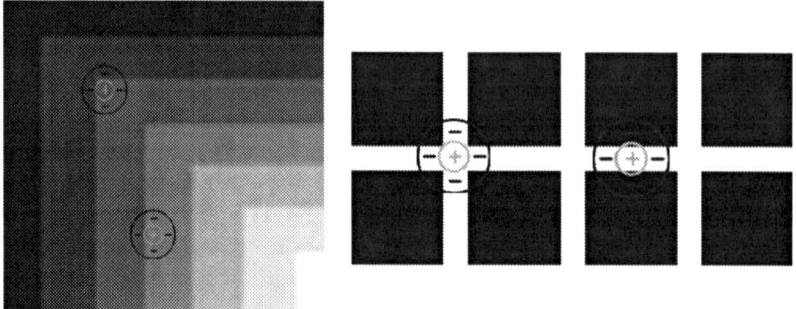

Figure 12. Lateral inhibition explanations of the pyramid illusion (left) and the Hermann-Hering grid illusion (right).

intersection results in the illusory darkness percept at the intersection. The explanation for the removal of the darker patch when the viewer fixates on a particular intersection is that the size of the receptive fields are smaller at the fovea and so there is no difference in the amount of inhibition at the intersections and the street.

The receptive field explanation for the illusory percepts only explains why the contrasts at the edges of different surfaces may appear different and provides no explanation of how absolute surface brightness is generated, the implication being — perhaps — that the luminance information is somehow passed to the visual cortex. The results showing collective perception of simple brightness illusions by the PixieDust framework are shown in Fig. 13. The results show the original image, the emergent trail patterns, and cross-section intensity plots showing cross section data plots of the original data (smooth lines) and the emergent trail patterns (more noisy lines). The cross section plots are included because of the possibility that the agent population could simply be accurately perceiving the original image, and that the observer would simply be re-experiencing the illusion when viewing the emergent trail patterns.

The results in Fig. 13 show that not only does the population collectively perceive the original image, but that it perceives the illusory brightness percept in a similar manner to the HVS. The activity of the agents — specifically their flux across brightness interfaces — is responsible for the emergent percept and a schematic example is given in Fig. 14 using the Hermann-Hering grid illusion. Note that only the behaviour of the light-attracted agents is shown for clarity.

The macroscopic behaviour in a single grid block is an expansive outward flux, initiated at the interface, which results in the sides of the block being perceived as brighter than the corners (see 3d plot of single block). The combined effect of the regularly arranged city blocks is a concentration of agent movement from the blocks into the sides of the streets. This greater concentration of agents at the streets, rather than at the intersections, results in greater accretion of trail at the street sides, and the perception of darker intersections (see 3d plot of intersection). Unlike the conventional lateral inhibition account of the Hermann-Hering grid illusion, the agent population perceives the illusion not as an active darkening of the intersections but rather as a consequence of the enhanced brightness of the streets.

If the agent flux across brightness interfaces is responsible for the illusory perception in the PixieDust framework, what exactly do the agents themselves represent? The agents cannot represent neurons - the agents are mobile and neurons are fixed networks. It may be more accurate to say that the agents are analogous to the signals passed from neuron

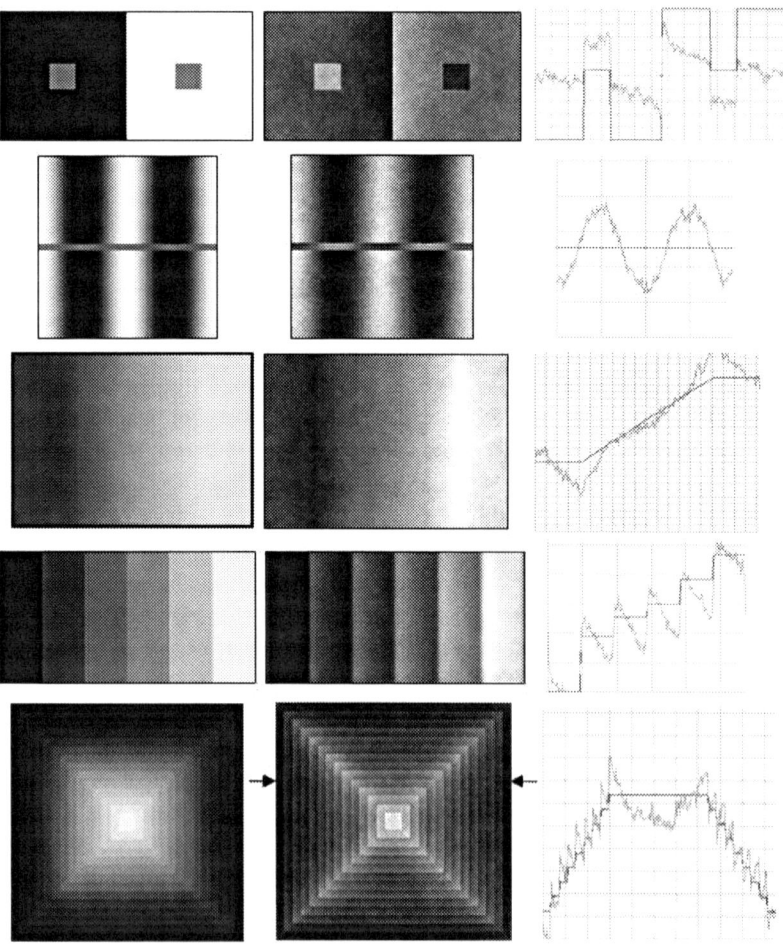

Figure 13. Collective perception of simple brightness illusions. Left: Original image. Middle: Emergent trail pattern. Right: Cross section data plot — Original data (smooth) and emergent trail (noisy). Cross section data is horizontal plot across screen sampled vertically halfway down image except where arrowed.

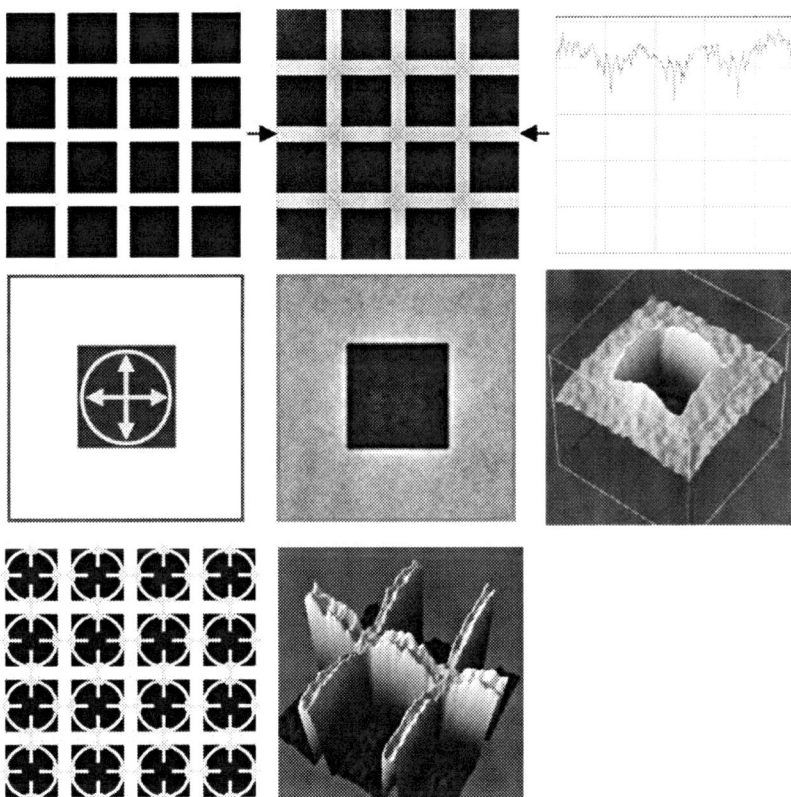

Figure 14. Collective perception of the Hermann-Hering grid illusion by the PixieDust population via population flux across brightness interfaces. Top row: Original image, emergent trail pattern, cross section plot through middle street. Middle row: Expansive diffusion across brightness interface by light-attracted agents, trail pattern and 3d representation. Middle row: combined pattern of expansive diffusion for entire Hermann-Hering grid and 3d plot of two intersections.

to neuron. The light-attracted agents represent the flow of brightness contrast information and the dark-attracted agents represent the flow of darkness contrast information. (The notion of darkness flow is not as unusual as it may first appear. The separation of ON and OFF fields in the retina and up to the LGN is well known and psychophysical evidence has shown the existence of darkness flow [29]). At the local level (at the point of stimulus) the agents resemble the behaviour of simple lateral inhibition and indeed exhibit the corresponding overshoot and undershoot characteristics in the stimulus patterns. The local movement of the agents results, over time, in distortions in the population density (and the resulting trail deposition / excavation patterns) and it is the summation of this activity that generates the global absolute brightness percept. That is not to suggest, of course, that this is the mechanism at work in the HVS but the global behaviour of the agent population does at least suggest possible mechanisms by which low-level local activity based upon the propagation of activity from relative cues can result in global absolute perception.

3.2   Perception of complex brightness illusions

In the early 1990s Adelson introduced a new wave of brightness illusions that exhibited effects much stronger than simple simultaneous brightness contrast [30]. These illusions were carefully crafted so as to rule out low level explanations. Examples of these complex constructions are the 'snake' illusions (Fig. 15). The snake comes in two variants: the strong snake, and the weak snake. The illusions consist of two rows of equally bright target diamonds set in complex backgrounds. The diamonds on the top row of the strong snake appear to be much brighter than the diamonds on the bottom row. The weak snake is a slightly rearranged version of the strong snake but the target diamonds and their relative local backgrounds are identical to those in the strong version. However for the weak snake the illusory brightness difference is much reduced. The identical backgrounds of the strong and weak version, according to Adelson, ruled out a low-level explanation. Suggested mechanisms proposed for these and later complex brightness illusions invoke higher-level unconscious inferences such as apparent transparency effects, the effect of particular junction types, the influence of grouping by similarity, and three dimensional cues.

The agent population collective perception also successfully perceived the difference in strength between the two versions of the illusion (Fig. 16). The population perceived that mean intensity values for the target diamonds as shown in Table 1.

Figure 15. The snake illusions. Left: The strong snake, Right: The weak snake.

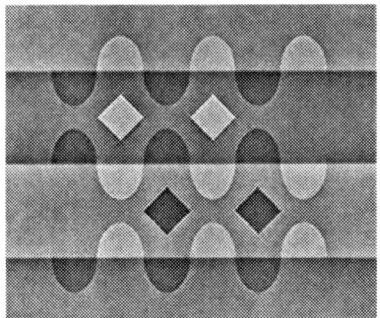 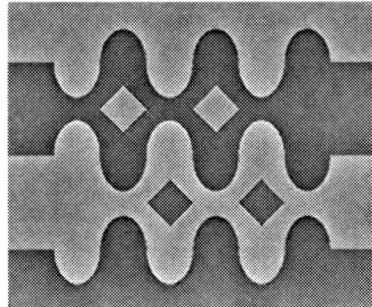

Figure 16. Collective perception of the strong and weak snake illusions - emergent trail patterns. The population correctly perceives the strong snake (left) as having a greater difference in the target diamond brightness levels than the weak version (right).

	Strong Snake	Weak Snake
Top Diamonds	168	159.5
Bottom Diamonds	82	91
Difference Between Top and Bottom Rows	86 pixel units	68.5 pixel units
Difference in illusion strength	17.5 pixel units	

Table 1. Difference in agent population perception of the snake illusions in mean target diamond brightness levels.

Why does the agent population perceive a difference in the two snakes when their targets and local surrounds are the same? The distribution of the agents on the image landscape (Fig. 17), at the end of the experiment, is telling (for clarity, only light-attracted agents are shown).

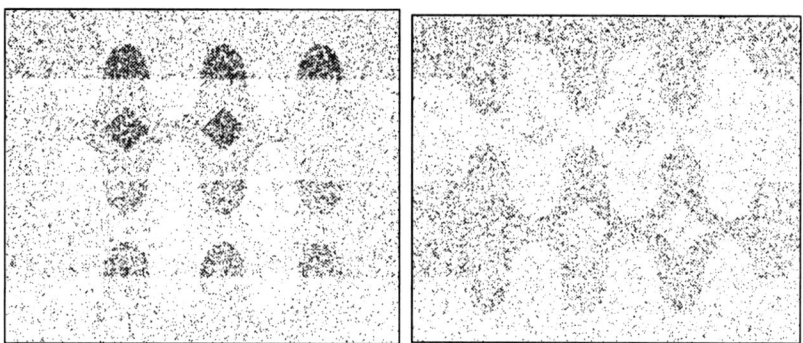

Figure 17. Light-attracted agent distribution at the end of the snake experiments. Flecks represent individual agents. Strong snake (left) and weak snake (right).

Figure 18 shows that despite having identical surrounds around the targets, the strong and weak snake landscapes end up with very different distribution of agents. The difference is caused by the mobility of the agents across brightness interfaces. A closer examination of the local, and non-local, surrounds explains why this is so.

The top target in the strong snake is brighter than its local surround and brighter than much of its non-local surround. The top target of the weak snake is also brighter than its local surround, but is not brighter than it's non-local surround. The global pattern of agent distribution for the strong snake shows that the light-attracted agents are attracted and corralled into the top target surround area, but in the weak snake the agents in the top target surround are drawn away from the local surround into further outlying areas. The relative depopulation in the weak snake top target surround ensures that, over time, fewer agents will be transported across the brightness interface into the target area, ultimately resulting in a less bright percept of the target area. A similar situation occurs with the bottom targets. In the strong snake, light-attracted agents present on the target area are likely to be drawn into the brighter local surround, but also likely to be later drawn out of the surround area into further distant areas, resulting in relatively little

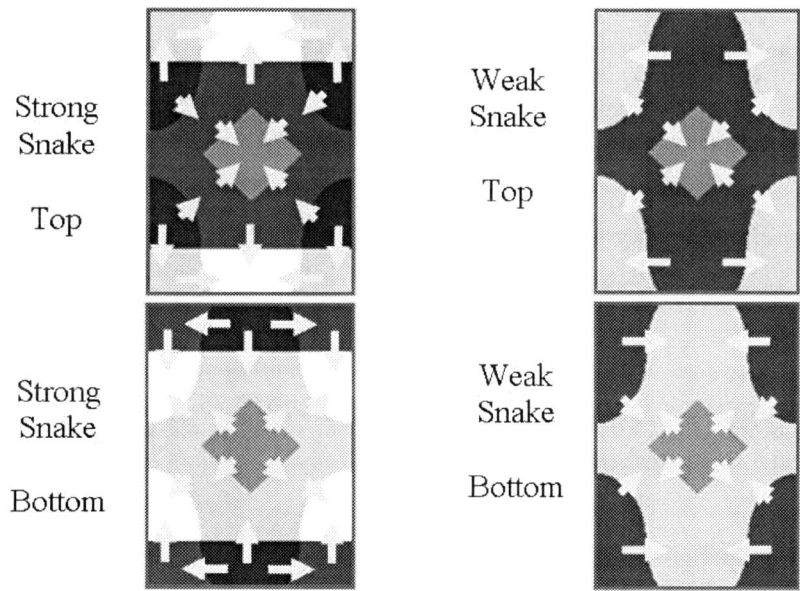

Figure 18. Direction of flux across brightness interfaces of light-attracted agents for the strong snake (left column) and the weak snake (right column) illusions.

trail deposition in the bottom target. In the weak snake, any agents present in the bottom target are also likely to leave the target area for the brighter surround but the agents in the local surround are not likely to migrate out of the local surround area because the more distant surround is darker in the weak snake. The corralling of agents into the lower surround increases the likelihood that agents will randomly cross into the surround, making the target appear brighter than in the strong snake version.

The collective illusory perception of the snake illusion suggests that low-level processes may in fact be able to account for at least some complex brightness illusions without recourse to complex, higher level inferences. The difference between the multi-agent collective perception and conventional lateral inhibition mechanisms is not a question of behavioural locality, since the agents sense and move in a local manner, but a question of distance - perhaps more easily stated as: 'how local is local?'. The long range effects in the PixieDust framework do not arise through large scale filtering, as with other models of brightness perception, but locality is breached by the massively iterative nature of the system — small local individual steps can ultimately become relatively large distances.

Adelson also created the 'Corrugated Plaid' illusions (Fig. 19), another pair of brightness illusions that is even more intriguing because in the two versions, the vertical plaid and the horizontal plaid, all of the panels in the 5×5 plaid are identical in brightness to the corresponding panels in the other version.

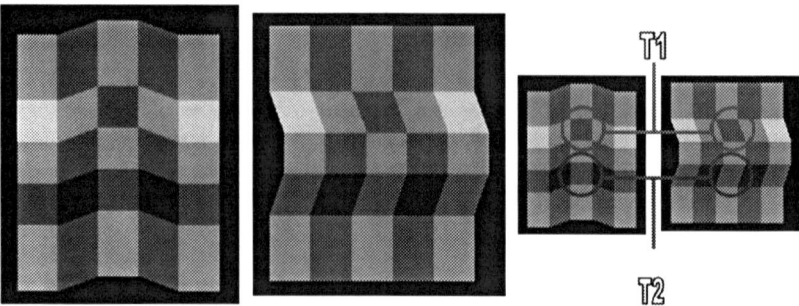

Figure 19. The Corrugated Plaid illusions. Vertical plaid (left) and horizontal plaid (right). Both versions of the illusion are composed of the same array of 5×5 panels, each panel being the same brightness as the corresponding panel in the other version. The targets in the middle column (2nd row and 4th row down) are of identical brightness.

The target panels (T1 and T2) described in the figure caption are actually the same brightness but to most observers the top target appears darker than the bottom target. The difference between the two targets may be explained in terms of the simple lateral inhibition explanation for SBC. However in the horizontal version of the plaid illusion, Adelson's subjects reported the illusory difference between the top and bottom targets is much stronger. The configuration of the two versions apparently rules out low level lateral inhibition type explanations for the differences in illusory strength: the target panels in both the vertical and horizontal plaid have the same surrounds as the corresponding version. The only difference between the two versions is the angles joining the panels together. Unlike the snake illusions no apparent transparency or layering effects may be inferred.

After ruling out low-level processes Adelson initially suggested that a 3d interpretation and an unconscious inference about lighting were the cause of the differences between the two versions. He suggested that in the vertical plaid, the centre column containing both targets appeared to be higher than its flanking sides and background and therefore more brightly lit. In the horizontal plaid, the middle row now appears to be higher and more brightly illuminated. The lower target appears to be in shadow and the perception of shadow generates an unconscious inference that the target is actually brighter than it appears. The three dimensional interpretation was questioned by Todorovic [31].Gilchrist also suggested that the different alignment of the panels may affect their perceptual grouping [32] and the two versions would be considered as different parts of the anchoring frameworks that (according to anchoring theories of brightness perception) is used to attribute global brightness values.

The collective perception of the plaid illusions by the agent population suggests that higher-level processes are not required to explain the differences in illusion strength (Fig. 20).

Like the human respondents, the agent population perceives a stronger illusory effect in the horizontal plaid than in the vertical plaid. The top target is perceived as darker in the horizontal plaid and the bottom target is brighter, increasing the difference in the perception of the two targets.

How does the agent population perceive the difference in illusion strength? The only difference between the two versions is in the geometry. Some squares have been replaced with parallelograms, resulting in changes of junction angles. T1 (on both versions) is bounded on all four sides by lighter areas. The difference between the two versions is that in the vertical plaid T1 is a square, whereas in the horizontal plaid T1 is

Mean Brightness	Vertical Plaid	Horizontal Plaid
Top Target (T1)	95	91
Bottom Target (T2)	122	128
Difference	27	37

Figure 20. Collective perception of the corrugated plaid illusions by the PixieDust framework. Emergent trail patterns of the vertical plaid (left) and the horizontal plaid (right).

a parallelogram. The same change occurs for the T2 patch: a square in the vertical plaid and a parallelogram in the horizontal plaid. A clue to understanding the different illusory strengths is the change in contrast from the vertical plaid to the horizontal plaid. The panel T1, which was perceived as darker in the vertical plaid than T2, now is perceived as even darker as a parallelogram. T2, which was perceived as lighter than T1 in the vertical plaid, is even lighter as a parallelogram in the horizontal plaid.

The difference between the perception of the horizontal and vertical plaid is an increase in contrast between the two illusions. PixieDust perceives differences between the T1 panel and T2 panel in both versions of the illusion. The cause of the misperception for both illusions is an imbalance between the net efflux of agents from T1 (to the four touching lighter side panels) and a disruption of the relative equilibrium of influx/efflux of agents at the T2 panel (two sides of the panel are darker and two are lighter). For the illusion to be perceived more strongly in the horizontal plaid (i.e. a higher contrast between the two panels: an even darker T1 and an even lighter T2), there must be both an increase in the net efflux from T1 (so the panel appears darker) and also a change of the equilibrium of the T2 panel, shifted towards a net influx.

Since we are only concerned with differences in the net efflux from T1 and the combined influx/efflux equilibrium from T2, the agent flow diagram shown in Fig. 21 only considers those possible flows. The flows relate to light-attracted agents and stop at the points where the agents cannot move to a lighter area. The only arrows from T1 are those to squares that are lighter. Flow arrows from T2 show influx into T2 from darker areas (sides) and efflux from T2 into lighter areas (top and bottom). Note that diagonal flow to neighbouring panels is not shown since this influence is relatively minor (recall that the pattern of flux at the

Hermann-Hering grid blocks was far stronger at the sides of the blocks than at their corners).

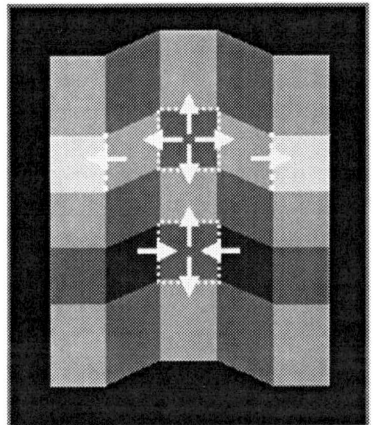 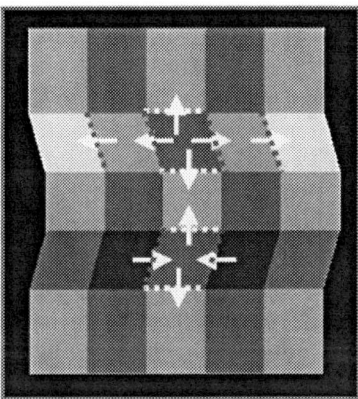

Figure 21. Net flow direction of light-attracted agents into and out of local and extra-local target areas.

As Fig. 21 shows, the net flow directions at the T1 and T2 target areas are exactly the same for the vertical and horizontal plaid. The only features on the figure that could possibly affect the flow of agents are the interfaces where attraction and migration occur. The notable change pertaining to the targets T1 and T2 (and related flow panels) is the border between panels that is changed due to the squares becoming parallelograms by rearranging the geometry of the figures. The diagonal borders of the parallelograms present a larger surface border for agent attraction and migration than the vertical borders of the squares. Both T1 and T2 will be considered separately for clarity.

- T1 – Vertical plaid: There is a 4 way efflux from T1. The panels directly to each side of T1 are also subject to an efflux of agents to the outer squares of the plaid since these are lighter. This represents a double increment step away from T1.
- T1 – Horizontal plaid: There is still a 4 way efflux from T1. The top and bottom exits have the same interface length but the two side exits, transformed by the rearrangement, are now diagonals of parallelograms that present a brighter interface length for attraction and migration. Furthermore the outermost side panels still present a double increment efflux and this border is also now represented

by a parallelogram diagonal — again with increased interface length
for attraction and migration across the brightness interface. The net
result of the changes regarding the T1 area is an increase in the net
efflux from T1 and thus less trail accretion and less brightness in the
emergent trail pattern.

- T2 – Vertical plaid: There is a net equilibrium of agent flux. The
  top and bottom panel interfaces present an increase in lightness and
  a resultant efflux from T2. The left and right bordering panels are
  darker and this interface presents an influx of agents into T2.

- T2 – Horizontal plaid: The rearrangement of geometry introduces
  two different border types into T2. The left and right side panels
  bordering T2 are now the diagonal sides of parallelograms, whose
  longer line length presents an increase in the interface length for
  attraction and migration. Note that the top and bottom borders of
  T2 are unchanged. The increase in interface size occurs at the borders
  of darker panels, so an increase in agent influx occurs. Because there
  is no increase in agent efflux from the T2 top and bottom panels
  the result is that the previous equilibrium of agent flux will now
  be shifted towards a net influx of agents into T2. The net result of
  changes to T2 results in a net influx of agents, more trail accretion
  and thus a brighter perception of T2 in the emergent trail pattern.

4   Conclusions and scope for further research

A multi-agent system framework was presented that uses very simple,
reactive agents residing on a topographic landscape representing a digi-
tised greyscale image. The cognitive behaviour of the mixed agent population is a very simple innate preference for locally higher/lower areas
(ON/OFF agents respectively). The agent population is decentralised
and uses local sensory and motor behaviours, responding to significant
changes in local contrast stimuli by changing their direction, and leaving marks on a corresponding stimulus map. The agents modify their
environment by modulating the trail level as they move (by accretion
or excavation) on a separate map that corresponds to the landscape.
The system evolves in a massively iterative manner and collectively the
population exhibits mass behaviour effects, resulting in global emergent
patterns being generated. The patterns reflect the emergence of population disparities at different parts of the landscape and the trail level
is interpreted as pixel brightness. The baseline grey level is deformed
by the individual actions of ON (upward deformation) and OFF agents
(downward deformation). The emergent trail patterns form a collective
perception of the original landscape's absolute brightness values using

reference to only local contrast levels. The absolute brightness percept reflects the historical activity of agents at the landscape sites.

The population is collectively able to accurately perceive 'real-world' imagery and responds in the same way as the human visual system (HVS) to illusory brightness phenomena. At a local level the agents behave in a manner analogous to simple lateral inhibition in the HVS. Collectively, the migration of the agents to different parts of the image landscape corresponds to the filling-in processes that are considered by some to be the mechanism for absolute surface brightness perception in the HVS. The distortion of the base grey level is similar to the models of eigengrau distortion by ON and OFF channels of the HVS and is similar in behaviour, if not implementation, to previous neural network mechanism that instantiate filling-in behaviour. Unlike the standard models of filling-in, however, there is no requirement that the boundary contour perception be completed before filling-in diffusion can occur. Indeed in this framework the collective perceptual filling-in behaviour is simply an emergent and collective consequence of independent agent activities. The mechanism of the population migration has been shown to be a shift from pseudo-random walk isotropic diffusion at locations without stimuli, to anisotropic movement across brightness interfaces where stimuli occur. The macroscopic effects of the shift towards anisotropic movement are an expansive or contractile wave of movement depending upon the orientation of the stimuli. The emergent differences in population density help to resolve difficulties contained in images containing transitive brightness such as luminance staircases that contain identical contrast levels but very different absolute brightness levels.

Whilst the framework is obviously not a model of the HVS neural behaviour, we have demonstrated that simple, local, low-level processes involving the flux of activity across brightness interfaces can explain the perception of classical brightness illusions. Furthermore the collective perception provides a parsimonious account of the perception of, and differences in the perception of, complex brightness illusions that have previously been explained only by complex higher-level unconscious inferences about the structure or luminance of the scene.

There is much scope for further work into collective perception by emergent pattern formation. Other illusory phenomena such as brightness assimilation (that appear to contradict the 'rules' of simultaneous brightness contrast effects [33]), and long range illusory phenomena (such as the Craik-O'Brien-Cornsweet effect) have shown positive preliminary results, based upon the mechanisms of random modulation of agent sensitivity.

References

[1] Kuffler, S., Discharge patterns and functional organization of mammalian retina. Journal of Neurophysiology, 1953. 16: p. 37-68.

[2] Kaplan, E., K. Purpura, and R. Shapely, Contrast affects the transmission of visual information through the mammalian lateral geniculate nucleus. Journal of Physiology, 1987. 391: p. 267-288.

[3] Pessoa, L., E. Mingolla, and H. Neumann, A contrast and luminance-driven multiscale network model of brightness perception. Vision Research, 1995. 35(2201-2223).

[4] Hong, S. and S. Grossberg, A neuromorphic model for achromatic and chromatic surface representation of natural images. Neural Networks, 2004. 17(5-6): p. 787-808.

[5] Blakeslee, B., W. Pasieka, and M.E. McCourt, Oriented multiscale spatial filtering and contrast normalization: a parsimonious model of brightness induction in a continuum of stimuli including White, Howe and simultaneous brightness contrast. Vision Research, 2005. 45(5): p. 607-615.

[6] Mariani, A., Biplexiform cells: ganglion cells of the primate retina that contact photoreceptors. Science, 1982. 216: p. 1134-1136.

[7] Berson, D., F. Dunn, and M. Takao, Phototransduction by retinal ganglion cells that set the circadian clock. Science, 2002. 295: p. 1070-1073.

[8] Li, C.-Y., et al., Role of the extensive area outside the X-cell receptive field in brightness information transmission. Vision Research, 1991. 31(9): p. 1529-1540.

[9] Li, C.-Y., et al., Extensive disinhibitory region beyond the classical receptive field of cat retinal ganglion cells. Vision Research, 1992. 32(2): p. 219-228.

[10] Knau, H. and L. Spillman, Brightness fading during Ganzfeld adaptation. Journal of the Optical Society of America A., 1997. 14(6): p. 1213-1222.

[11] Spillmann, L., et al., Perceptual filling-in from the edge of the blind spot. Vision Research, 2006. 46(25): p. 4252-4257.

[12] Zur, D. and S. Ullman, Filling-in of retinal scotomas. Vision Research, 2003. 43(9): p. 971-982.

[13] Paradiso, M. and K. Nakayama, Brightness perception and filling-in. Vision Research, 1991. 31(7-8): p. 1221-1236.

[14] Davey, M., T. Maddess, and M. Srinivasan, The spatiotemporal properties of the craik-o'brien-cornsweet effect are consistent with 'filling-in'. Vision Research, 1998. 38: p. 2037-2046.

[15] Gerrits, H. and A. Vendrik, Simultaneous contrast, filling-in process and information processing in a man's visual system. Experimental Brain Research, 1970. 11: p. 411-430.

[16] Arrington, K.F., The temporal dynamics of brightness filling-in. Vision Research, 1994. 34(24): p. 3371-3387.

[17] Grossberg, S. and F. Kelly, Neural dynamics of binocular brightness perception. Vision Research, 1999. 39(22): p. 3796-3816.

[18] Mingolla, E., W. Ross, and S. Grossberg, A neural network for enhancing boundaries and surfaces in synthetic aperture radar images. Neural Networks, 1999. 12(499-511).

[19] Grossberg, S. and D. Todorovic, Neural Dynamics of 1-d and 2-d brightness perception: A unified model of classical and recent phenomena. Perception and Psychophysics, 1988. 43: p. 241-277.

[20] Arrington, K., Directional filling-in. Neural Computation, 1996. 8: p. 300-318.

[21] Neumann, H., L. Pessoa, and T. Hansen, Visual filling-in for computing perceptual surface properties. Biological Cybernetics, 2001(85): p. 355-369.

[22] Keil, M.S., et al., Recovering real-world images from single-scale boundaries with a novel filling-in architecture. Neural Networks, 2005. 18: p. 1319-1331.

[23] Ramos, V. and F. Almeida. Artificial ant colonies in digital image habitats - A mass behaviour effect study on pattern recognition. in Ants 2000 2nd Int. Workshop on Ant Algorithms. 2000. Brussels, Belgium.

[24] Smolka, B. and K.W. Wojciechowski, Random walk approach to image enhancement. Signal Processing, 2001. 81(3): p. 465-482.

[25] Toffoli, T., Programmable matter methods. Future Generation Computer Systems, 1999. 16(2-3): p. 187-201.

[26] Jones, J. and M. Saeed, Image enhancement - an emergent pattern formation approach via decentralised multi-agent systems. Multi-Agent and Grid Systems, 2007. 3(1): p. 105-140.

[27] Hartline, H.K. and F. Ratliff, Inhibitory interaction in the retina of Limulus, in Handbook of Sensory Physiology, M.G.F. Fuortes, Editor. 1972, Springer-Verlag: Berlin. p. 381-447.

[28] Spillman, L., The Hermann Grid Illusion: a tool for Studying Human Perceptive Field Organization. Perception, 1994. 23: p. 691-708.

[29] Rudd, M.E. and K.F. Arrington, Darkness filling-in: a neural model of darkness induction. Vision Research, 2001. 41(27): p. 3649-3662.

[30] Adelson, E., Perceptual organization and the judgment of brightness. Science, 1993. 262: p. 2042-2044.

[31] Todorovic, D., Lightness and junctions. Perception, 1997. 26: p. 379-394.

[32] Gilchrist, A., et al., An Anchoring Theory of Lightness Perception, ,. Psychological Review, 1999. 106(4): p. 795-834.

[33] Bindman, D. and C. Chubb, Brightness assimilation in bullseye displays. Vision Research, 2004. 44(3): p. 309-319.

Evaluation of a multi-path maze-solving cellular automata by using a virtual slime-mold model

Masayuki Ikebe and Yusuke Kitauchi

Graduate School of Information Science and Technology, Hokkaido University,
Kita 14 Nishi 9, Sapporo 060-0814, Japan

Abstract We describe cellular-automata that will enable solving a multi-path maze. When a slime mold and a food are put in an entrance and a goal of a maze, a shape of the slime mold indicates a shortest path from the entrance to the goal. We propose CA algorithm in which a slime mold changes its body to obtain food effectively. We simulated the virtual slime-mold model and the simulation results indicated obtaining shortest path of a multi-path maze.

1 Introduction

We developed a cellular automaton to solve mazes and find the shortest path based on the behavior of a slime mold by using topology preservation. LSI computing, such as von the Neumann machine and Boolean logic, has recently experienced extraordinary development and is currently indispensable in mainstream information-processing hardware.

However, current LSI computing cannot satisfy all of the requirements of newly emerging areas (such as high-speed image processing and network analysis in which pixels or nodes affect each other) because of the development of the multimedia society and the tendency for various pieces of information to subdivide. LSI computing must therefore be further developed. Operations can be processed effectively by using a suitable algorithm in a parallel-distributed architecture.

CA (Cellular automata) are information-processing systems that have both a parallel and a distributed architecture. For example, a 2D CA with a Moore neighbourhood are shown in Fig. 1. They are configured as matrices of unit cells that interact with each other. Various spatial

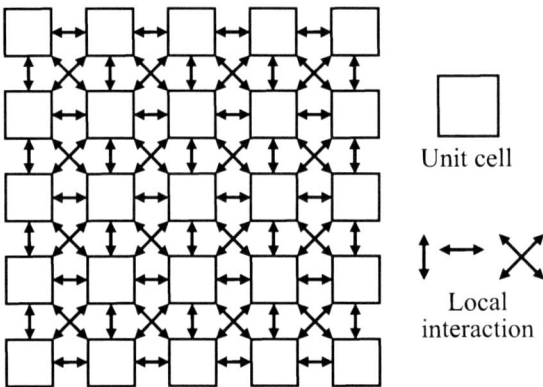

Figure 1. Structure of 2D CA with a Moore neighbourhood.

patterns can be generated by setting up a suitable interaction between the cells [1–8].

For the computation of shortest path in cellular automata, methods of using reaction diffusion system and leaving directions of movement in a maze [9] are known. About the former, based on a chemical reaction, collisions of waves are checked by multi-state expression, and the shortest path is formed [10, 11]. Or traveling of a wave is memorized for every definite period of time, and the shortest path is formed by the alteration of a wave [12]. However, in the above-mentioned method, the arrival to the goal is checked from the outside, and then the process which analyzes the shortest path from the goal to a start is needed. Therefore, the rules of maze search and the shortest path analysis will differ greatly, and the number of cell states will increase.

In this research, we focus on the behavior of slime molds and propose a 2D CA algorithm for shortest path maze solving in the five cell states by a object modification. The cellular automata check itself the arrival to the goal, branches are contracted by object modification, and the shortest path is formed. The rule which performs dilation and erosion with topology preservation was applied to modification of an object. In the above-mentioned rule, collisions of the objects was expressed by using rotated template matching.

The remainder of this paper is organized as follows. We describe the topology-preservation rule for the cellular automaton in section 2. In section 3, we discuss the transition rule for finding the shortest path. In section 4, we discuss the simulation results, which demonstrate the capability of the algorithm to solve the maze by finding the shortest path.

2 Topology-preservation rule for the cellular automaton

When we change the form of an object by using ordinary dilation or an erosion rule, the topology of the object will be disrupted because of parts of the object dividing or disappearing. Therefore, we developed a rule that preserves topology. In preserving the topology of an object, the features of the object are preserved when dilation or an erosion rule is used. In this study, we consider "erosion" as applied to the interrelations between the eight neighboring cells. Because all the cells in an automaton operate in parallel, it is difficult to transform objects whose width is an even number of pixels into one-pixel-size central points. (If an object is erased evenly from both sides, it will eventually disappear.) Therefore, erosion operations must be carried out by successively changing the templates. To preserve the topology, we used one-directional templates that rotate at each step.

The topology-preservation rule is shown in Fig. 2a. Using this rule requires three templates. The cell state is binary. Each template represents an inward contraction that is slanted and convex. Furthermore they do not disrupt the topology. If the templates match, the cell state changes from 1 to 0. We can gradually contract objects from four directions by rotating the templates at 90 degrees for each step. Unlike in the case of a conventional isotropic rule, the form of an object does not disappear (Fig. 2b). The topology of an object without a hole converges to one point. Examples of the application of this rule are shown in Fig. 2c. They show that applying the contraction operation preserves the topology of the object. By reversing the state of a template, a dilation operation is also possible.

3 Transition rule for finding the shortest path by using virtual slim-mold model

3.1 Maze solution by slime mold

Recently, the report that the capability to solve mazes is in a slime mold was made, and the high level information processing capability of low organisms was proved [13].

The maze solution by slime mold consists of two steps. After slime mold spreads in the maze whole route, when foods are put on an entrance

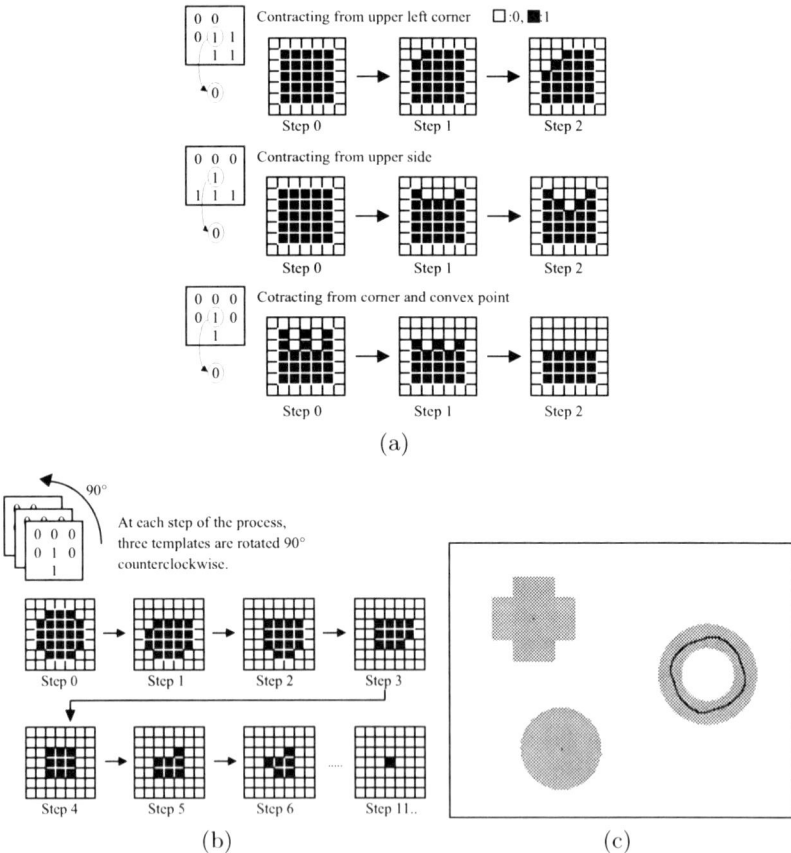

Figure 2. Topology-reservation erosion rule. (a) Templates for proposed erosion. (b) Rotation operation of templates. (c) Simulation results obtained with proposed rule.

and an exit (1) the body in the maze of the dead end is shrinking and the slime mold leaves pipes to all the routes that connect an entrance and an exit, (2) the pipe of the shortest distance is chosen and, finally the slime mold leaves one thick pipe. Thus, the food in two distant places of the entrance and the exit is crowded with slime mold. And, by the pipe of the shortest path formed, the slime mold can take in food efficiently, with one body maintained. Under the complicated situation like a maze, slime mold is able to discover the optimal answer to the problem how to connect two places with food.

The maze solution by the slime mold can be explained based on the physical property of the substance which constitutes a cell. Everywhere inside of the slime mold, the parts which repeats contraction movement spontaneously exists. They affect each other and form time/spatial patterns, such as a wave of contraction movement, in the whole cell. This pattern formation is connected with modification of the slime mold, and the maze can be solved.

3.2   Modeling of virtual slime mold

In this work, we focus on the behavior of slime mold instead of the strict physical characteristic of slime mold. Moreover, we assumed as follows. (1) In the initial state, slime mold is arranged at an entrance and food is arranged to an exit. (2) The slime mold spreads the inside of the maze and looks for food. The differences between true slime mold and virtual one are shown in Tab. 1.

Table 1. Comparison between true slime mold and virtual model

	True slime mold	Virtual model
Initial state	Filled in the maze	Put on the entrance
Selection of the shortest path	Disappearance of the path	Interception of the path
Behavior in the dead end	Contraction	Contraction

Then, the behavior of the slime mold was decomposed into the following three states (Fig. 3):

− Scanning for food
− Securing food

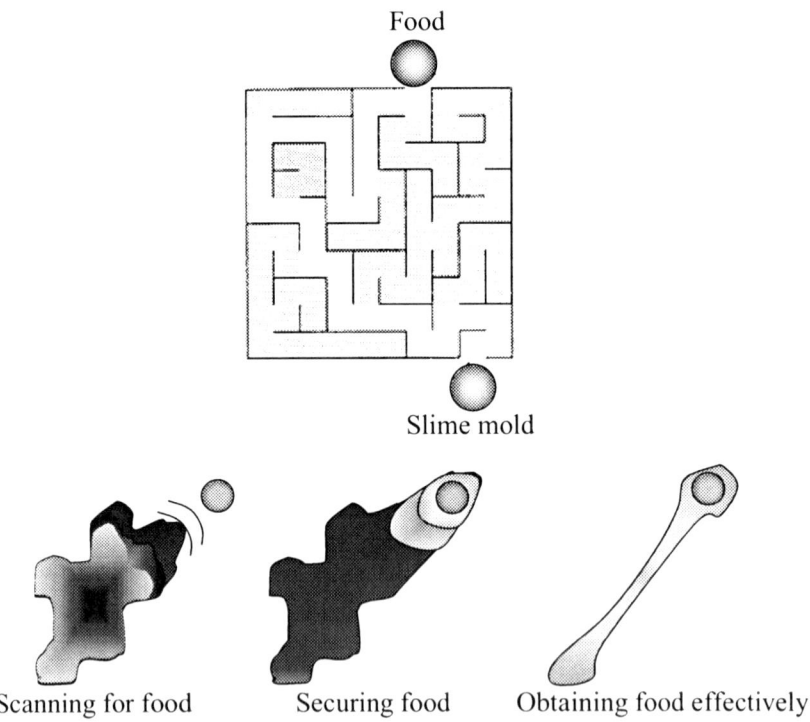

Figure 3. Modeling of slime-mold behavior.

– Obtaining food effectively

In the state of the scanning for food, preserving own topology, the slime mold spreads the inside of a maze and looks for food. In the state of the securing food, If the slime mold contacts food, the information will spread the inside of slime mold. In the state of the obtaining food effectively, the slime mold mutates its body, in order to obtain food efficiently (the slime mold indicates the shortest path of the maze). The slime mold chooses the above-mentioned states by itself automatically.

3.3   Transition rule for solving the maze

The transition rule for finding the shortest path is shown in Fig. 4. Using this rule requires five cell states transitions:

– 0: Road
– 1: Gel1
– 2: Wall
– 3: Gel2
– 4: Food

For example, state "2" cells are located on a wall. In state "2"(Wall) and "4"(Food), the cell states are constant. In the other three states, the cells have transition rules and imitate the behavior of a slime mold.

The transition rule of "0" (Road) represents the interaction between "0"(Road) and "1"(Gel1). Rotated templates for contraction are used. If the current states, the templates match, and the center cell ("0") changes its state. With this rule, the state of the center cell "0" changes to "1", if the templates match. Gel1 dilates and decreases the Road area (Road contraction) by using this rule. This state transition represents Gel1 moving along the Road.

The transition rule of "1"(Gel1) represents the interaction between "1"(Gel1) and "3"(Gel2). With this rule, an ordinary dilation template is used instead of the rotating templates. The template matching uses the values of the changed states "1"(Gel1), "3"(Gel2), and "4"(Food). The state of the center cell "1" changes to "3", if the templates match. Gel1 (which touches the food) mutates into Gel2 by using this rule, and the mutation is transmitted through Gel1. This represents times when a virtual slime mold acquires food, and the constitution of virtual slime molds varies.

The transition rule of "3"(Gel2) represents the interaction between "3"(Gel2) and "0"(Road). The rotating templates are used and the rule uses the values of "0"(Road), "2"(Wall), and "3"(Gel2). The state of the center cell "3" changes to "0", if the templates match. However, Gel2

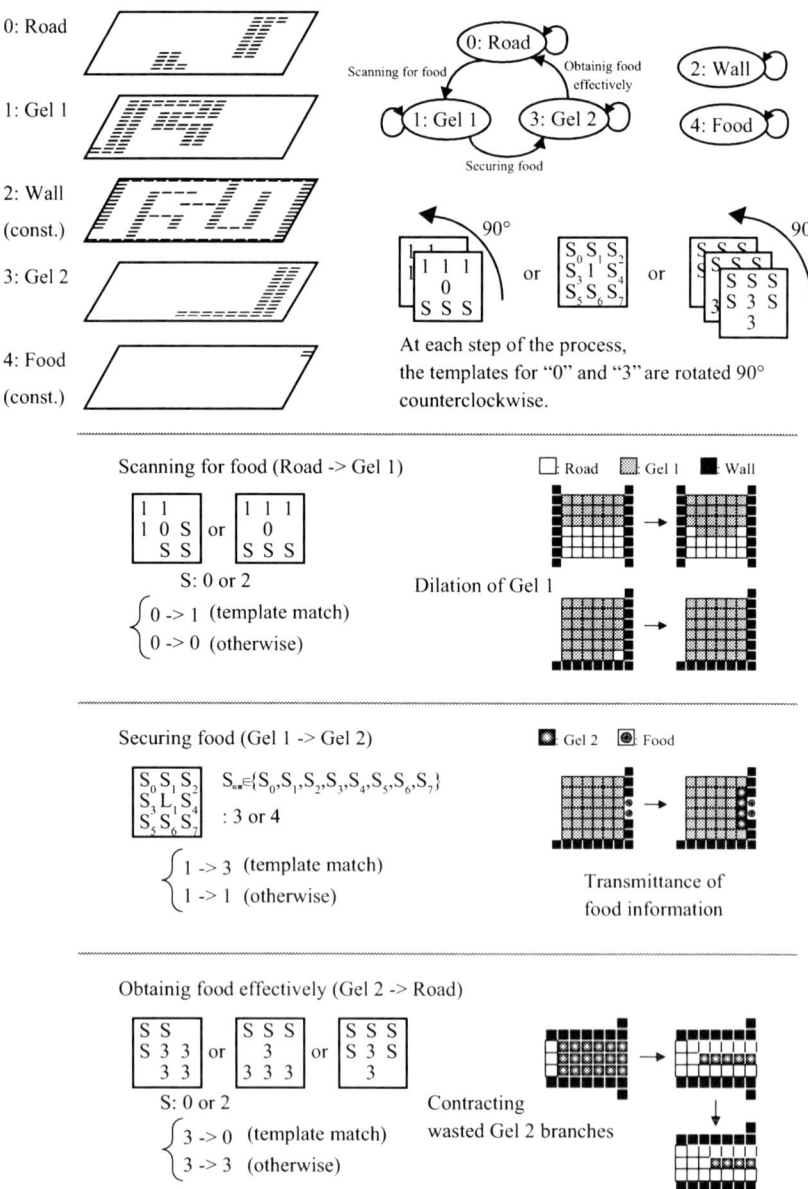

Figure 4. Transition rule for finding shortest path.

is in contact with Food and does not change. The wasteful branches of Gel2 contract, and Gel2 mutates into an effective body for obtaining food by using this rule. Gel2 represents the shortest path to the food after mutation.

When two or more paths exist toward Food, matched rotating templates control the contraction of "0"(Road) and the blocks of Gel1 for the shortest path and Gel1 for the other path together. By applying the Gel2 transition rule in this state, Gel2 (which represents the shortest path) will remain. Other maze-solving algorithms for cellular automata have previously been proposed. In conventional algorithms, a cell needs 14 states [14]. However, in this work, a cell can be used to solve a maze problem with only five states.

4   Simulation results

The simulation results of the cellular automaton that searches for the shortest path are shown in Figs.5 and 6. We set routes like mazes and simulated a virtual slime mold that searched for the shortest path toward the goal of the food in the mazes. The cellular automaton consisted of 120×120 cells, and we set Gel1 as the start and Food as the goal. First, when there is only one path toward the goal, in the initial state, step 0. Gel1 moves along the road: step 1670 and step 2340. Then, Gel1 touches the food and mutates into Gel2, step 2600 and step 2900. Gel2 started shrinking before Gel1 filled the maze because the dilation speed of Gel1 was slower than its mutation speed. After 3864 steps, the shrinking of Gel2 made the shortest path to the food.

Second, when there are several possible paths toward the goal, the initial state in, step 0, the rotating templates worked effectively. When Gel1 of the shortest path and other paths contact, the wall (Road) by the topology preservation effect is generated. At the time of transition of Gel2, the shortest path was formed efficiently: step 2340.

5   Applications and future work

The ad hoc network of a short-distance radio terminal can be considered as a cellular automaton, and the terminal can be considered as a unit cell. An efficient algorithm for searching for the shortest path to the terminal is important for connecting the terminal to the desired network. Our transition rule only works based on the states of neighboring cells, so it is suitable for communications-range networks with narrow short-distance wireless connections. Furthermore, parallel operation, like

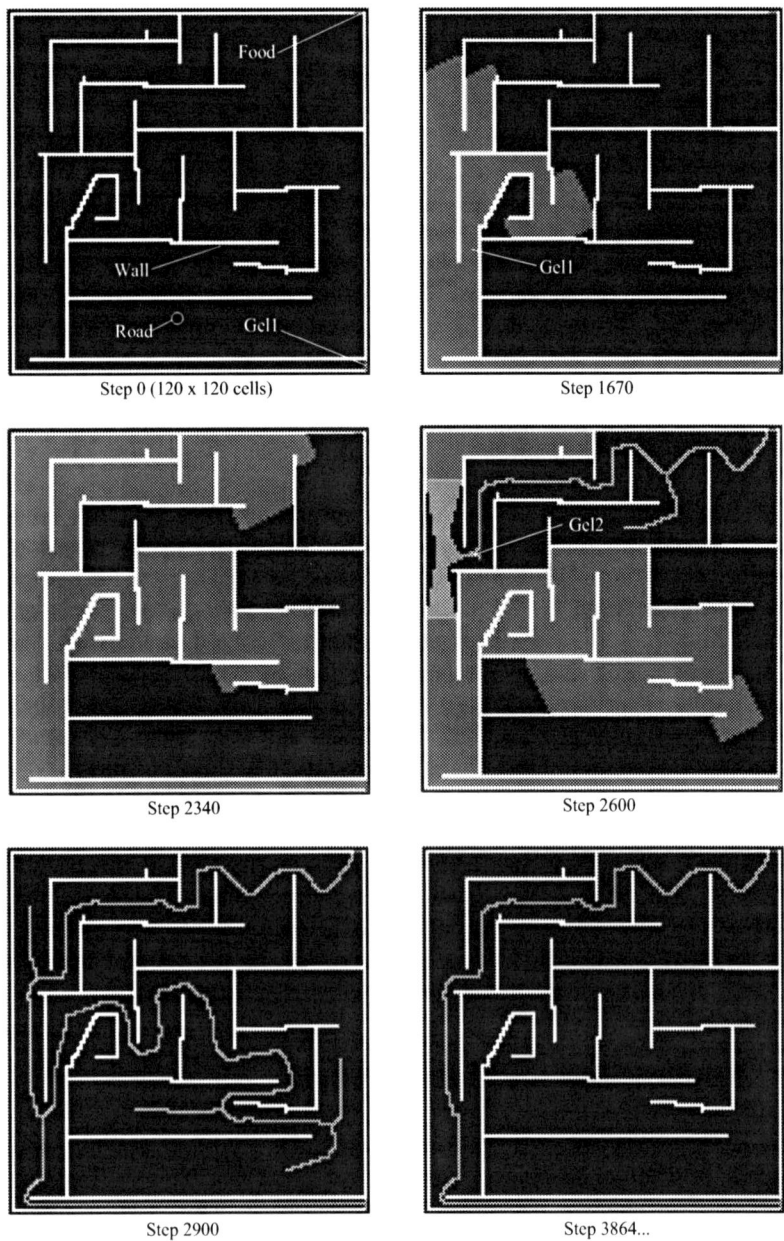

Figure 5. Results of simulation to find shortest path (1).

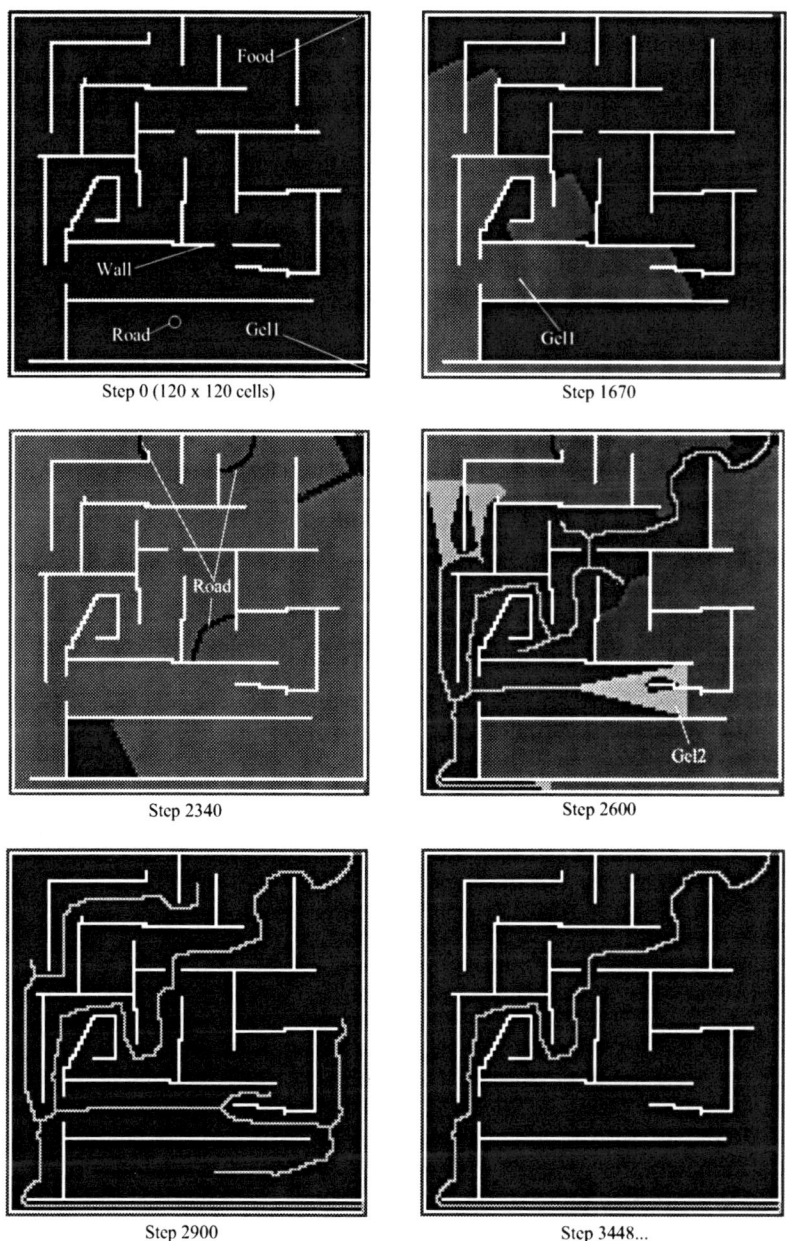

Figure 6. Results of simulation to find shortest path (2).

that of a cellular automaton is expected because all terminals are operating simultaneously. Because network composition is not a perfect 2-dimensional mesh, our transition rule cannot be used in its current state. However, improving the transition rule would make it correspond to an irregular mesh structure.

We developed a transition rule for a cellular automaton that imitates the behavior of a virtual slime mold and found that this rule can be used to find the shortest path from two or more paths to a target point. By further developing this algorithm, we should be able to apply it to search for the shortest path in a network field.

References

[1] Preston, K., Duff, M., "Modern cellular automata," Plenum Press, 1984

[2] Toffoli, T., Margolus, N., "Cellular automata machines," MIT Press, 1987

[3] Golay, M.J.E., "Hexagonal parallel pattern transformations," IEEE Trans. Comput. C 18, pp. 733−740, 1969

[4] Arcelli, C., Cordella, L., Levialdi, S., "Parallel thinning of binary pictures," Electron Lett. 11, pp. 148−149, Dec. 1975

[5] Hilditch, C.J., "Linear skeletons from square cupboards," In: Melter, B., Michie, D., editors, "Machine Intelligence" Edinburgh Univ. Press, 403, 1969

[6] Stefanelli, R., Rosenfeld, A., "Some parallel thinning algorithms for digital pictures," J. ACM 18, pp. 256−264, 1971

[7] Hwang, F.K., "An O(nlogn) algorithm for rectilinear minimal spanning trees," Journal of the ACM, Vol. 26, pp. 177−182, 1979

[8] Lee, D.T., "Two dimensional Voronoi Diagram in the Lp-metric" Journal of the ACM, Vol. 27, pp. 604−618, 1980

[9] Adamatzky, A., "Cellular automaton labyrinths and solution finding," Computers and Graphics, 21, 519?−522, 1997

[10] Adamatzky, A., "Computation of shortest path in cellular automata," Mathematical and Computer Modeling, 23, 105−11, 1996

[11] Adamatzky, A. and B. De Lacy Costello, "Reaction-diffusion path planning in a hybrid chemical and cellular-automaton processor," Chaos, Solitons and Fractals, Vol. 16, No. 5, pp. 727−736, 2003.

[12] Ito, K. Hiratsuka, M. Aoki, T. and Higuchi. T, "A Shortest Path Search Algorithm Using an Excitable Digital Reaction-Diffusion System,". IEICE Transactions on Fundamentals of Electronics, Communications and Computer Sciences,E89-A(3), 735-743, 2006

[13] Nakagaki, T., Yamada, H., and AGOTA, T., "Intelligence: Maze-solving by an amoeboid organism " Nature Vol. 407, 470, 2000

[14] Stefania, B., Giancarlo, M., "Solving routing problems with cellular automata", Proceedings of the Second Conference on Cellular Automata for Research and Industry, Milan, Italy, 16−18 October 1996

A cellular-automaton-based anisotropic diffusion algorithm for subjective contour generation and its digital VLSI implementation

Takashi Morie and Takahiro Yamamoto

Graduate School of Life Science and Systems Engineering,
Kyushu Institute of Technology,
2-4, Hibikino, Wakamatsu-ku, Kitakyushu, 808-0196, Japan
morie@brain.kyutech.ac.jp
http://www.brain.kyutech.ac.jp/~morie

Abstract Subjective contours are imaginary contours that humans generate in the visual systems of their brains to perceive incomplete natural scene images. Therefore, generating subjective contours is important to realize human-like high-level vision functions. This paper proposes a cellular automaton (CA) based anisotropic diffusion algorithm for subjective contour generation. The diffusion state at each pixel is binary and it is updated using the neighboring pixel's states. By using two-step updating rules, gradually spreading diffusion is achieved, and it leads to successful subjective contour generation. For its VLSI implementation, we have designed a digital pixel circuit and verified the correct diffusion operation in the pixel circuit array by logic circuit simulation.

Key words: cellular automaton, anisotropic diffusion, subjective contour, VLSI implementation, FPGA

1   Introduction

In natural-scene images obtained by image sensors, the edge information is often lacking due to the illumination conditions and the poor contrast with the background. Even in such cases, a human can recognize an object by generating imaginary contours, which are known as subjective contours. A typical figure that induces subjective contours is the Kanizsa triangle shown in Fig. 1(a), in which a human can perceive a white triangle in the center region [1].

Regarding subjective contour generation, physiological, psychophysical, and computational models have been proposed. Some physiological experiments have revealed that neurons in the primary visual cortex (V1 or V2) respond to subjective contours as well as to real contours [2]. This may suggest that subjective contours are generated by some low-level simple neuronal network mechanisms, although there should be some feedback from a higher-level cortex. On the other hand, in some computational models, physically existing points are combined with a spline function to realize a subjective contour [3]. These models are useful but it is difficult to apply them to VLSI implementation because of their high complexity.

Thus, we use a model that causes subjective contours to grow from the physically existing points. To implement such a model, we proposed a pixel-parallel directional (anisotropic) diffusion algorithm (DDA) for subjective contour generation [4], and also proposed a VLSI circuit using a merged analog-digital circuit approach [5]. The DDA generates analog ridge-like functions of the diffusion state by propagating the internal state of pixel units in the given direction. The DDA is suitable for VLSI implementation because it is based on the framework of cellular neural networks (CNNs) with nearest-neighbor connections.

The DDA requires analog calculations, and an analog dedicated VLSI chip has to be designed for implementing it. However, for practical applications, digital VLSI implementation is preferable to analog one because the former is easy to design and is realized by an FPGA (field-programmable gate array) chip, which can be programmed in the field by the users.

In this paper, we propose a cellular automaton (CA) based anisotropic diffusion algorithm that can generate subjective contours, in which only binary signals are used, and therefore, it is suitable for digital VLSI implementation. We show the pixel circuit design result and verify the diffusion operation by logic circuit simulation.

2   CA-based anisotropic diffusion algorithm

In the 2-dimensional space corresponding to a given image, let us define x- and y-axes, discrete coordinates (m, n) indicating a pixel position, and angle θ indicating the diffusion direction, as shown in Fig. 2. Each pixel has a binary diffusion state, and at the initial state only diffusion

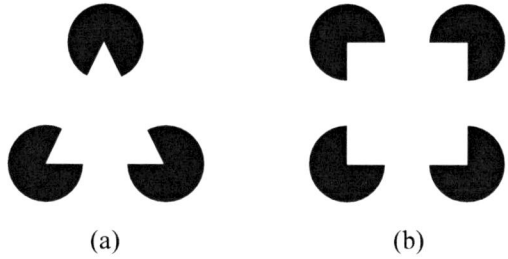

Figure 1. Kanizsa figures: (a) triangle and (b) square.

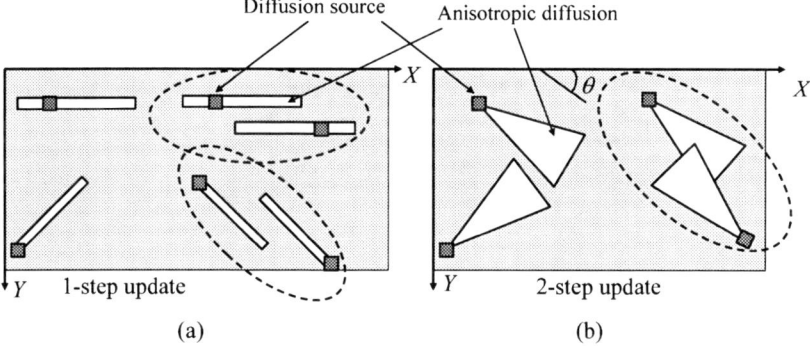

Figure 2. Definition of parameters and diffusion states.

starting pixels (diffusion sources) have a state '1'. Each pixel state is updated only using the neighboring pixel's states.

However, if we use the pixel's states only one-time-step before the present (one-step updating), diffusion along the horizontal, vertical, or diagonal directions is only achieved; i.e., $\theta = n\pi/4$, where n is an integer. Moreover, the one-step updating algorithm only realizes narrow diffusion, which forms a diffusion line with a one pixel width. Such diffusion often fails to cross the two diffusion lines generated from confronting diffusion sources, as shown in Fig. 2(a).

On the other hand, our proposed algorithm is more controllable in diffusion orientation, and realizes gradually spreading diffusion, as shown in Fig. 2(b). To achieve these advantages, we use two-step updating processes, that is the neighboring pixel's states two time-steps ago are used for calculation. Let us define $A(m,n)$, $B(m,n)$ and $C(m,n)$ as state values at pixel (m,n) at $t = t-2, t-1, t$, respectively. The updating

Table 1. Index sets for various diffusion orientations θ.

Case	Diffusion orientation	i	j	k	l
1	$\tan^{-1}(\frac{2}{3}) \leq \theta \leq \frac{\pi}{4}$	-1	-1	-1	0
2	$\frac{\pi}{4} \leq \theta \leq \tan^{-1}(\frac{3}{2})$	-1	-1	0	-1
3	$\pi - \tan^{-1}(\frac{3}{2}) \leq \theta \leq \frac{3}{4}\pi$	$+1$	-1	0	-1
4	$\frac{3}{4}\pi \leq \theta \leq \pi - \tan^{-1}(\frac{2}{3})$	$+1$	-1	$+1$	0
5	$\pi + \tan^{-1}(\frac{2}{3}) \leq \theta \leq \frac{5}{4}\pi$	$+1$	$+1$	$+1$	0
6	$\frac{5}{4}\pi \leq \theta \leq \pi + \tan^{-1}(\frac{3}{2})$	$+1$	$+1$	0	$+1$
7	$2\pi - \tan^{-1}(\frac{3}{2}) \leq \theta \leq \frac{7}{4}\pi$	-1	$+1$	0	$+1$
8	$\frac{7}{4}\pi \leq \theta \leq 2\pi - \tan^{-1}(\frac{2}{3})$	-1	$+1$	-1	0

rule consists of two sets according as the diffusion orientation. One rule is as follows:

$$\textbf{if } \ C(m,n) = 0 \ \textbf{ and } \ C(m+i, n+j) = 1 \ \textbf{ then } \ B(m,n) \leftarrow 1,$$
$$\textbf{if } \ B(m,n) = 0 \ \textbf{ and } \ B(m+k, n+l) = 1 \ \textbf{ then } \ A(m,n) \leftarrow 1,$$
$$C(m,n) \leftarrow A(m,n) + B(m,n), \tag{1}$$

where indices i, j, k, and l are determined by the diffusion orientation, as shown in Table 1. An example of diffusion state update is shown in Fig. 3 and diffusion results for these cases are shown in Fig. 4.

The other rule is as follows:

$$\textbf{if } \ C(m,n) = 0 \ \textbf{ and } \ C(m+i, n+j) = 1 \ \textbf{ then } \ B(m,n) \leftarrow 1,$$
$$\textbf{if } \ B(m,n) = 0 \ \textbf{ and } \ [B(m+k, n+l) = 1 \ \textbf{ or } \ B(m+p, n+q) = 1]$$
$$\textbf{then } \ A(m,n) \leftarrow 1, \tag{2}$$

where indices i, j, k, l, p, and q are determined by the diffusion orientation, as shown in Table 2. An example of diffusion state update is shown in Fig. 5 and diffusion results for these cases are shown in Fig. 6.

By combining Eqs. (1) and (2), all orientations are covered.

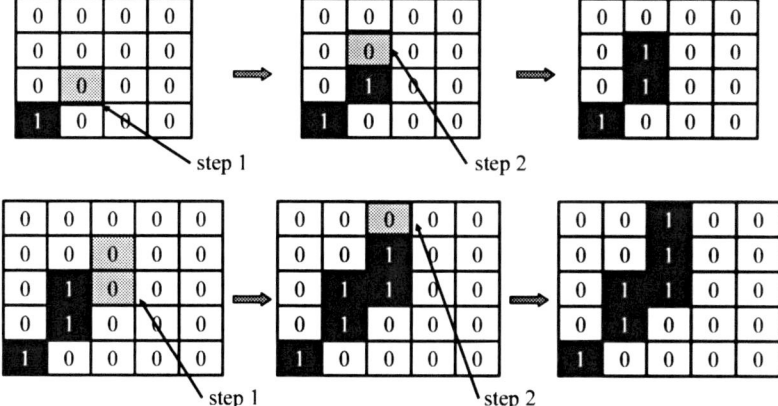

Figure 3. Diffusion state update for case 7.

Table 2. Index sets for various diffusion orientations θ.

Case	Diffusion orientation	i	j	k	l	p	q
9	$-\tan^{-1}\left(\frac{2}{3}\right) \leq \theta \leq \tan^{-1}\left(\frac{2}{3}\right)$	-1	0	-1	0	-1	± 1
10	$\tan^{-1}\left(\frac{3}{2}\right) \leq \theta \leq \pi - \tan^{-1}\left(\frac{3}{2}\right)$	0	-1	0	-1	± 1	-1
11	$\pi - \tan^{-1}\left(\frac{2}{3}\right) \leq \theta \leq \pi + \tan^{-1}\left(\frac{2}{3}\right)$	$+1$	0	$+1$	0	$+1$	± 1
12	$\pi + \tan^{-1}\left(\frac{3}{2}\right) \leq \theta \leq 2\pi - \tan^{-1}\left(\frac{3}{2}\right)$	0	$+1$	0	$+1$	± 1	$+1$

3   Generation of subjective contours

The algorithm described above has been applied to subjective contour generation in Kanizsa figures shown in Fig. 1.

In order to obtain the diffusion source positions from the given image, orientation-selective Gabor-filtering is performed. Figure 7 shows an example of diffusion source position extraction. One of the authors has already proposed a pixel-parallel algorithm for Gabor filtering [6], and also has developed a VLSI chip for Gabor filtering [7].

Using the obtained diffusion source positions and orientations, the anisotropic diffusion algorithm is executed, and then skeletonization, which is a well-known CA-based image processing algorithm, is performed. If the skeletonized lines generated from two confronting diffusion

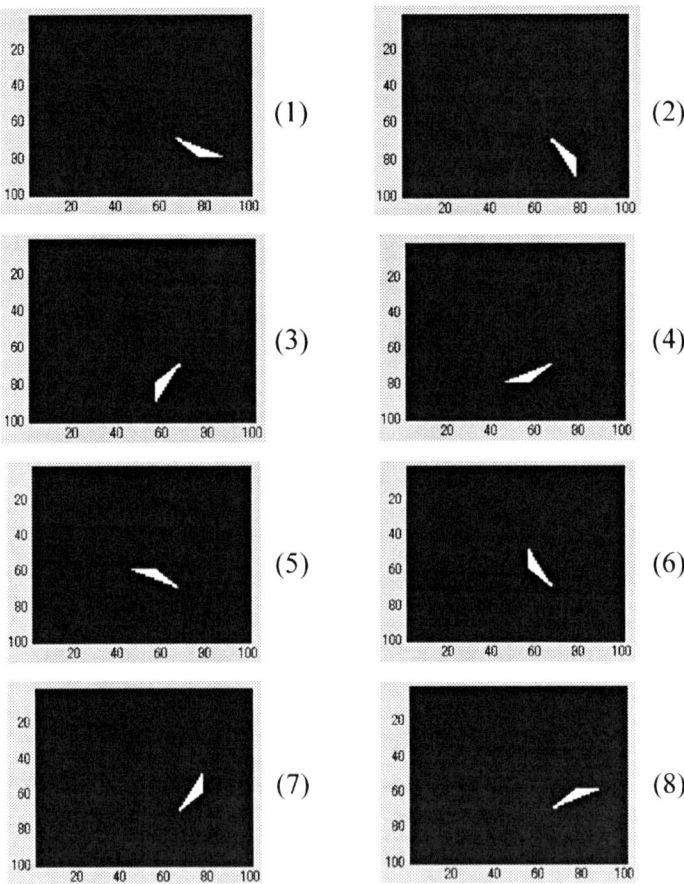

Figure 4. Diffusion results for cases 1 to 8.

sources are connected, the pair of lines are considered as a subjective contour.

After the above processing is performed for various orientations, if a closed region is formed by a set of the generated subjective contours, perception of Kanizsa figures is achieved. Using the proposed algorithm, the Kanizsa figures have successfully been extracted as shown in Figs. 8 and 9.

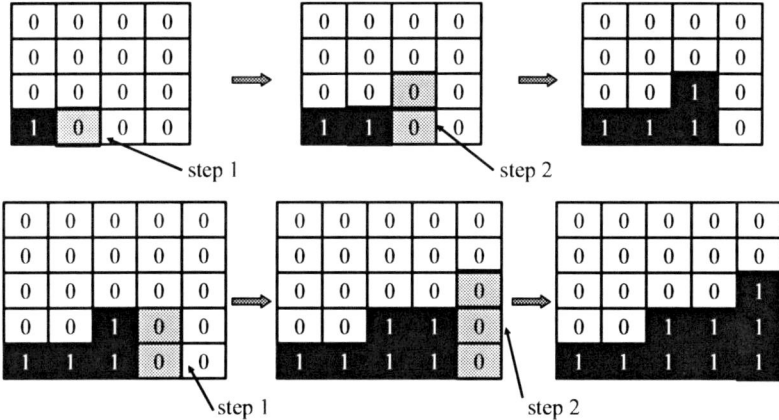

Figure 5. Diffusion state update for case 9 ($q = -1$).

4   Pixel circuit for CA-based anisotropic diffusion algorithm

We designed a pixel circuit for the proposed CA-based anisotropic diffusion algorithm. Because the algorithm is based on binary logic, a digital pixel circuit was designed as shown in Fig. 10. We constructed a 11×11 array of the pixel circuits. The logic simulation results of the circuit is shown in Figs. 11 and 12. It has been verified that the circuit successfully performs correct anisotropic diffusion processes.

According to the estimation of FPGA implementation using Altera Apex or Stratix II devices, an image of about 24×24 pixels can be treated if EP20K200C is used, and an image of more than 100×100 pixels can be treated if EP2S180 is used. Even in the former case, processing time for an image of 100×100 pixels was estimated to be less than $100~\mu$sec, where the input image is divided into 5×5 image parts each of which has 20×20 pixels.

5   Conclusion

An anisotropic diffusion algorithm for subjective contour generation was proposed. The proposed algorithm is based on a cellular automaton model and it can be executed by pixel-parallel operation. The diffusion state at each pixel is binary and its updating is performed using the neighboring pixel's states. By using two-step updating rules, gradually spreading diffusion can be achieved. We designed a digital pixel circuit

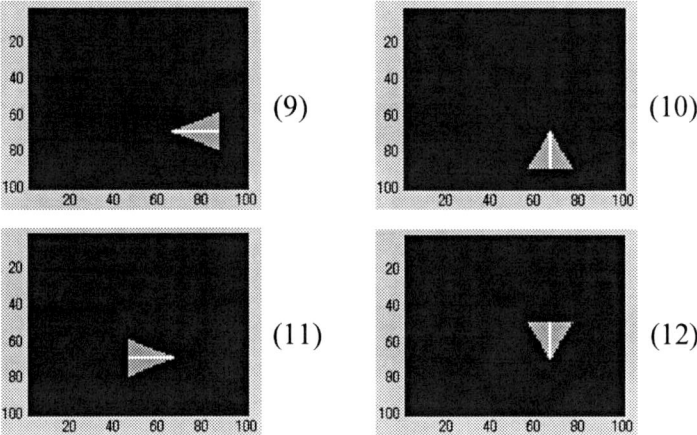

Figure 6. Diffusion results for cases 9 to 12.

and verified the correct diffusion operation by logic circuit simulation. We also estimated the performance of the FPGA implementation, and showed that we can treat an image of more than 100×100 pixels within 100 μsec using a commercial FPGA chip.

Acknowledgment

This work was partly supported by a COE program (center #J19) granted to Kyushu Institute of Technology by MEXT, Japan.

References

[1] Kanizsa, G.: Subjective contours. Scientific American **234** (1976) 48–52
[2] von der Heydt, R. and Peterhans, E.: Mechanisms of Contour Perception in Monkey Visual Cortex. I. Lines of Pattern Discontinuity. J. Neurosci. **9** (1989) 1731–1748
[3] Yasuda, H., Ando, K., Ohnish, N., and Sugie, N.: Extracting Physically Nonexistent Contours. Trans. IEICE **J73-D-II** (1990) 906–913
[4] Kim, Y. and Morie, T.: A Pixel-parallel Anisotropic Diffusion Algorithm for Subjective Contour Generation. IEEE Proc. of Int. Symp. Circuits and Systems (ISCAS) (2005) 4237–4240
[5] Kim, Y. and Morie, T.: A Pixel Circuit Implementing an Anisotropic Diffusion Algorithm for Subjective Contour Generation Using Merged Analog-Digital circuit Approach. J. Signal Processing **10** (2006) 259–262

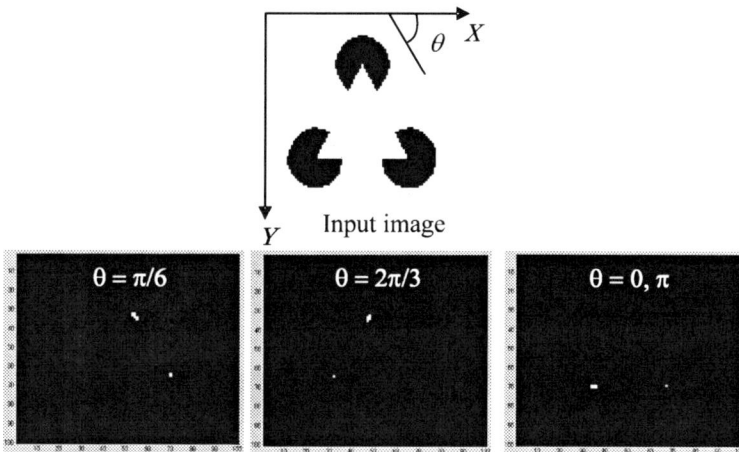

Figure 7. Detection of diffusion start points using Gabor filtering.

[6] Morie, T., Umezawa, J., and Iwata, A.: Gabor-Type Filtering Using Transient States of Cellular Neural Networks. Intelligent Automation and Soft Computing **10** (2004) 95–104

[7] Morie, T., Umezawa, J., and Iwata, A.: A Pixel-Parallel Image Processor for Gabor Filtering Based on Merged Analog-Digital Architecture. Symposium on VLSI Circuits, Digest of Technical papers (2004) 212–213

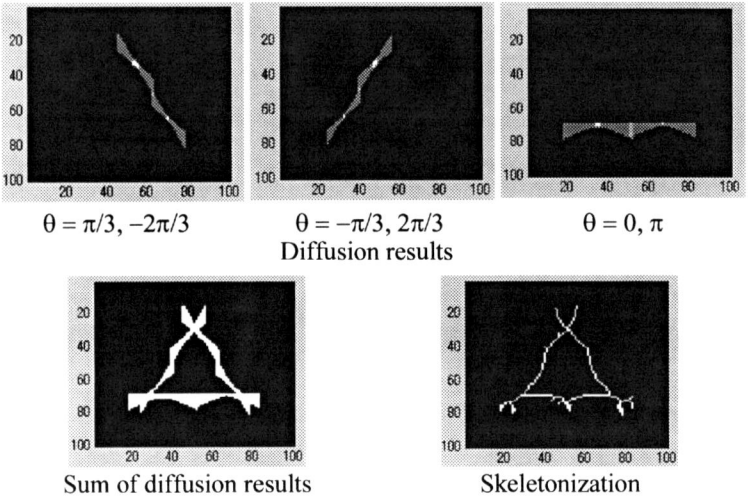

Figure 8. Subjective contour generation for Kanizsa triangle.

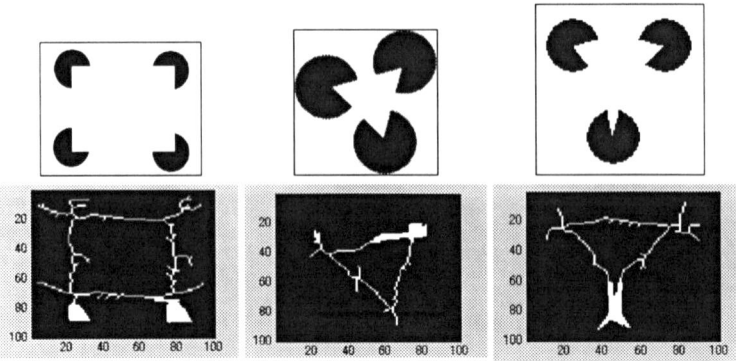

Figure 9. Subjective contour generation for other Kanizsa figures.

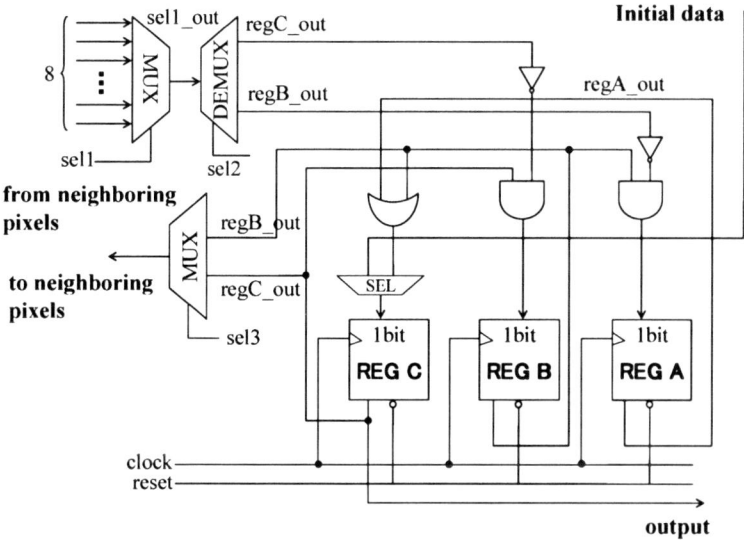

Figure 10. Pixel circuit for anisotropic diffusion.

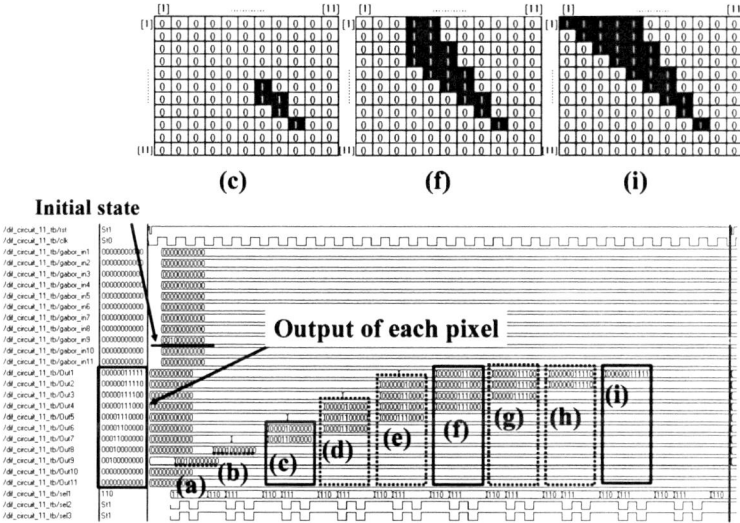

Figure 11. Logic simulation results of the diffusion circuit for case 6.

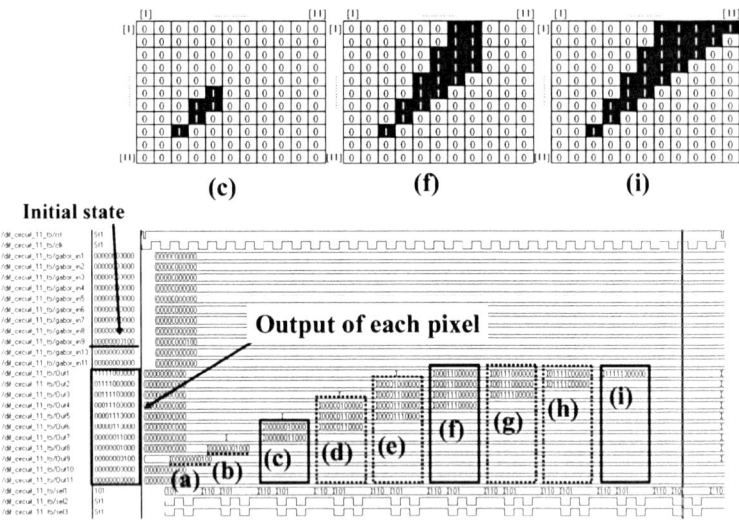

Figure 12. Logic simulation results of the diffusion circuit for case 7.

Cellular automata networks

Xin-She Yang* and Young Z. L. Yang

Department of Engineering, University of Cambridge
Trumpington Street, Cambridge CB2 1PZ, UK

Abstract A small-world cellular automaton network has been formulated to simulate the long-range interactions of complex networks using unconventional computing methods in this paper. Conventional cellular automata use local updating rules. The new type of cellular automata networks uses local rules with a fraction of long-range shortcuts derived from the properties of small-world networks. Simulations show that the self-organized criticality emerges naturally in the system for a given probability of shortcuts and transition occurs as the probability increases to some critical value indicating the small-world behaviour of the complex autamata networks. Pattern formation of cellular automata networks and the comparison with equation-based reaction-diffusion systems are also discussed.

1   Introduction

Theory and computation about complex networks such as the bacterial colonies, interacting ecological species, and the spreading of computer virus over the Internet are becoming very promising and they may have important applications in a wide range of areas. The proper modelling of these networks is a challenging task and the studies in this area are still at very early stage. However, various techniques and applications have been investigated, especially in the area of computational logic, the Internet network, and application of bio-inspired algorithms [1-7]. Since the pioneer work of Watts and Strogatz on small-world networks, a lot of interesting studies on the theory and application of small-world networks [7-9,12-18] have been initiated. More recently, the automata networks have been developed by Tomassini and his colleagues [14, 15] to study the automata network in noise environment. Their study shows that small-world automata networks are less affected by random noise.

---

* Corresponding Author

The properties of complex networks such as population interactions, Internet servers, forest fires, ecological species and financial transactions are mainly determined by the way of connections between the vertices or occupied sites.

Network modelling and formulations are essentially discrete in the sense that they deal with discrete interactions among discrete nodes of networks because almost all the formulations are in terms of the degrees of clustering, connectivity, average nodal distance and other countable degrees of freedom. Therefore, they do not work well for interactions over continuous networks and media. In the later case, modelling and computations are usually carried out in terms partial differential equations (PDEs), however, almost all PDEs (except those with integral boundary conditions) are local equations because the derivatives and the dependent variable are all evaluated at concurrent locations. For example, the 3-D reaction-diffusion equation

$$u_t = D\nabla^2 u + r(u; x, y, z, t), \tag{1}$$

describes the variation of $u(x, y, z, t)$ such as temperature and concentration with spatial coordinates (x, y, z) and time t. While the diffusion coefficient D can be constant, but the reaction rate r may depends on u and location (x, y, z) as well as time t. This equation is local because u, u_t, r and $\nabla^2 u$ are all evaluated at the same point (x, y, z) at any given time t. Now if we introduce some long-distance shortcuts (e.g., a computer virus can spread from one computer to another computer over a long-distance, not necessary local computers), then the reaction rate can have a nonlocal influence in a similar manner. We can now modify the above equation as

$$u_t = D\nabla^2 u + q\Big(u(x, y, z, t), u(x_*, y_*, z_*, t); x, y, z, t\Big), \tag{2}$$

where q depends on the local point (x, y, z) and another point (x_*, y_*, z_*) far away. Obviously, q can be any function form. As a simple example in the 1-D case: $u_t = Du_{xx} + u(x, t) * [1 - u(x, t)] + \beta u(x - s, t)$, where s is simply a shift and $\beta \in [0, 1]$ is a constant. This equation is nonlocal since the reaction rate depends the values of u at both x and $x - s$ at t. This simple extension makes it difficult to find analytical solutions. Even numerical solutions are not easy to find because the standard numerical methods do not necessarily converge due to the extra nonlocal term. This paper will investigate this aspect in detail using unconventional solution methods such as small-world cellular automata networks.

The present work aims to develop a new type of small-world cellular automata by combining local updating rules with a probability of long-range shortcuts to simulate the interactions and behaviour of a complex system. By using a small fraction of sparse long-range shortcut interactions together with the local interactions, we can simulate the evolution of complex networks. Self-organized criticality will be tested based on the results from the cellular automata. The important implications in the modelling and applications will also be discussed.

2 Small-world networks

Small-world networks are a special class of networks with a high degree of local clustering as well as a small average distance, and this small-world phenomenon can be achieved by adding randomly only a small fraction of the long-range connections, and some common networks such as power grids, financial networks and neural networks behave like small-world networks [10, 11]. The application of small-world networks into the modelling of infection occurring locally and at a distance was first carried out by Boots and Sasaki [7] with some interesting results. The dynamic features such as spreading and response of an influence over a network have also been investigated in recent studies [11] by using shortest paths in system with sparse long-range connections in the frame work of small-world models. The influence propagates from the infected site to all uninfected sites connected to it via a link at each time step, whenever a long-range connection or shortcut is met; the influence is newly activated at the other end of the shortcut so as to simulate long-range sparkling effect. These phenomena have successfully been studied by Newman and Watts model [12] and Moukarzel [10]. Their models are linear in the sense that the governing equation is linear and the response is immediate as there is no time delay in their models [20]. More recently, one of the most interesting studies has been carried out by De Arcaneglis and Herrmann [8] using the classic height model on a lattice, which implied the self-organized criticality in the small-world system concerned.

On the other hand, cellular automata have been used to simulate many processes such as lattice gas, fluid flow, reaction-diffusion and complex systems [18–22] in terms of interaction rules rather than the conventional partial differential equations. Compared to the equation-based models, simulations in term of cellular automata are more stable due to their finite states and local interacting rules [18]. In fact, in most cases, the PDE models are equivalent to rule-based cellular automata if the local rules can be derived from the corresponding PDE models [9,19], and thus both PDE models and CA rules can simulate the same process [22]. However, we will show that cellular automata networks are a better approach for solving nonlocal equation-based models.

The rest of the present paper will focus on: 1) to formulate a cellular automaton network on a 2-D lattice grid with sparse long-range shortcuts; 2) to simulate the transition and complexity concerning small-world nonlocal interactions; 3) to test the self-organized criticality of the constructed network systems; 4) to find the characteristics of any possible transition.

3   Cellular automata networks

Earlier studies on cellular automata use local rules updating the state of each cell and the influence is local. That is to say, the state at the next time step is determined by the states of the present cell concerned and those of its immediate surrounding neighbour cells. Even the simple rules can produce complex patterns [19]. The rule and its locality determine the characteristics of the cellular automata. In fact, we do not have to restrict that the rules must be local, and in general the influence can be either local or nonlocal. Thus, we can assume the rules of cellular automata can be either local or nonlocal or even global. The state of a cell can be determined by m cells consisting of m_i immediate neighbour cells and $m_o = m - m_i$ other cells at longer distance. In the case of local rules only, $m = m_i$ and $m_o = 0$. If $m_o \neq 0$, then the rules are nonlocal. If m is the same order of the total cells $N \times N$ of the cellular automaton, then rules are global. Nonlocal interactions rule for lattice-gas system was first developed by Appert and Zaleski in the discussion of a new momentum-conserving lattice-gas model allowing the particles exchange momentum between distant sites [2]. Some properties of local and nonlocal site exchange deterministic cellular automata were investigated by researchers [5, 15]. As the nonlocal rules are different from the local rules, it is naturally expected that the nonlocal rules may lead to different behavior from conventional local rule-based cellular automata. Furthermore, self-organized criticality has been found in many systems in nature pioneered by Bak and his colleagues [3, 4]. One can expect that there may be cases when self-organized criticality, cellular automata, and small-world phenomena can occur at the same time. More specifically, if a finite-state cellular automaton with a small-fraction of long-range shortcuts is formulated, a natural question is: Do the self-organized criticality exist in the small-world cellular automaton? Is there any transition in the system?

3.1   Local cellular automata

A cellular automaton is a finite-state machine defined on a regular lattice in d-dimensional case, and the state of a cell is determined by the current state of the cell and the states of its neighbour cells [18,19]. For simplicity, we use 2-D in our discussions. A state $\phi_{i,j}$ of a cell (i,j) at time step $n+1$ can be written in terms of the previous states

$$\phi_{i,j}^{n+1} = \sum_{k,l=-r}^{r} c_{k,l}\phi_{i+k,j+l}^{n}, \quad i,j = 1,2,...,N \tag{3}$$

where summation is over the $4r(r+1)$ Moore neighbourhood cells. In the d-dimensional case, there are $(2r+1)^d - 1$ Moore neighbourhood cells. N is the size of the 2-D automaton, and $c_{k,l}$ are the coefficients. For the simplest and well-known 2-D Conway's game of life $c_{k,l} = 1$ for 8 neighbour cells ($r = 1$). Now let us introduce some nonlocal influence from some sparse long-range cells (see Fig. 1) by combining small-world long-range shortcuts and conventional cellular automata to form a new type of cellular automata networks.

3.2 Small-world automata networks

For simplicity, we define a small-world cellular automaton network as a local cellular automaton with an additional fraction or probability p of sparse long-range nonlocal shortcuts (see Fig. 1). For $m_i = 4r_m + 1$ immediate Neumann neighbours and $m_o = 2r_o$ nonlocal cells, the updating rule for a cell becomes

$$\phi_{i,j}^{n+1} = \sum_{k,l=-r_m}^{r_m} c_{k,l}\phi_{i+k,j+l}^{n} + \delta(p) \sum_{s,q=-r_o}^{r_o} c_{s,q}\phi_{i+d_i+s,j+d_j+q}^{n}, \tag{4}$$

where $\delta(p)$ is a control parameter that can turn the long-range cells on ($\delta = 1$) or off ($\delta = 0$) depending on the probability p.

The probability p is the fractions of long-range shortcuts in the total of every possible combinations. For $N \times N$ cells, there are $N^2(N^2 - 1)$ possible connections. The simplest form of δ can be written as $\delta(p) = pH(p - p_0)$ where H is a Heaviside function. p_0 is a critical probability, and p_0 can be taken as to be zero in most simulations in this paper. The updating rules are additive and thus form a subclass of special rules. We can extend the above updating rules to a generalized form, but we are only interested in additive rules here because they may have interesting properties and can easily be transformed to differential equations. In addition, the neighourhood can be either extended Moore neighourhood or Neumann neighbourhood. For Moore neighbourhood, $m_i = 4r_m(r_m + 1)$ and $m_o = 4r_o$. Our numerical experiments seem to indicate that Moore neighbourhood is more sensitive for avalanche and Neumann neighbourhood is more stable for pattern formation. A simple case for a small-world cellular automaton in 2-D case is $r_m = 1$ and $r_0 = 5$ (or any $r_0 > r_m$) so that it has 5 immediate Neumann neighbour cells and 2 shortcuts.

The distance between the nonlocal cells to cell (i, j) can be defined as

$$S = \sqrt{s_i^2 + s_j^2}. \tag{5}$$

The nonlocality requires

$$\min(|s_i|, |s_j|) > r_m. \tag{6}$$

The nonlocal influence can also be introduced in other ways. Alternatively, we can use the conventional local rule-based cellular automaton and adding the long-distance shortcuts between some cells in a random manner. The probability p of the long-range shortcuts in all the possible connections is usually very small. Under certain conditions, these two formulations are equivalent. More generally, a finite-state cellular automaton with a transition rule $G = [g_{ij}]$ $(i, j = 1, 2, ..., N)$ from one state $\Phi^n = [\phi_{ij}^n]$ $(i, j = 1, 2, ..., N)$ at time level n to a new state $\Phi^{n+1} = [\phi_{ij}^{n+1}]$ $(i, j = 1, 2, ..., N)$ at time level $n + 1$ can be written as

$$G : \Phi^n \mapsto \Phi^{n+1}, \quad g_{ij} : \phi_{ij}^n \mapsto \phi_{ij}^{n+1}, \tag{7}$$

where g_{ij} takes the same form as equation (4) for small-world cellular automata.

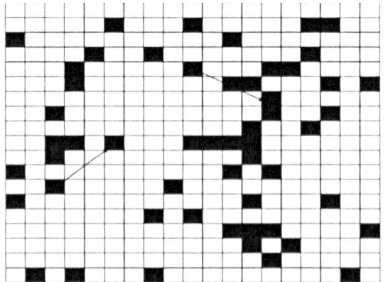

Figure 1. A cellular automaton network with a probability p of long-range shortcuts

The state of each cell can be taken to be discrete or continuous. From simplicity, we use n_v-valued discrete system and for most of the simulations in the rest of the paper, we use $n_v = 2$ (thus, each cell can only be 0 or 1) for self-criticality testing, and $n_v = 1024$ for pattern formation. Other numbers of states can be used to meet the need of higher accuracies.

4   Simulations and results

By using the small-world cellular automaton formulated in the previous section, a large number of computer simulations have been carried out in order to find the statistic characteristics of the complex patterns and behaviour arising from cellular automata networks with different probabilities p of long-range shortcuts. Numerical simulations are carried out on an $N \times N$ lattice in 2-D setting, and usually, $N \geq 40$, or up to 5000. Different simulations with different lattice size are compared to ensure the simulated results are independent of the lattice size and time steps. In the rest of the paper, we present some results concerning the features of transition and self-organized criticality of small-world cellular automata.

4.1   Self-organized criticality

For a lattice size of $N = 2000 \times 2000$ with a fixed p, a single cell is randomly selected and perturbed by flipping its state in order to simulate an event of avalanche in 2-D automata networks with the standard Moore neighbourhood and Game-of-life updating rules, but a probability p is used to add long-range shortcuts to the cellular automaton. A shortcut forces the two connecting cells having the same state. Figure 2 shows the avalanche size distribution for two different values of $p = 0.05$ and $p = 0.2$, respectively. The avalanche size is defined as the number of cells affected by any single flipping perturbation.

In the double logarithmic plot, the data follows two straight lines. It is clearly seen that there exists a power law in the distribution, and the gradient of the straight line is the exponent of the power-law distribution. A least-square fitting of $N \propto s^{-\gamma}$, leads to the exponents of $\gamma = 1.06 \pm 0.04$ for $p = 0.05$ and $\gamma = 1.40 \pm 0.05$ for $p = 0.2$. Although a power-law distribution does not necessarily mean the self-organized criticality. Self-organized criticality has been observed in other systems [1-4,26]. The pattern formed in the system is quasi-stable and a little perturbation to the equilibrium state usually causes avalanche-like readjustment in the system imply the self-organized criticality in the evolution of complex patterns of the cellular automaton. This is the first time of its kind by using computer simulations to demonstrate the feature of self-organized criticality on a *cellular automaton network*. We can also see in Figure 2 that different probabilities p will lead to different values of exponents. The higher the probability, the steeper the slope.

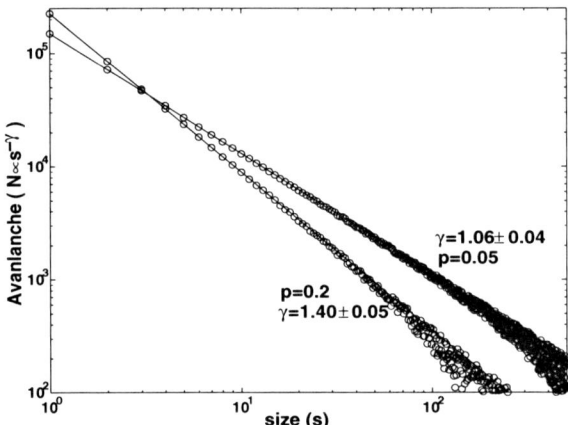

Figure 2. Avalanche size distribution for a small-world cellular automaton.

4.2 Transition of small-world systems

For a fixed grid of 2000×2000 cells, we can vary the probability p to
see what can happen. For a single event of flipping state, the fraction
of population affected is plotted versus p in a semi-logarithmic plot as
shown in Figure 3 where the fraction of population is defined as the
number (N_a) of cells affected among the whole population N^2, that
is N_a/N^2. The sharp increase of the fraction versus the probability p
indicates a transition in the properties of cellular automata networks.
For a very small probability $p < 0.004$, the influence of the event mainly
behaves in a similar way as the conventional local cellular automata.
As the probability increases, a transition occurs at about $p = 0.01$. For
$p > 0.05$, any event will affect the whole population. This feature of
transition is consistent with the typical small-world networks [12, 21].

Comparing with the local rule-based cellular automata, the transition in small-world cellular automata is an interesting feature. Without
such shortcuts, there was no transition observed in the simulations. However, self-organized criticality was still observed in finite-state cellular automata [1, 21] without transition. In the present case, both self-organized
criticality and transition emerge naturally. Thus, the transition in cellular automata networks suggests that this transition may be the result of
nonlocal interactions by long-range shortcuts.

This feature of transition may have important implications when applied to the modelling of real-world phenomena such as the Internet and
social networks. For a system with few or no long-range interactions,
there is no noticeable change in its behavior in transition. However, as

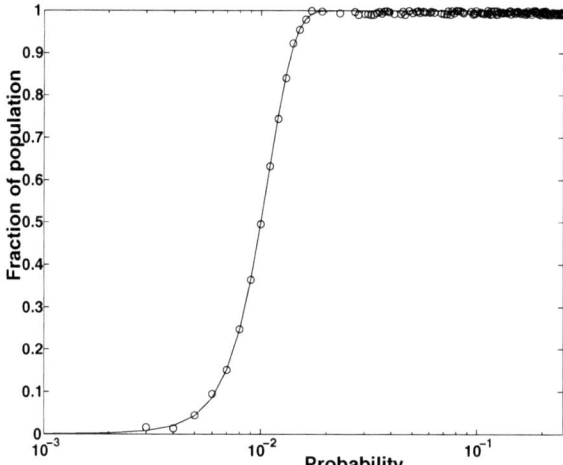

Figure 3. Transition of influence with different probabilities p of long-range shortcuts.

the long-range shortcuts or interacting components increase a little bit more, say to $p = 0.01$, then a transition may occur and thus any event can affect a large fraction of the whole population. For example, to increase the speed of finding information on the Internet, a small fraction of long-range shortcuts in terms of website portals and search engine (e.g., Google) and high-capacity/bandwidth connections could significantly increase the performance of the system concerned. In addition, the self-organized criticality can also imply some interesting properties of the Internet and other small-world networks, and these could serve as some topics for further research.

4.3 Nonlocal partial differential equations

The evolution of a system can usually be described by two major ways: rule-based systems and equation-based systems. The rule-based systems are typically discrete and use local rules such as cellular automata or finite difference system. As discussed by many researchers [18, 22], the finite difference systems are equivalent to cellular automata if the updating rules for the cellular automata are derived directly from their equation-based counterpart. On the other hand, the equation-based systems are typically continuous and they are often written as partial differential equations. Sometimes, the same system can described using these two different ways. However, there is no universal relationship between a rule-based system and an equation-based system [22]. Given differential equation, it is possible to construct a rule-based cellular automaton by discretizing the differential equations, but it is far more complicated to formulate a system of partial differential equations for a given cellular automaton. For example, the following 2-D partial differential equation for nonlinear pattern formation for $u(x, y, t)$

$$\frac{\partial u}{\partial t} = D(\frac{\partial^2 u}{\partial x^2} + \frac{\partial^2 u}{\partial y^2}) + \gamma f(u, x, y, t), \tag{8}$$

can always be written as an equivalent cellular automaton if the local rules of cellular automaton are obtained from the PDE. Conversely, a local cellular automaton can lead to a local system of partial differential equation (PDE), if the construction is possible [22]. A local PDE can generally be written as

$$\mathcal{F}(u, u_x, u_y, ..., x, y, t) = 0. \tag{9}$$

A nonlocal PDE can be written as

$$\mathcal{F}(u, u_x, u_y, ..., x, y, t, x + S(x, y, p), y + S(x, y, p)) = 0, \tag{10}$$

where $S(x, y, p)$ is the the averaged distance of long-range shortcuts and p is the probability of nonlocal long-range shortcuts. In order to show what a nonlocal equation means, we modify the above equation for pattern formation as

$$u_t = D\nabla^2 u(x, y, S, p, t) + \gamma u(x, y, S, p, t)[1 - u(x, y, S, p, t)]. \tag{11}$$

This nonlocal equation is far more complicated than equation (8). For the proposed cellular automaton networks, a system of nonlocal partial differential equations will be derived, though the explicit form of a generic form is very difficult to obtain and this requires further research.

4.4   Pattern formation

Even for simple nonlinear partial differential equations, complex pattern formation can arise naturally from initially random states. For example, the following nonlinear partial differential equation for pattern formation $u(x, y, t)$

$$\frac{\partial u}{\partial t} = D(\frac{\partial^2 u}{\partial x^2} + \frac{\partial^2 u}{\partial y^2}) + \gamma u(1 - u) + \beta u(x - S), \tag{12}$$

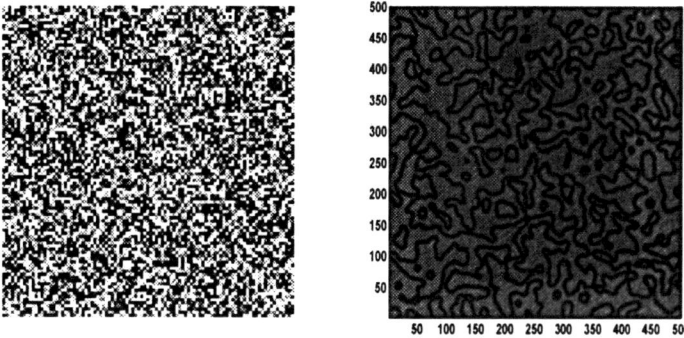

Figure 4. a) Initial random configuration on a 500×500 grid at $t = 0$; b) 2-D Pattern formation and distribution displayed at $t = 100$.

can be discretized using central finite difference scheme in space with $\Delta x = \Delta y = \Delta t = 1$. Then, it is equivalent to

$$u_{i,j}^{n+1} = \sum_{k,l=-r}^{r} a_{k,l} u_{i+k,j+l}^{n} + \gamma u_{i,j}^{n}(1 - u_{i,j}^{n}) + \beta u_{i-S,j}^{n}, \qquad (13)$$

where $r = 1, a_{0,0} = 1 - 4D, a_{-1,0} = a_{+1,0} = a_{0,-1} = a_{0,+1} = D$. It is a cellular automaton with the standard Neumann neighbourhood for this PDE.

The formed patterns and their distribution resulting from the system on a 500×500 grid are shown in Fig. 4 where $D = 0.2$, $S = 5$, $\gamma = 0.5$ and $\beta = 0.01$. We can see that stable patterns can be formed from initial random states.

The formed patterns using the Neumann neighbourhood ($r = 1$) are very stable and are almost independent of the initial conditions. In fact, the initial state does not matter and the only requirement for the initial state is some degree of randomness. If we run the same program using a photograph or the UC2007 conference logo, similar patterns can also form naturally as shown in Figure 5. In our simulations, we have used $D = 0.2$ and $\gamma = 0.5$. Other parameters S and β can vary. For the case shown in Fig. 5b, $S = 5$ and $\beta = 0.05$, while $S = 50$ and $\beta = 0.2$ are used in Fig. 5c. This means that the initial state does not affect the characteristics of pattern formation and this is consistent with the stability analysis [22].

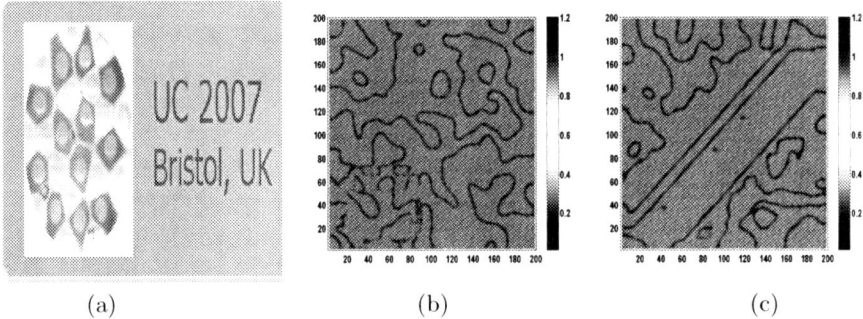

(a)   (b)   (c)

Figure 5. a) The UC2007 logo has been mapped onto a 200×200 grid as the initial condition at $t = 0$; b) 2-D pattern distribution with a short-cut probability $p = 0.0025$ and $S = 5$ (or $\beta = 0.05$) displayed at $t = 200$; c) 2-D pattern with a short-cut probability $p = 0.01$ and $S = 50$ (or $\beta = 0.2$) at $t = 200$.

5 Discussions

Small-world cellular automata networks have been formulated to simulate the interactions and behaviour of multi-agent systems and small-world complex networks with long-range shortcuts. Simulations show that power-law distribution emerges for a fixed probability of long-range shortcuts, which implies self-organized criticality in the avalanche and evolving complex patterns. For a given size of cellular grid, the increase of the probability of long-range shortcuts leads to a transition, and in this case, a single even can affect a large fraction of the whole population. In this sense, the characteristics of small-world cellular automota are very different from the conventional locally interacting cellular automata.

The nonlocal rule-based network systems in terms of cellular automata can have other complicated features such as its classifications compared with the conventional automata and its relationship its partial differential equations. In addition, cellular automota networks could provide a new avenue for efficient unconventional computing for simulating complex systems with many open questions such as the relationship between cellular automata networks and nonlocal PDEs, and the potential implication on the parallelism of these algorithms. These are open problems to be investigated in the future research.

References

[1] Adami, C, Self-organized criticality in living systems, *Phys. Lett. A*, **203**, 23 (1995).

[2] Appert C. and Zaleski S., Lattice gas with a liquid-gas transition, *Phys. Rev. Lett.*, **64**, 1-4 (1990).

[3] Bak P, Tang C and Wiesenfeld K, Self-organized criticality: an explanation of 1/f noise, *Phys. Rev. Lett.*, **59**, 381-384 (1987).

[4] Bak P, *How nature works:the science of self-organized criticality*, Springer-Verlag, (1996).

[5] Boccara N. and M. Roger, Some properties of local and nonlocal site exchange deterministic cellular automata, *Int. J. Modern Phys.*, **C5**,581-588 (1994).

[6] Bollobas B., *Random graphs*, Academic Press, New York, (1985).

[7] Boots M., Sasaki A., Small worlds and the evolution of virulence: infection occurs locally and at a distance, *Proc Roy Soc Lond*, B **266**, 1933-1938 (1999).

[8] De Arcaneglis, L. and Herrmann, H. J., Self-organized criticality on small world networks, *Physica A*, **308**, 545-549 (2002).

[9] Guinot V., Modelling using stochastic, finite state cellular automata: rule inference from continuum model, *Appl. Math. Model.*, **26**, 701-714(2002).

[10] Moukarzel C. F., Spreading and shortest paths in systems with sparse long-range connections, *Phys. Rev. E*, **60**, 6263-6266, (1999).

[11] Newman M. E. J., Moore C., Watts D. J., Mean-field solution of the small-world network model, *Phys. Rev. Lett.*, **84**, 3201-3204 (2000).

[12] Newman M. E. J., Watts D. J., Scaling and percolation in the small-world network model, *Phys. Rev. E*, **60**,7332-7342 (1999).

[13] Pandit S. A., Amritkar R. E., Characterization and control of small-world networks, *Phys. Rev. E*, **60**, 1119-1122 (1999).

[14] Tomassini M., *Generalized automata networks*, 7th Int. Conference on Cellular Automata for Research and Industry, ACRI 2006, France, *Lecture Notes in Computer Sciences*, **4173**, 14-28 (2006).

[15] Tomassini M., Giacobini M., Darabos C., Evolution and dynamics of small-world cellular automata, *Complex Systems*, **15**, 261-284 (2005).

[16] Watts D. J., *Small worlds: The dynamics of networks between order and randomness*, Princeton Univ. Press, (1999).

[17] Watts D. J., Strogatz S. H., Collective dynamics of small-world networks, *Nature* (London), **393**, 440-442 (1998).

[18] Weimar J. R., Cellular automata for reaction-diffusion systems, *Parallel computing*, **23**, 1699-1715 (1997).

[19] Wolfram, S., *Cellular automata and complexity*, Reading, Mass: Addison-Wesley (1994).

[20] Yang X. S., Chaos in small-world networks, *Phys. Rev. E.*, **63**, 046206 (2001).

[21] Yang X. S. and Young Y., Cellular automata, PDEs, and pattern formation, in: Handbooks of Bioinspired Algorithms and Applications (eds. Olarius S. and Zomaya A. Y.),Chapman & Hall /CRC Press, 271-282 (2005).

[22] Yang X. S., Computational modelling of nonlinear calcium waves, *Applied Maths. Modelling*, **30**, 200-208 (2006).

Wholeness based on gluing of incomplete information

Eugene S. Kitamura[1] and Yukio-Pegio Gunji[2]

[1] Graduate School of Science and Technology
Kobe University, Nada, Kobe 657-8501, Japan
[2] Department of Earth and Planetary Sciences, Faculty of Science,
Kobe University, Nada, Kobe 657-8501, Japan
esk@luna.email.ne.jp

Abstract The cause of emergence is described as a constant collapse and restoration of intent-extent relationships in robust systems. We use a cellular automata (CA) model as a platform with conditions such that 1) each cell is not fully aware of the reference CA rule (indeterminacy), therefore it must make up its own rule by partial disjunctive normal form (PDNF), and consequently, 2) every cell obeys a different CA rule. By incorporating such conditions to a CA model, Class 4 "complex" behavior is seen more ubiquitously, unrestricted to "the edge of chaos." Standard deviation of entropy is examined to quantitatively measure the Class 4 behavior. When internal-perturbation is applied to the CA (map-perturbation), the system shows even greater robustness.

Key words: Internal measurement; cellular automata; edge of chaos; robustness; intent-extent; partial disjunctive normal form.

1   Introduction

Cellular automata (CA) are a computer modeling platform used in numerous kinds of research, from biological behavior to image recognition. It is a spatially and temporally discrete model, and its "cells" take on finite discrete states. This model was first introduced by von Neumann and Ulam [1–3]. Studies were made later on statistical and universality properties of CA by Stephen Wolfram [4–6]. Wolfram found that CA can be grouped into four categories: Class 1, orderly patterns that dissipate immediately; Class 2, orderly patterns that settle in localized repetitive structures; Class 3, chaotic patterns; and Class 4, complex patterns that contain both orderly and chaotic patterns. These four groupings are analogous to behaviors in the field of dynamic systems; dissipative, repetitive, chaotic, and complex dynamics.

It is said that complex dynamics occur at the "edge of chaos," a region between orderly and chaotic dynamics [7,8]. With this notion, the occurrence of complex dynamics and resulting systems such as evolution and biological systems or social systems becomes a rarity, only capable of flourishing in a limited parameter range. Naturally, one could question what the mechanism is in nature to choose such particular parameters in systems to emerge and evolve. This question is yet to be answered.

Robust systems with emergent properties can be seen abundantly in our living environment. At every moment, adaptation or formation of structure and functionality takes place on all levels of an emergent system. The CA model used to derive the "emergence at the edge of chaos" is rigid and does not include important features of emergent systems in the model itself, that is, robustness and the indeterminacy of states.

Robustness is the ability of a system to retain its function while tolerating perturbations to its structure [9]. Indeterminacy is assumed when the states of the subcomponents of a system cannot be detected accurately [10–13]. Indeterminacy is a characteristic in a model that makes emergent phenomena possible. This notion originates from the internal measurement perspective [14–16] or endo-physics [17,18]. Indeterminacy occurs when the information that can be acquired by an observer is compromised because the observer is located within the system. An internal observer is restricted from the view of the whole system. However, this constraint does not prevent the observer from modeling the whole system. Indeterminacy forces the observer to "speculate" based on what it already knows (or has) to fill in the unknown states of subcomponents. In fact, choices for such "speculation" are partially open to possibilities outside of the normally expected states. This partial openness creates a discrepancy between intent, which is the formula (or map) that generates the output (or state), and extent, which is the collection of states to which the formula applies. The relationship between intent and extent collapses. Yet this collapse is immediately restored by the internal dynamics of the system's subcomponents. This process is what is commonly known as "emergence."

2   Collapse and restoration of adjoint functors

In a categorical theory, when there is a transformation relationship between two categories, this transformation is called a functor. Intent is defined as a collection of attributes of a concept and extent is defined as a collection of objects to which the concept is applied. The world of intent and extent can be described as a category and the relationship between them is described as a functor. When intent and extent fully match each other, the pair of functors is called adjoint functors.

We here illustrate an example of intent-extent relationships that match each other (adjoint) by using a simple 2-neighbor 2-state CA. A category of Intent, Int, consists of a formula expressed as a disjunctive normal form (DNF) and inclusion relation between formulas. DNF is a Boolean logic formula consisting of $\wedge$ (and) and $\vee$ (or). Since each DNF is expressed as a $\vee$-bind consisting of $\wedge$-term such as $x \wedge y$, the inclusion stands for the inclusion with respect to $\wedge$-terms. A category of Extent, Ext, consists of a truth table (T, a relation of input-output) representing 2-state CA and the inclusion between truth tables with respect to the (input, output) relation such as $((x, y), 1)$. The x and y represent a two-neighboring cell input.

A functor $Exp_1 : Ext \rightarrow Int$ sends a truth table T to a DNF and the inclusion in Ext to the inclusion in Int by the following. For a truth table $T = \{((i, j), k) \mid i, j, k \in \{0, 1\}\}$ where T never contains both $((i, j), 0)$ and $((i, j), 1)$,

$$Exp_1(T_1) = \vee \left(x^{e(p)} \wedge y^{e(q)} \right) , \tag{1}$$

where $x^e(0) = x^c, x^e(1) = x$. $y^e(q)$ is defined in the same manner, and (p, q) satisfies $((p, q), 1)$. Given two truth tables $T_1 = ((0,0), 1)$, $((0, 1), 0)$, $((1, 1), 0)$ $T_2 = ((0, 0), 1), ((0, 1), 1)$, we have

$$Exp_1(T_1) = x^c \wedge y^c \subseteq Exp_1(T_2) = (x^c \wedge y^c) \vee (x^c \wedge y) , \tag{2}$$

standing for the preservation of inclusion.

A functor $Val_1 : Int \rightarrow Ext$ sends DNF to the truth table T and the inclusion in Int to the inclusion in Ext by the following. For a DNF $= f(i, j)$,

$$Val_1(\text{DNF}) = \bigcup_{i,j \in \{0,1\}} \{(i, j), f(i, j)\} . \tag{3}$$

It is easy to see that Val_1 preserves inclusion. The adjunction between Exp_1 and Val_1 stands for

$$Ext(Val_1(\text{DNF}), T) \simeq Int(\text{DNF}, Exp_1(T)) . \tag{4}$$

Note that the formula in Intent is based only on a known input-output relation that outputs 1 in Extent (i.e. $((p, q), 1)$). This formula

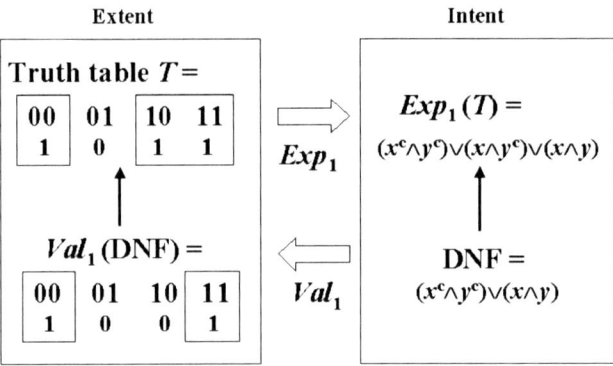

$$Ext(Val_1(DNF), T) \cong Int(DNF, Exp_1(T))$$

Figure 1. An example of intent-extent relationship and adjoint functors that satisfy $((x, y), 1)$ Known input-output relations in the truth table T are outlined. A pair made of 1 or 0 (i.e. $00, 01, \cdots$) is the input. It outputs a single number, 1 or 0. The vertical arrows suggest inclusive relationship.

in Intent outputs the same (or "correct") values for both the known and unknown input-output relations in Extent. See Fig. 1.

When a mismatch occurs between intent and extent, the adjoint functors collapse. Fig. 2 shows an example of a process where the adjoint relation between intent and extent collapses. Given a truth table $T' = \{((0, 0), 1), ((0, 1), 0), ((1, 0), 0),$ $((1, 1), 0)\}$, $Exp_1(T) = x^c \wedge y^c$. Now we replace DNF by DNF with redundant terms such that, for $T = \{((i, j), k) \mid i, j, k \in \{0, 1\}\}$

$$Exp(T) = (\vee(x^{e(p)} \wedge y^{e(q)})) \vee (\vee(x^{e(p')} \wedge y^{e(q')}))^c \qquad (5)$$

where $\{((p, q), 1), ((p', q'), 0)\} \in T$. For T' we get

$$Exp(T') = (x^c \wedge y^c) \vee ((x \wedge y^c) \vee (x^c \wedge y) \vee (x \wedge y))^c \quad . \qquad (6)$$

It is clearly seen that $Val(Exp(T')) = T'$. Since in the sense that $Val(Exp(T')) = Val(Exp_1(T))$, the term $((x \wedge y^c) \vee (x^c \wedge y) \vee (x \wedge y))^c$ is redundant. If one accepts DNF with redundant terms, the outputs of this new formula in Intent may be different from the actual input-output relation of the Extent. A mismatch may occur between intent and extent. However, it does not matter whether the output from the formula is "correct" compared with the original input-output relation of Extent. Because of indeterminacy, the "correct" states cannot be known. The point is, the relationship between intent and extent of a system collapses, but is perpetually restored. It is restored by filling in the gaps with new variety of states due to partial openness of one of the categories. The dynamics of the adjoint relationship collapsing, or "the bottom (of the limit to possibilities) falling out," and the adjoint being restored again with expanded possibilities of states is the source of emergent phenomena [19, 20].

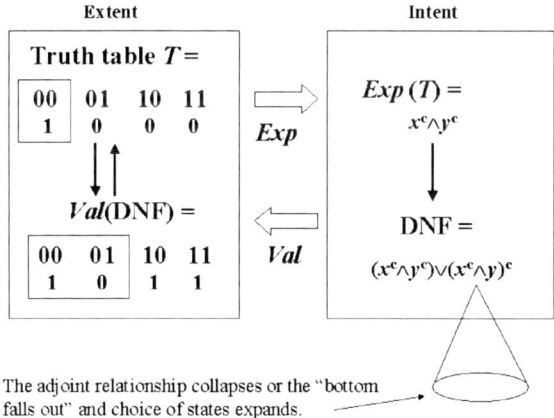

Figure 2. Example of intent-extent relationship with collapsed adjoint functor Intent is "partially opened" to the outside of its corresponding Extent. The formula in Intent is based on known input-output relations that output 1 and 0 in Extent. Such an equation outputs the same (or "correct") result for the known input-output relation in Extent, but not necessarily the "correct" values for the unknown input-output relation.

Emergence is the dynamics of constant collapse and restoration of intent and extent in robust systems. When an observer takes an internal measurement perspective, Class 4-like dynamics are more likely to be observed. We use a 3-neighbor 2-state (CA) model as a platform with conditions such that 1) each cell is not fully aware of the reference CA rule table due to indeterminacy, therefore it must make up its own rule, and consequently, 2) every cell obeys a different CA rule. By incorporating such conditions to a CA model, Class 4 "complex" behavior is seen more ubiquitously, regardless of the "edge of chaos." It shows that indeterminacy contributes much to models with robustness.

3 Partial disjunctive normal form fashioned cellular automata

3.1 Disjunctive normal form

The rule for elementary cellular automata (ECA) can be described by using DNF. The three neighboring cells, the input triplets, $(x_{i-1}^t, x_i^t, x_{i+1}^t)$ can be grouped together by $\wedge$. To represent a complete rule table, each of the rules written in the form $(x_{i-1}^t \wedge x_i^t \wedge x_{i+1}^t)$ is connected by $\vee$. The following equation is an example of the rule table, represented as a disjunctive normal form equation when the output for 000 and 001 are 1.

$$x_i^{t+1} = (f(0,0,0) \wedge x_{i-1}^c \wedge x_i^c \wedge x_{i+1}^c) \vee (f(0,0,1) \wedge x_{i-1}^c \wedge x_i^c \wedge x_{i+1}) \ . \ (7)$$

In this equation, "t" is omitted. When 000 or 001 is the input, the equation will output 1. For other input states $(010 \cdots 111)$ the equation will output 0.

In order to interrupt the adjunction between intent and extent of our CA model, we add redundant terms to the disjunctive normal form, as described in Fig. 2. The additional terms are based on three neighboring cells that output 0. For terms based on cells that output 0, we take the complement of the entire group of terms that output 0 in order to obtain the correct results for the known triplets. For example, in addition to the above equation, if we know that 010 and 011 outputs 0, the equation becomes

$$\begin{aligned} x_i^{t+1} = \ & (f(0,0,0) \wedge x_{i-1}^c \wedge x_i^c \wedge x_{i+1}^c) \vee (f(0,0,1) \wedge x_{i-1}^c \wedge x_i^c \wedge x_{i+1}) \vee \\ & \{(f(0,1,0) \wedge x_{i-1}^c \wedge x_i \wedge x_{i+1}^c) \vee (f(0,1,1) \wedge x_{i-1}^c \wedge x_i \wedge x_{i+1})\}(8) \end{aligned}$$

3.2 Partial disjunctive normal form

When only a portion of the rule table is available due to indeterminacy, the general equation of the rule table is made up from the available triplets by using the partial disjunctive normal form (PDNF). The formulated equation gives the same outputs for the known triplet inputs on which the equation was based. For the other triplet inputs for which the outputs are unknown, the equation does not necessarily output the "correct" values that correspond to the unknown triplet inputs. For triplets with unknown outputs, the equation outputs the following. If the equation was based only on triplet inputs that output 1, such as $101 \to 1$, the equation always outputs 0. This type of equation is called a 1-term equation. If the equation was based only on triplet inputs that output 0, such as $010 \to 0$, the equation always outputs 1. This type of equation is called a 0-term equation. If the equation was based on triplets that output 1 and on triplets that output 0, such as $101 \to 1$ and $010 \to 0$, the equation always outputs 1. The equation behaves like a 0-term equation. Therefore, a 0-term equation which outputs 1 is more likely to occur. Fig. 3 describes the PDNF CA's process in a diagram. The outputs are generated by the partial differential normal form. As a result, different output values may be created for each input triplet compared to the original target rule table.

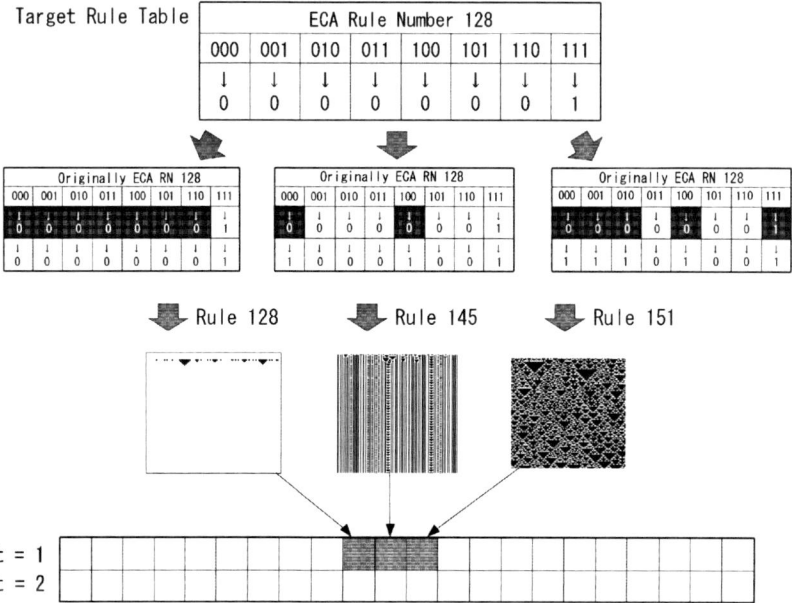

Figure 3. The PDNF CA process. The top table represents target rule number 128. The three tables in the second row are examples of tables with parts of the target table hidden. The three images in the third row are the resulting patterns due to the new rule table. The fourth row is an image of cells extracted from a one dimensional PDNF CA.

4 Results

As a result of the above process, we obtain a Class 4 image that is neither periodic, nor chaotic, as shown in Fig. 4. Cluster patterns may emerge from various locations, grow for some time steps, and eventually fade away. The mechanism that generates such patterns is the strong presence of $111 \rightarrow 1$, which forms a black upside-down triangle, since 1's are marked with black dots. Rule numbers such as 128, 129, 131 contain $111 \rightarrow 1$. Additionally, the PDNF method is more likely to produce $111 \rightarrow 1$, since 0-term equations are more likely to occur.

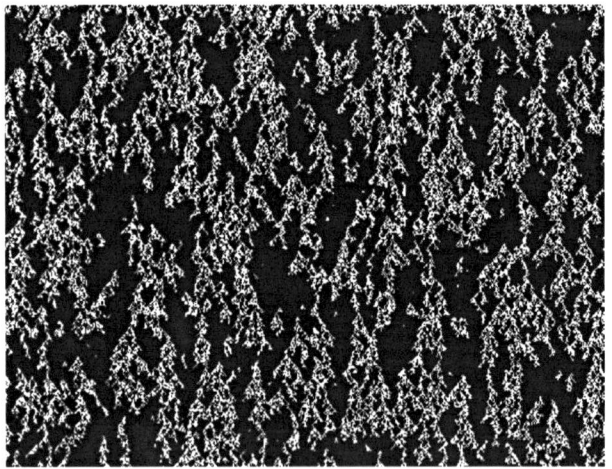

Figure 4. An example of CA pattern created with PDNF process. The target rule used is the ECA equivalent rule number 128

We measured the standard deviation of entropy and the Lyapunov exponent of the PDNF CA. The standard deviation of entropy was compared with the probability controlled CA and the Lyapunov exponent was compared with the ECA RN110 with randomly assigned initial conditions.

4.1 Comparing standard deviation of entropy of PDNF CA with probability controlled CA

We use the Shannon entropy,

$$S = -k \sum_{i=1}^{8} (p_i \times \log p_i) \; , \tag{9}$$

where S is the average entropy of the different triplets that occur in the system, i is each triplet in decimal, and p_i is the occurrence frequency of each input triplet. k is a normalization factor, $k = 1/n$, where n is the number of cells in the neighborhood. The data used to calculate the entropy is collected 100 to 300 time steps after the program has started, when the system reaches a steady state in order to eliminate the noise from the initial values.

The standard deviation of entropy reflects the variation in the occurrence of triplets. The standard deviation σ is,

$$\sigma = \sqrt{\frac{1}{N-1} \sum_{i=1}^{N} (S_i - S_{avr})^2} \; , \tag{10}$$

where S_i is the entropy of the occurred triplet, S_{avr} is the average entropy of S_i, and N is the number of entropy used to calculate the standard deviation. If the model is Class 4 type, the standard deviation of entropy is relatively high [21]. This means that there is a bias in the occurrence of triplets in the rule table.

The probability controlled CA is a model where the occurrence of each triplet in a rule table is controlled by a probability. With a strong presence of certain triplets, one can obtain Class 4 behavior. Here we focus on probability controlled CA with rule numbers varying from 128 to 135, since the PDNF CA uses these rule numbers. See Fig. 5. The transition from rule number 128 to 129 is achieved by gradually increasing the occurrence probability of 000 → 1. From 129 to 131, we increase 001 → 1, etc. With the method introduced above to measure the standard deviation of entropy, we can compare the standard deviation of entropy of probability controlled CA and that of PDNF CA. The graph below compares the standard deviation squared of entropy of the PDNF CA and the probability controlled CA. The dark circles indicate the standard deviation squared of entropy of the PDNF CA and the diamond marks are that for the probability controlled CA. The vertical arrows show the rise in the standard deviation squared of entropy due to the effect of PDNF. The standard deviation squared of entropy rises significantly from the probability controlled CA to the PDNF CA at target rule tables of 128, 129, 131, and 135. The rise is 0 to 0.014658 for target rule table 128, 0.001981 to 0.013733 for 129, 0.001651 to 0.009734 for 131, and 0.000268 to 0.005821 for 135. This suggests that higher stan-

dard deviations of entropy can be obtained more ubiquitously with the
PDNF CA method.

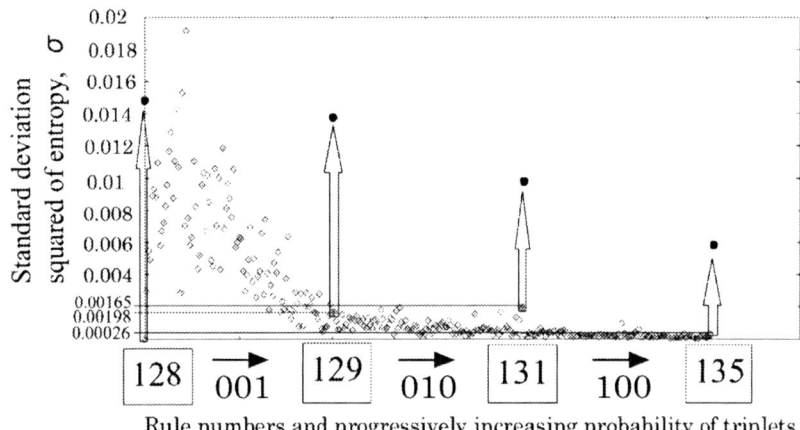

Figure 5. Standard deviation squared of entropy for PDNF CA and probability CA, as the probability of specific triplets is increased along the x-axis. The
dark circles indicate the standard deviation squared of entropy of the PDNF
CA and the diamond marks are that for the probability controlled CA. The
vertical arrows show the rise in the standard deviation squared of entropy. The
standard deviation squared of entropy rises significantly from the probability
controlled CA to the PDNF CA at target rule tables of 128, 129, 131, and 135.

4.2   Comparing the Lyapunov exponent of PDNF CA with elementary cellular automata

To test the robustness of a system to perturbation, we can study the
Lyapunov exponent of the system. The Lyapunov exponent measures the
time development of an infinitesimal difference in the initial states in the
model. To find the Lyapunov exponent for CAs, we measure how quickly
the difference between the two CAs grows. The time steps, $T(\Delta 2^n)$,
for the difference to propagate over an exponentially growing horizontal
distance $\Delta 2^n$ is measured. Generally, when the inverse time $1/T(\Delta 2^n)$ is
plotted against the exponentially growing horizontal distance $\Delta 2^n$, the
slope λ would tell how rapidly the expansion is growing. λ is calculated
by the equation,

$$\lambda = \frac{\left(\frac{1}{T(\varDelta 2^n)}\right)}{\varDelta 2^n} \ .$$

(11)

The log of the above equation was taken so the log of the inverse time, $\log\left(1/T(\varDelta 2^n)\right)$, was plotted against the exponent n. The horizontal distance $T(\varDelta 2^n)$ is measured on both directions from the center. The smaller $T(\varDelta 2^n)$ of the two directions was used for calculation.

Lyapunov exponent for ECA 110 ECA 110 is known to have a complex behavior. Fig. 6 (a1) shows an example of two CAs with random initial conditions (left and center) and the image of the difference between the two CAs (right). The only difference between the left and center CA is the state of the initial condition at the center of the system. For the image on the right, there is a black dot at a cell where the initial perturbation has propagated. The perturbation generally keeps expanding, even though there are occasional contractions and stability. Fig. 6 (a2) shows the value of the Lyapunov exponent at every n for ECA 110. The difference keeps expanding through greater n and grows steadily and rapidly. This graph is formulated by averaging 100 cases of ECA 110 with different randomly assigned initial conditions for every case.

Lyapunov exponent for PDNF CA with external-perturbation Here we present the Lyapunov exponent of CA generated by partial disjunctive normal form. The two CAs have the same randomly assigned initial conditions, except for the state of the cell at the center. Since the state, the output of the CA input-output relation, is perturbed, it is called external-perturbation or state-perturbation. Both CAs obey the same map (the input to output relation) constructed by the PDNF equation. The propagation of initial perturbation of the two CAs eventually disappears, as shown at the far right of Fig. 6 (b1). The perturbed difference of the two CAs is likely to disappear for the following reason. The maps created by PDNF are more likely to produce 1s, thus making it easier for the two CAs to result in common states. In some cases, the initial perturbation does not propagate at all; the difference disappears only after few time steps. In other cases, as shown in Fig. 6 (b1), the difference lasts longer. Unlike the Lyapunov exponent values for ECA 110, the difference between the two CAs dissipates after a limited time (Fig. 6 (b2)). On average, the initial perturbation propagates only as far as $2^4 = 16$ cells horizontally. This data is averaged over 100 PDNF CA systems. The target rule used is the ECA equivalent rule number 128.

Lyapunov exponent for PDNF CA with internal-perturbation
We now compare two CAs with the same, randomly selected initial
states, but with a different initial map for the center cell. A map is the
relationship between the input and the output of a CA. Since the map,
an internal property of the PDNF CA model, is perturbed, it is called
internal-perturbation or map-perturbation. The different map is chosen
randomly from the 256 possible maps. As time steps progress and initial
perturbation propagates, the maps for cells where initial perturbation
has propagated are also assigned randomly different maps between the
two CAs. (The maps for cells where initial perturbation has not propagated have the same maps.) An example image of two such CAs is shown
at the left and center in Fig. 6 (c1). The left image is the reference PDNF
CA. The center image is the PDNF CA with the initial map difference
at $t = 0$ at the center cell of the system. The image on the right is the
difference between the two CAs, indicated by black dots. The graph of
the Lyapunov exponent at every n is shown in Fig. 6 (c2). Similar to
PDNF CA with the same map and different initial states (PDNF CA
with state-perturbation), the initial perturbation does not keep propagating as in ECA 110. However, the initial perturbation does grow larger
than PDNF CA with state-perturbation (Fig. 6 (b2)). Additionally, the
initial perturbation keeps propagating through greater time steps (or
smaller $\log(1/T)$). On average, the difference seems to expand as far as
$2^6 = 64$ cells, greater than PDNF CA with the same maps and different
initial states. The target rule used is the ECA equivalent rule number
129.

**Lyapunov exponent for PDNF CA with maximizing external
perturbation** The last model (Fig. 6 (d1)) is a preparation for the
following section. There is an initial difference in the state at the center
cell. In this model, the same map is created for both CAs at every cell
such that the output states will always be different for the two CAs. For
example, if a map output is 1 (e.g. $111 \rightarrow 1$ in the center at $t = 0$) for one
CA, the same map output is 0 (e.g. $101 \rightarrow 0$ in the center at $t = 0$) for the
other CA. Consequently, the progression rate of perturbation is always
maximized. Since a PDNF equation is created such that a different state
will result in the two CAs, naturally the initial perturbation will expand
steadily over time, as seen in Fig. 6 (d2). The target rule used is the
ECA equivalent rule number 128.

4.3   Contributions of map-perturbation to robustness

By using PDNF CA, we are not only able to control or perturb the states
of cells, but also the maps of the PDNF CA model. We can analyze the

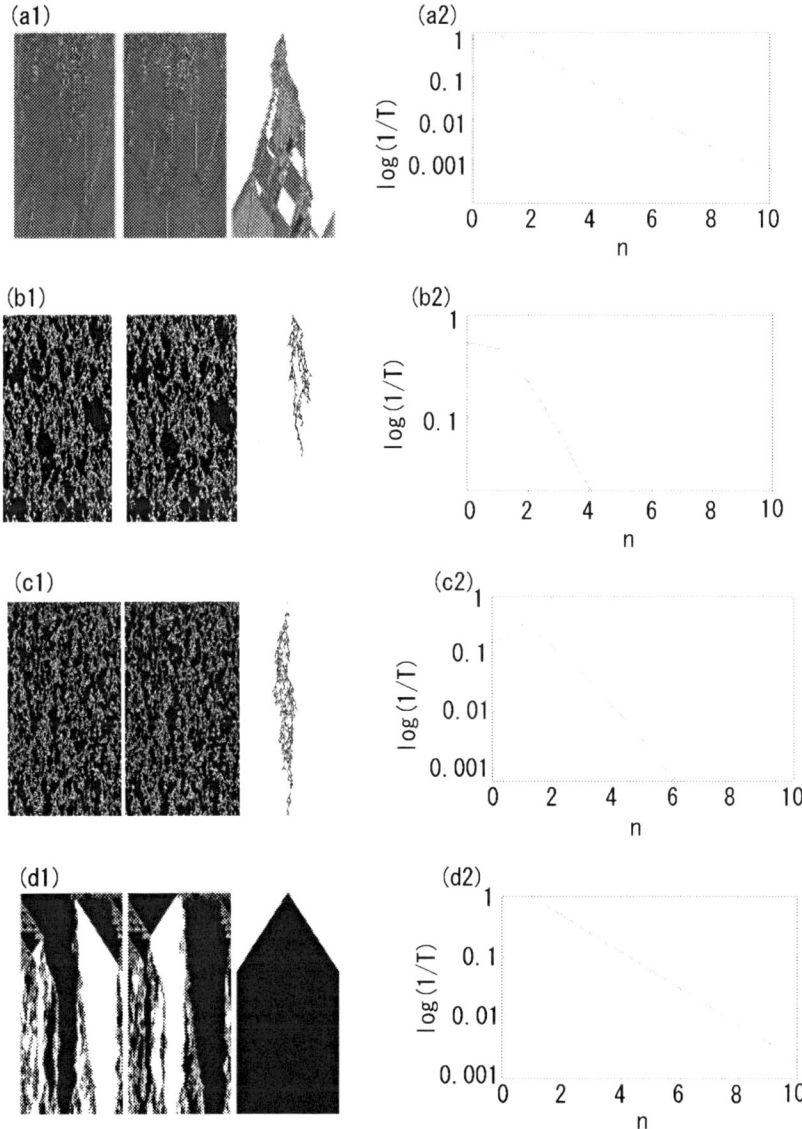

Figure 6. Images (a1)-(d1) show an example of two CAs (left and center image) and the difference between the two CAs (right image). Graphs (a2)-(d2) show the Lyapunov exponent at every n. Figures (a1-2) are for ECA 110; (b1-2) are for PDNF CAs with a same map for both CAs, but with a different initial state condition for the cell in the center; (c1-2) are for PDNF CAs with the same randomly chosen initial states, but with different maps at the center cell; (d1-2) are for PDNF CAs with different initial states at the center, with the same maps for both CAs such that it would always produce different states in the two CAs.

robustness of the model by varying the degree of perturbation applied
to the model.

First we start with a two CA system with identical maps that always output the opposite states as in Fig. 6 (d1) and (d2). The initial
state-perturbation expands at the fastest rate as time progresses. (Line
with cross markers in Fig. 7, at $x = 0$.) Next, instead of choosing maps
that produce opposite output states for every cell, we progressively introduce arbitrary maps that do not necessarily output opposite states.
Consequently, the initial perturbation expansion is not as strong as when
perturbation-maximizing maps are used. The result is plotted as a line
with cross markers in Fig. 7. The y-axis is the perturbation growth rate.
We normalize the maximum perturbation expansion to 1. The x-axis is
the probability of choosing an arbitrary map for every cell. The graph
shows that when the perturbation propagation is the largest (at $x = 0$),
the CA behaves like a chaotic Class 3 CA. As arbitrary maps are chosen
more frequently, the CA behaves like a Class 4 CA between orderly and
chaotic dynamics. When all maps are chosen randomly (at $x = 100$),
there is virtually no expansion of perturbation. The CA behaves like a
Class 1 or Class 2.

Now we introduce map-perturbation to one of the two CAs, in addition to state-perturbation, as in Fig. 6 (c1) and (c2). This means
that a portion of the two CAs would not have the same maps. As time
steps progress, cells where the initial perturbation has propagated have
different maps randomly chosen (map-perturbation is applied). Map-
perturbation is not applied to cells where the initial perturbation has not
propagated. (Cells in areas where initial perturbation has not propagated
have the same maps.) The result is plotted as a line with square markers. At $x = 0$, the effect of initial perturbation is damped significantly, to
almost half of the CAs with perturbation-maximizing maps. Randomly
choosing different maps between the two CAs (map-perturbation) interferes with the perturbation-maximizing maps created for the original
state-perturbation CAs. The PDNF CA with map-perturbation starts
out from $x = 0$ with a Class 4 type response to perturbation. As x
increases, the effect of perturbation is damped, but not as rapidly as
CA without map-perturbation (line with cross markers). The system's
response to initial perturbation remains robust. At $x = 90$, the initial
perturbation growth for CA with and without map-perturbation is about
the same. At $x = 100$, the influence of the initial perturbation continues in the CA with map-perturbation, whereas there is hardly any initial perturbation development in the CA with only state-perturbation.
While PDNF CAs with the same maps are likely to end up with the
same states, PDNF CAs with map-perturbation are less likely to have

the same states, therefore the initial perturbation remains in the system. The CA without map-perturbation settles in the Class 1 and Class 2-like stable behavior, however the CA with map-perturbation continues with a Class 4 behavior.

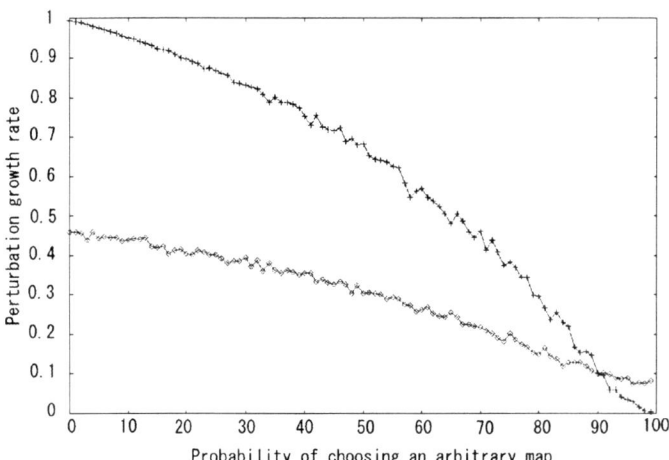

Figure 7. Comparing CAs with only state-perturbation (line with cross markers) and CAs with map-perturbation (line with square markers). Perturbation growth is normalized to one.

Fig. 8 shows various images of the initial perturbation propagation along the line of cross markers (CAs with state-perturbation) and square markers (CAs with state-perturbation and map-perturbation) from Fig. 7. Along the line with cross markers, at $x = 0$ the perturbation forms an isosceles triangle because maps are selected such that the spread of perturbation is maximized. The image of perturbation propagation becomes smaller as maps are selected randomly. When the CAs are completely dependent on arbitrary maps at $x = 100$, the image for the perturbation expansion barely exists. Along the line with square markers, map-perturbation interferes with the perturbation-maximizing map which produces the isosceles triangle image. At $x = 0$, the image of propagation of initial perturbation is the largest along the x-axis, however it is half the size of the CAs without map-perturbation. As x increases, the image of the propagation of initial perturbation gradually decreases (lower images in Fig. 8) and the size of the image becomes narrower, but at a slower rate compared to the rapidly decreasing perturbation propagation of CAs without map-perturbation (upper images in Fig. 8). At $x = 100$, the effect of map-perturbation persists, reflecting the behavior of robust or emergent systems.

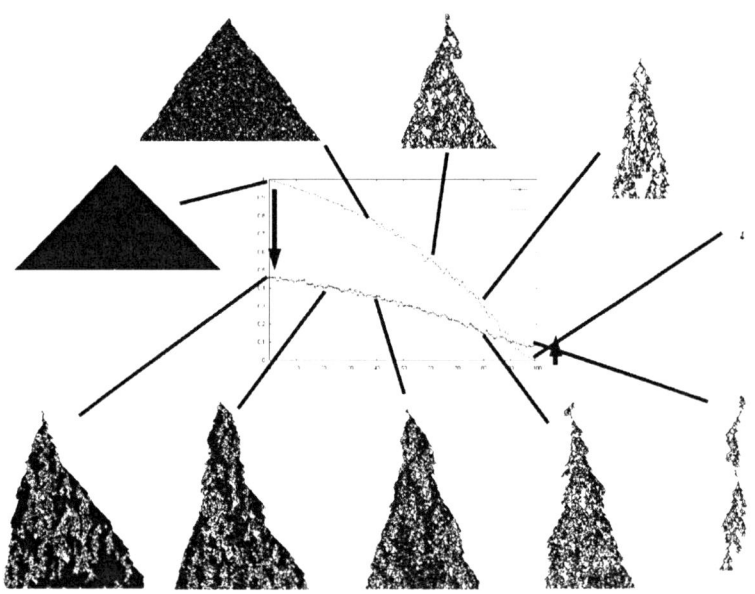

Figure 8. Images of propagation of initial perturbation for CAs with and without map-perturbation.

5   Discussion and conclusion

Emergence is a consequence of constant collapse and restoration of adjoint relationships between intent and extent. When intent or extent exchanges information (or matter) with the outside of its boundaries, the adjoint relationship collapses. However, with the restoration process, an observer perceives the system as an emergent system. For such description to be made, indeterminacy of the internal measurement perspective is required [14–18]. By incorporating such characteristics, emergence becomes a commonly occurring phenomenon.

When modeling complex systems with emergent or robust characteristics that consist of subcomponents, it ought to be captured as a conglomeration of subcomponents which are not fully aware of the dynamics of the whole system. When the condition of the whole system changes because of shifts in the environment, the subcomponents must adapt to the changing dynamics of the whole system. Each subcomponent adapts to the change based on the limited information available to it. When there is a lack of information on the directionality of the change, this lack of information does not halt the subcomponent from adaptation. To deal with the lack of information, the subcomponent would make up a new map to adapt to the current situation. The "correctness" of the made up map for a daptation is irrelevant. In a new situation there is no "correct" adaptation, and one could even choose paths that may seem out of context. This type of response is called emergence [22].

Our model systematically captures such functions of emergent systems. We used CA as a model structure. Unlike conventional CA that uses a rule table commonly referred to by all cells in the grid, our CA model uses a reference rule table in which randomly selected different parts are available to different cells. The unknown rules in the incomplete rule table are not arbitrarily generated, but are formulated based on the rules of PDNF. By using PDNF, this allows the adjunction between intent and extent to collapse. In this model, a given rule table is only a reference. It is not the system's goal to find the "ideal" rule table or to find the original rule table. The subcomponents make up their own map based on information already available. This means that each cell is partially open to options outside of the regular context or the adjoint relationship. Remarkably, instead of the system falling into utter chaos, it shows a complex Class 4 CA behavior that contains both orderly and chaotic characteristics. Additionally, the experiment compareing the standard deviation of an entropy for PDNF CA and probability controlled CA suggest that such complex dynamics can be seen more ubiquitously beyond the "edge of chaos."

Another experiment to test the robustness of our model is to observe the system's response to perturbation. Robustness is the ability of a system to retain its functionality while allowing structural changes within the system [9, 19]. Rosen suggests that when dealing with the robustness of a system, one must consider not only the perturbation of states $(x+\delta x)$ but also the perturbation of maps $(f + \delta f)$ of the system [23, 24]. Complex systems in nature experience not only perturbation to the states, but also perturbation to the internal structure of the system. In a human body, for example, cells are constantly renewed by metabolic activities. Yet a person basically continues to maintain oneself as the same person. A large company or business constantly experiences staff changes due to retirement or new hiring. In spite of such internal perturbation, the company maintains its functionality. Computational machines and robots today have little such robust traits. Such systems could be stable. However stability is a result of invariance in the structure of the system [19]. Typically, when a structural irregularity occurs in such systems, it ceases to function or outputs an error [25]. Such machines may be stable, but they are rigid. If we seek to construct emergent or robust machines, we must embed concepts such as "looseness" and "indeterminacy" in the design so that the adjoint between intent and extent can collapse and be restored.

Flexibility in controlling the robustness of the system is only possible when the model contains indeterminacy. With only external- (or state-) perturbation $(x + \delta x)$ in the PDNF CA model, it does result in Class 4 emergent behavior, however only under limited conditions. In the Lyapunov exponent experiment with state-perturbation, when the map for two CAs is chosen such that it would output opposite states, the propagation of the initial perturbation is maximized and the system shows chaotic behavior. On the other hand, when the same map for two CAs is arbitrarily chosen for all cells, the state-perturbation is minimized and the system shows a stable behavior. A Class 4 complex behavior can be seen only between the chaotic (Fig. 7, line with cross markers at $x = 0$) and orderly behaviors (Fig. 7, line with cross markers at $x = 100$). However, when internal- (or map-) perturbation $(f + \delta f)$ is introduced to the model, chaotic instability in the system is stabilized and orderly stability is destabilized, allowing the system to demonstrate an even stronger robust trait. Such flexibility in the model can only be obtained with the presence of indeterminacy.

We conclude that in order to model a robust and flexible system that is capable of adaptation, the model must contain indeterminacy. This allows the subcomponents of the system to be partially (not completely!) open to interactions with the "outside," beyond its expected behavioral boundaries. Such partial openness allows collapse and restoration of the adjunction between intent and extent. This mechanism is commonly referred to as emergence or robustness.

References

[1] von Neumann, J.: The general and logical theory of automata. In: von Neumann, J.: Collected Works, Edited by Taub, A. H. 5, 288. (1963)
[2] von Neumann, J.: Theory of Self-Reproducing Automata. Edited by Burks, A. W., University of Illinois, Urbana. (1966)
[3] Ulam, S.: Some ideas and prospects in biomathematics. Ann. Rev. Bio., 255 (1974)
[4] Wolfram, S.: Universality and complexity in cellular automata. Physica D 10 (1984) 1-35

[5] Wolfram, S.: Statistical mechanics of cellular automata, Rev. Mod. Phys. 55 (1983) 601-644

[6] Wolfram, S.: A New Kind of Science, Wolfram Media, (2002)

[7] Langton, C.G..: Computation at the edge of chaos: phase transitions and emergent computation. Physica D 42 (1990) 12-37

[8] Kauffman, S. A.: Antichaos and adaptation. Scientific American. (1991) 265(2): 78-84

[9] Jen, E.: Stable or robust? What's the difference?. Complexity 8, (2003) 12-18

[10] Gunji, Y.-P., Sadaoka, H., and Ito, K.: Inter- and intra-cellular computation models based on Boolean vs. non-Boolean inconsistency. Bio Systems 35 (1995) 213-218

[11] Ito, K., Gunji, Y.-P.: Self-organized criticality resulting from minimization of perpetual disequilibration. Physica D 102 (1997) 275-284

[12] Ito, K., Gunji, Y.-P.: Self-organization toward criticality in the Game of Life. Bio Systems 26 (1992) 135-138

[13] Kamiura, M., Nakajima, K., Gunji, Y.-P.: Generative Pointer: Dynamical System with a Fluctuant Parameter Motivated by Origin of Fraction. Physica D (submitted)

[14] Matsuno, K.: Protobiology: Physical Basis of Biology. CRC Press, Boca Raton, MI, (1989)

[15] Gunji Y-P. Global logic resulting from disequilibration process. BioSystems 1995; 35:33-62

[16] Gunji, Y-P., Ito, K., Kusunoki, Y.: Formal model of internal measurement: alternate changing between recursive definition and domain equation. Physica D 110 (1997) 289-312

[17] Rossler OE.: Explicit observers. In: Plath, PJ. (ed.): Optimal structures in heterogeneous reaction systems. Springer-Verlag, New York (1989)123-38

[18] Rossler OE.: Endophysics, the world as an interface. World Scientific, Singapore (1998)

[19] Gunji Y-P., Kamiura M.: Observational heterarchy enhancing active coupling. Physica D. 198 (2004) 74-105

[20] Kamiura M., Gunji, Y-P.: Robust and ubiquitous on-off intermittency in active coupling, Physica D 218 (2006) 122-130

[21] Wuensche, A.: Classifying cellular automata automatically: finding gliders, filtering, and relating space-time patterns, attractor basins, and the Z parameter. Complexity. Volume 4, Issue 3, (1999) 47-66

[22] Gunji Y-P.: Protocomputing and Ontological Measurement. Tokyo University Press, Tokyo (2004) (in Japanese)

[23] Rosen, R.: Theoretical Biology and Complexity, Three Essays on the Natural Philosophy of Complex Systems, Academic Press, Orlando (1985)

[24] Rosen, R.: Life Itself. Columbia University Press, New York (1991)

[25] Bickhard, M. H.: Why Children Don't Have to Solve the Frame Problems: Cognitive Representations Are Not Encodings. Developmental Review, Volume 21, Issue 2 (2001, 224-262)

Abstracts of invited talks presented at Unconventional Computing 2007, Bristol, United Kingdom

1 Tetsuya Asai (Sapporo, Japan): Neuromorphic VLSIs: past, present and future

"Neuromorphic Engineering" introduced by Carver Mead is a new research field based on the design and fabrication of neural systems whose architecture and design principles are based on biological nervous systems. I will briefly review the past and present works including my developments on neuromorphic VLSIs, and will discuss the future prospectives.

2 Julian Miller and Simon Harding (York, UK): Evolution in Materio: evolving computation in materials

In 1958 Gordon Pask described a method of manipulating a physical system (a series of electrodes in a chemical chamber) so that it would carry out a new function. This work was largely forgotten for many years. In 1996, Adrian Thompson demonstrated that when artificial evolution is sufficiently unconstrained it is possible for it to exploit physical properties for computation. He demonstrated this using an re-programmable electronic device called an Field Programmable Gate Array (FPGA). This inspired the idea that artificial evolution might be able to used to discover hitherto unknown ways of configuring matter to carry out computation. This led to a research project that demonstrated that artificial evolution can be used to manipulate a liquid crystal device so that it can do computations that solve a number of tasks (frequency discrimination, robot control, Boolean logic). We describe this work and discuss further prospects for evolving computation in materials.

3 Jonathan Mills (Indiana, USA): Natural Computing

Drawing on over a decade of experience with Rubel's extended analog computer (EAC), some simple but important questions are addressed.

What is a natural computer? How does a natural computer differ from a digital computer? How does a natural computer perform computation? What is an analogy, for those machines that are configured by analogy? How is a natural computer programmed? For what classes of applications are natural computers the most efficient choice? The relationships between the laws of physics, important mathematical principles, and the technology and architecture of natural computers are presented to answer these questions. Applications with a real natural computer, Rubel's EAC, are used to illustrate our growing understanding of the principles of natural computing. Participants will be able to explore these ideas for themselves using several EACs that will be available during the conference.

4   Kenichi Morita (Hiroshima, Japan): Computation-universality in simple reversible systems

Reversible computers that are defined as "backward deterministic" systems are unconventional models of computation having an analogous property of physical reversibility. Until now, many kinds of reversible models, such as reversible Turing machines, reversible logic circuits, and reversible cellular automata, have been proposed and investigated. In this talk, we mainly discuss reversible cellular automata and their computation-universality. We can see that even very simple reversible cellular automata have universal computing ability. We first give a brief survey on this topic, and show several simple models of 2-dimensional universal reversible cellular automata [1]. Then, we show that also in the 1-dimensional case there are very simple universal reversible cellular automata [2]. Universality issues on simple reversible logic elements and Turing machines are also discussed.

[1] Morita, K., Tojima, Y., Imai, K., and Ogiro, T. Universal computing in reversible and number-conserving two-dimensional cellular spaces, in Collision-based Computing (ed. A. Adamatzky), 161-199, Springer-Verlag, (2002).

[2] Morita, K., Simple universal one-dimensional reversible cellular automata, Journal of Cellular Automata, (in press).

5 Toshiyuki Nakagaki, Tetsu Saigusa, Atsushi Tero, Ryo Kobayashi (Sapporo, Japan): Solving networking problems by amoeba: dynamics and computation

We demonstrate that an amoeboid organism, a plasmodium, of the true slime mold can solve certain computational tasks while foraging in a complex environment. Thus when presented with two food sources separated one from each other by a maze of obstacles the plasmodium calculates a shortest path connecting the sources through the maze by forming a tubular structure of protoplasm so that absorptions of nutrients and intracellular communication are maximised. We propose a mathematical model for the plasmodium-based problem-solving, based on physiological mechanism of morphogenesis of the tubular structure. The mathematical model indicates prospective ways of designing unconventional computing architectures on biological substrates.

6 Susan Stepney (York, UK): The Neglected pillar of material computation

Biological organisms and processes are often touted as information processing systems, and then analysed in computational terms. But their properties differ in many important ways from our "classical" mathematical-logical computational system formalisms, and the way these are implemented "in silico". In particular, they have an extra important feature: their operation is deeply entwined with the physical and chemical properties of the substrates of which they are composed. Those properties both impose constraints on, and provide capabilities to, the computations being performed. Here I discuss the "missing pillar", of "in materio" computation, that is needed to complement classical computational models, before we can understand biological information processing in full.

7 Christof Teuscher (Los Alamos, USA): Computation in self-assembled Avogadro-scale systems — challenges and opportunities

By using chemical self-assembly and self-organizing principles at the cellular, molecular, or atomic scale, it is nowadays relatively straightforward to "build" functional assemblies in a bottom-up way that involve an Avogadro number (1023) of interconnected components. The hope in using such unconventional fabrication paradigms is to eventually be able to go beyond Moore's law by creating highly complex information processing devices in a simple and cheap way. Avogadro-scale engineering is not only concerned with building such devices, but also addresses the challenges of how one can efficiently and reliably compute and how the component's functionality can be programmed and controlled. Since such large-scale nano, bio, and chemical systems often don't work according to the principles we know from silicon-based digital computers, there is a clear lack of design principles and programming paradigms. Usual irregularities, inhomogeneities, and unreliabilities in the physical substrate make the entire undertaking even more challenging. In this talk, I will review the challenges and opportunities of Avogadro-scale engineering, identify possible show-stoppers, and delineate a roadmap of both the issues that need to be addressed and the routes to avoid.

8   Klaus-Peter Zauner (Southampton, UK): Biological computing substrates

A crucial difference sets apart present computing technology from information processing mechanisms utilised by organisms: The former is based on formalisms which are defined in disregard of the physical substrate used to implement them, while the latter directly exploit the physico-chemico properties of materials. There are many advantages to isolating the operation from implementation as is the case in current computers—but theses come at the cost of low efficency. In applications where size and energy-consumption is tightly restricted, or where real-time response to ambiguous data is required organisms cope well, but existing technology is unsatisfactory. Taking heed of the clues from biology the question arises how the realm of computer science can be extended from formal to physical information paradigms. The aim is to arrive at a technology in which the course of computation is driven by the physics of the implementation substrate rather than arbitrarily enforced. The traditional tools and approaches of computer science are ill suited to this task. In particularly it will be necessary to orchestrate autonomously acting components to collectively yield desired behaviour without the possibility of prescribing individual actions. Bio-electronic hybrid systems can be serve as a starting point to explore approaches to computing with autonomous components. We take a two-pronged approach in which we recruit both molecules and complete cells as biological computing substrate. Molecules offer reproducible nonlinearity, self-assembly, and high integration density of complex input-output mappings. Cells, on the other hand, provide cheap and fast nano-engineering through self-reproduction, build in quality-assurance through testing at the point of assembly, self-reconfiguration, and self-repair. Molecules, however, require infrastructure and cells are typically too complex for efficient computation. Our expectation therefore is that in the long term practical biological computing substrates will be situated at the sub-pramolecular and subcellular level, i.e., at the interface between inanimate and animate matter.

318

Index

Lightning Source UK Ltd.
Milton Keynes UK
UKOW031529080212

186903UK00012B/167/A